Springer Complexity

Springer Complexity is an interdisciplinary program publishing the best research and academic-level teaching on both fundamental and applied aspects of complex systems—cutting across all traditional disciplines of the natural and life sciences, engineering, economics, medicine, neuroscience, social and computer science.

Complex Systems are systems that comprise many interacting parts with the ability to generate a new quality of macroscopic collective behavior the manifestations of which are the spontaneous formation of distinctive temporal, spatial or functional structures. Models of such systems can be successfully mapped onto quite diverse "real-life" situations like the climate, the coherent emission of light from lasers, chemical reaction-diffusion systems, biological cellular networks, the dynamics of stock markets and of the internet, earthquake statistics and prediction, freeway traffic, the human brain, or the formation of opinions in social systems, to name just some of the popular applications.

Although their scope and methodologies overlap somewhat, one can distinguish the following main concepts and tools: self-organization, nonlinear dynamics, synergetics, turbulence, dynamical systems, catastrophes, instabilities, stochastic processes, chaos, graphs and networks, cellular automata, adaptive systems, genetic algorithms and computational intelligence.

The three major book publication platforms of the Springer Complexity program are the monograph series "Understanding Complex Systems" focusing on the various applications of complexity, the "Springer Series in Synergetics", which is devoted to the quantitative theoretical and methodological foundations, and the "Springer Briefs in Complexity" which are concise and topical working reports, case studies, surveys, essays and lecture notes of relevance to the field. In addition to the books in these two core series, the program also incorporates individual titles ranging from textbooks to major reference works.

Indexed by SCOPUS, INSPEC, zbMATH, SCImago.

Series Editors

Understanding Complex Systems

Diego Rybski
Editor

Compendium of Urban Complexity

Springer

Editor
Diego Rybski
Leibniz Institute of Ecological Urban
and Regional Development (IOER)
Dresden, Germany

ISSN 1860-0832 ISSN 1860-0840 (electronic)
Understanding Complex Systems
ISBN 978-3-031-82665-8 ISBN 978-3-031-82666-5 (eBook)
https://doi.org/10.1007/978-3-031-82666-5

This Springer imprint is published by the registered company Springer Nature Switzerland AG
The registered company address is: Gewerbestrasse 11, 6330 Cham, Switzerland

Foreword

In the last 25 years, theories of how cities develop and evolve have rapidly gained ground after many years of speculation but with little sense of explanation as to how they form. Over 100 years ago, Geddes [1], amongst others, argued that cities largely grow from the bottom up and evolve, despite our best efforts to improve the quality of life that they offer through top-down planning. Our quest to provide formal theories of how cities develop has largely assumed that explanation begins by articulating the city as being in some form of equilibrium, a 'statis' that remains stable through negative feedback. We assume that cities in their most obvious form, look much the same in visual and cartographic terms from generation to generation. However, this picture of the city has been questioned quite strongly from the middle of the last century. Commentators such as Jacobs [2] argue that the diversity of cities is continually changing due to the fact that they are places of continuous and rapid innovation, diverse places which act as the motors of successive revolutions in culture, science and technology: crucibles of innovation that manifest a complexity that was and continues to be hard to explain and even harder to theorise about.

Much of this recent theorising has been woven into the science of complexity which emerged to deal with systems for which the traditional systems approach whose focus was on how systems could be organised from the top down as nested hierarchies of sub-systems, proved to be inadequate. This systems approach dominated the applied sciences until the last quarter of the twentieth century. However, it was unable to handle the complex nonlinear dynamics, such as discontinuities, bifurcations and catastrophes that characterised the way most social and biological systems actually evolve. In fact, the focus on cities as one of the key exemplars of complexity theory mirrored many of these limitations. Although a rudimentary science of cities focusing on cities in equilibrium had begun to develop in areas such as location theory, urban and regional economics and transportation modelling, it took the development of complexity theory to cast all these ideas into a wider framework of explanation which could embrace qualitative change in city systems as they evolved.

In this book, Diego Rybski has collected together a series of papers that provide the most comprehensive set of ideas about this field so far. In fact, the field has now developed to the point where it is possible to provide a rich range of ideas that are slowly being blended together to produce a basic theory of how city systems are structured and developed in terms of their key determinants. Scale, size and shape provide the basic dimensions of this new science [3] while key ideas involving agglomeration, mobility, movement, diffusion and migration, as well as urban morphology define the forces that knit these issues together. New ways of measuring urban structure and dynamics involving self-organisation, emergence, diversity and entropy as well as hierarchy and fractality are all central ideas introduced here in the various contributions that now define this wider science.

Besides defining a series of dimensions such as those we have just noted, it is possible to provide a conception of city systems that embody the key features of complexity theory. Complex systems are composed of many interacting parts that scale up and down, often in a nested fashion, to a hierarchy or set of overlapping groups. Connections and flows that provide the cement that relates these parts together change as these systems get bigger and as they scale in size, the flows within them, can change qualitatively. The process of change is often referred to as agglomeration, sometimes allometry, and it implies that as systems and their components grow in number and size, their functions also begin to change. The simplest example of this is the value of the components in question. If the components are elements of population, then one of the best examples is that the wealth of the population can increase more than proportionately as the city itself grows. The implication of this is that bigger cities are richer cities, all other things being equal, of course.

This is key to the notion that cities and their components scale and it is due to the fact that as the number of components increases linearly, their interactions scale exponentially. If we have a city of say 1000, there are 1000^2 or 1 million possible interactions. Of course, as individuals we cannot realise ('know') all of these interactions; more likely, we know only a tiny fraction of these, something like $1000^{1.1}$ which gives us about 2000. But as we add components one at a time, the number of interactions increases more than proportionality to the number so far, and it is this interaction that provides opportunities for social interaction. It is this that generates the potential for innovations. This, as we will see in various contributions to the book that follow, is reflected in ideas about scaling, power laws and economies of agglomeration. These ideas weave their way into all the chapters and although there are specific chapters dealing with individual perspectives as we have identified, the seeds of a much more integrated theory of cities lie between the lines of the chapters that follow.

There are 13 chapters in all which the editor has implicitly grouped into cognate areas. The first three chapters deal with city size distributions that follow rank size laws and reflect ideas about the costs and benefits of agglomeration. The next three chapters group together mobility, migration and human traffic flow using gravitational models. Power laws and their approximations are used to describe all these themes. The next set of three chapters deals with urban morphology, in particular segregation, polycentricity in cities and climate change specifically heat islands, while the final grouping includes an overview of complexity science and self-organisation as well as fractality and entropy, two key statistical concepts that are important in measuring degrees of complexity in cities at different scales.

To provide some sense of how all these ideas begin to fit together, power laws or scaling laws are resonant notions in this new urban science, or a science of cities, however, you wish to refer to it. As cities get bigger, there are more than proportionately less of them. These follow what Zipf [4] called 'the rank-size rule' which in essence is a power law that, in its purest form, suggests that the population size of a city is equal to the population of the largest city in the system divided by the rank of the city in question. This is one of the simplest power laws, but it is universal in form in that it appears to apply to any system of cities, west and east, in the Global South or the Frozen North. The law seems to work at many different scales from the distribution of neighbourhoods in a city to the entire world system of cities. There is plenty of noise around these kinds of relationships and it is clear that if we define the spatial system—the system of cities and cities within it—differently in physical terms, these relations change in terms of their parameterization but are still robust and appear to hold over many intervals of space and time.

New data sources that define cities and their components at scales from the individual (often generated using individual-based mobility and data) to city and regional aggregates are now beginning to enable new models to be built in the context of mobility, migration and gravitation and these models that are rapidly being further developed link to wider questions of city size. Some of these data sources suggest how new forms of segregation and hot spots in cities can be identified and how they scale. Methods for scaling are then developed in more detail using measures such as fractal dimension, Shannon entropy and variations thereof.

All the elements needed to turn many of these models into applications are contained in the pages that follow but there are still important elements needed in these kinds of translation. It is thus important to link the contributions here to the wider field, to new developments in data, to ideas about spatial and temporal representation and to ways in which this kind of science can be used to help urban planners and policymakers in their quest to design more sustainable cities. The contributions gathered here are the best of the state of the art at present, but continued effort is required to improve these theories, to continue to enrich the field with new ideas, and to integrate them through exploiting systematic ideas from a variety of disciplines, philosophies and technologies. The keys to this kingdom are implicit in the contributions that follow. Read on and enjoy.

London, UK
June 2024

Michael Batty

References

1. P. Geddes, *Cities in Evolution* (Williams and Norgate, London, 1915)
2. J. Jacobs, *The Death and Life of Great American Cities* (Random House, New York, 1961)
3. M. Batty. "The size, scale, and shape of cities". In: *Science* 319.5864 (2008), pp. 769–771. https://doi.org/10.1126/science.1151419
4. G.K. Zipf, *Human Behavior and the Principle of Least Effort: An Introduction to Human Ecology*. (Reprint of 1949 Edition) (Manfield Centre, CT: Martino Publishing, 2012)

Preface

Cities are the habitat of humans. Almost every person lives in a city, town, village or some sort of settlement. Cities have a remarkable endurance [1]—they survive plagues, fires and storms [2]. If they don't, they get rebuilt. In other words, 'Cities are seen around the world and through history, suggesting "universal" reasons for their existence' [3].

Interactions are fundamental to cities and urban systems. For example, urbanites interact with one another and infrastructure interacts with urban networks. In a general sense, it has been shown, that depending on the character of such interactions, respective systems can be additive or non-additive [4, 5]. Non-additive means that an additional unit interacts—weakly but not negligibly—with all other units so that a surplus emerges, making the contribution of an extra unit larger than expected from the plain count. This is precisely what is characteristic of complex systems, i.e. 'it [a complex system] is a system composed of many interacting parts, such that the collective behavior of those parts together is more than the sum of their individual behaviors' [6]. Disadvantages resulting from extreme densities of people and infrastructure in cities are outweighed by the surplus due to the interactions with others. An urbanite has a non-negligible chance to meet a friend on the other side of the city. In other words, the size of the city is limited by the time it takes to travel through it [7].

As suggested above, there is a considerable intersection between urban systems and complex systems [6, 8]. *Urban complexity* research is dedicated to phenomena of emergence in and around cities. The urban phenomenon can be described by what cities have in common instead of what distinguishes a specific one.

What would be a 'counterfactual' to cities? Living in isolated houses that are spread homogenously in natural and agricultural lands would be inefficient in regards to the mobility between buildings. Moreover, the houses and transport ways would represent obstacles and fragment fields. Thus, it is plausible that buildings are clustered spatially into settlements. But how big should the settlements be? Another counterfactual could be the normal distribution. Why do poeple not organise themselves into towns of more or less equal (optimal) size? Maybe actual city-size distributions represent sort of optimal configurations. Even if not optimal, it is plausible that the configuration of few large cities and many small ones brings some advantage. Metaphorically, we are searching for the *Galton board of power-law distributions* [9].

Science is about gaining insights. We want to understand cities—and their role in the world. Are cities a resolution to the dystainable realities? For example, until now it remains unclear if urbanisation drives or mitigates urban carbon emissions [10, 11]. Which policies can modify the above-mentioned city-size distributions? How would such modification affect carbon emissions? Understanding city systems should not serve as an end to itself. Rather, our understanding should be a step towards improving them.

This textbook aims at a comprehensive introduction to the exciting field of urban complexity. The chapters give an overview of selected topics around cities, from the perspective of physics, economics and planning. As a reference book on urban complexity, it addresses scholars from both city science and complex systems and has a quantitative emphasis. The compendium can also be of interest for statistical physics illustrating an interdisciplinary application outside conventional physics. As methods and insights from physics, economics and planning are combined, it can serve all those studying cities as complex systems.

The idea of a book on urban complexity emerged during the symposium *Cities as Complex Systems—Structure, Scaling, and Economics* (CTCS2016)[1] which took place from July 13th to 15th, 2016 in Hanover (Germany) and was generously funded by VolkswagenFoundation. Most of the 45 participants had a background in physics, economics, geography or urban planning. The symposium addressed major questions faced today towards constructing a (new) science of cities [12].

I would like to thank A. Bunde whose books inspired me to edit one myself. F. L. Ribeiro, as a close colleague, supported this project at various stages. I appreciate the kind Foreword by M. Batty. I am very grateful to the authors, without whom this book would not have been possible. I also appreciate the efforts of reviewers. These are not only internal ones, i.e. authors of other chapters, but also external ones, notably M. Roos, B. Zhou, Y. Li and R. L. Fagundes. I owe gratitude to the Complexity Science Hub (CSH) Vienna for hosting the meeting CTCS2022 in June 2022. I would like to thank the Wuppertal Institute (M. Fischedick, O. Lah) for hosting me from 11/2022 till 01/2024. Additionally, I am thankful for the financial support I received from the German Research Foundation (DFG; Gropius-Project #511568027) and the Alexander von Humboldt Foundation (AvH). Last but not least, I would like to thank the staff of Springer-Verlag, most importantly Zach Evenson.

Dresden, Germany
May 2024

Diego Rybski

[1] 'Sustainable cities: Researchers discuss urban complexity' (last accessed 23.5.2024): https://www.pik-potsdam.de/en/news/latest-news/sustainable-cities-researchers-discuss-urban-complexity.

References

1. M.E. Smith et al., "The persistence of ancient settlements and urban sustainability". In: *Proc. Natl. Acad. Sci. USA*. 118.20 (2021), e2018155118. https://doi.org/10.1073/pnas.2018155118
2. D.R. Davis, D.E. Weinstein, "Bones, bombs, and break points: the geography of economic activity". In: *Am. Econ. Rev.* 92.5 (2002), pp. 1269–1289. https://doi.org/10.1257/000282802762024502
3. M. Barthelemy, "The statistical physics of cities". In: *Nature Reviews Physics* 1.6 (2019), pp. 406–415. https://doi.org/10.1038/s42254-019-0054-2
4. A. Campa, T. Dauxois, S. Ruffo, "Statistical mechanics and dynamics of solvable models with long-range interactions". In: *Phys. Rep.* 480.3-6 (2009), pp. 57–159. https://doi.org/10.1016/j.physrep.2009.07.001
5. F.L. Ribeiro et al., "Analytical solution for the long- and short-range every-pair-interactions system". In: *Chaos Soliton. Fract.* 183 (2024), p. 114771. https://doi.org/10.1016/j.chaos.2024.114771
6. M.E.J. Newman, "Complex systems: A survey". In: *arXiv.org e-Print archive* arXiv:1112.1440 [cond-mat.stat-mech] (2011). https://doi.org/10.48550/arXiv.1112.1440. https://arxiv.org/abs/1112.1440
7. M. Batty, P. Ferguson. "Defining city size". In: *Env. Plan. B* 38.5 (2011), pp. 753–756. https://doi.org/10.1068/b3805ed
8. F. Heylighen, P. Cilliers, C. Gershenson, "Philosophy and complexity". In: *Complexity,science and society*. Ed. by J. Bogg and R. Geyer. London: CRC Press, 2017. Chap. 8, pp. 117–134. isbn: 9781315383132
9. E. Limpert, W.A. Stahel, M. Abbt, "Log-normal distributions across the sciences: keys and clues". In: *BioScience* 51.5 (2001), pp. 341–352. https://doi.org/10.1641/0006-3568(2001)051[0341:LNDATS]2.0.CO;2
10. R. Louf, M. Barthelemy, "Scaling: lost in the smog". In: *Env. Plan. B* 41.5 (2014), pp. 767–769. https://doi.org/10.1068/b4105c
11. R. Gudipudi et al., "The efficient, the intensive, and the productive: Insights from urban Kaya scaling". In: *Appl. Energ.* 239 (2019), pp. 155–162. https://doi.org/10.1016/j.apenergy.2018.11.054
12. J. Lobo et al., "Urban science: Integrated theory from the first cities to sustainable metropolises". In: (2020). Report submitted to the NSF on the Present State and Future of Urban Science. https://doi.org/10.2139/ssrn.3526940

Contents

Contributors

Laura Alessandretti DTU Compute, Technical University of Denmark, Lyngby, Denmark

Efrat Blumenfeld-Lieberthal The Azrieli School of Architecture, Tel Aviv University, Tel Aviv, Israel

Carmen Cabrera Department of Geography and Planning, University of Liverpool, Liverpool, UK

Caio Cacholas Graduate Programme of Architectural and Urban Studies (PPGAU), Federal Fluminense University (UFF), Niterói, Brazil

Clémentine Cottineau-Mugadza Centre National de la Recherche Scientifique (CNRS), Paris, France;
Technische Universiteit Delft (TU-Delft), Faculty of Architecture and the Built Environment, Department of Urbanism, Delft, The Netherlands

Michail Fragkias Department of Economics, Boise State University, Boise, ID, USA

Laetitia Gauvin IRD, UMR 215 PRODIG, Aubervilliers cedex, France;
ISI Foundation, Turin, Italy

Anat Goldman The Azrieli School of Architecture, Tel Aviv University, Tel Aviv, Israel

Andres Gomez-Lievano Analysis Group, Boston, MA, USA;
Growth Lab, Harvard University, Cambridge, MA, USA

Rémi Lemoy IDEES Laboratory UMR 6266 CNRS, University of Rouen, Mont-Saint-Aignan, France

Gabriele Manoli Laboratory of Urban and Environmental Systems, École Polytechnique Fédérale de Lausanne, Lausanne, Switzerland

Vinicius M. Netto Research Centre for Territory, Transports and Environment (CITTA), Faculty of Engineering (FEUP), University of Porto, Porto, Portugal

Otavio M. Peres Faculty of Architecture and Urbanism (FAUrb), Federal University of Pelotas (UFPel), Pelotas, Brazil

Rafael Prieto-Curiel Complexity Science Hub, Vienna, Austria

Julien Randon-Furling Centre Borelli, ENS Paris-Saclay, Gif-sur-Yvette, France;
MSDA, UM6P, Ben Guerir, Morocco

Fabiano L. Ribeiro Department of Physics, Universidade Federal de Lavras (UFLA), Lavras, Brazil

Diego Rybski Leibniz Institute of Ecological Urban and Regional Development (IOER), Dresden, Germany;
Complexity Science Hub, Vienna, Austria

Hadrien Salat Alan Turing Institute, British Library, London, England, UK

Michael Szell NEtwoRks, Data, and Society (NERDS), ITU Copenhagen, Copenhagen, Denmark

Chapter 1
City Size Distributions

Clémentine Cottineau-Mugadza

Abstract Despite high levels of diversity and contingency in the creation and historical development of cities, the distribution of their sizes in a given area presents a surprisingly robust and simple pattern, known as Zipf's law or the rank-size curve. This chapter offers an overview of this regularity in city size distribution. Starting from a short history of early findings, it goes on to present Zipf's contribution, the plethora of empirical material which can be interpreted as a confirmation of the law and some models generating such a regular distribution. The deviations from the law highlighted in the literature are presented along with the theories to explain them. The last section of the chapter reviews recent propositions to advance the study of urbanisation "beyond Zipf's law", namely by using more sophisticated statistical tools or by reintegrating space in the analysis of urban population.

1.1 Introduction

Cities are complex organisations which have emerged independently at several points in human history. Where there was sufficient surplus in agriculture to fund the activity of non-farmers, the concentration of people has allowed a positive feedback loop between the specialisation and complementarity of social groups to produce larger populations and a growing array of functions performed at city level. Depending on their access to resources (from food to tax payers), cities have been

C. Cottineau-Mugadza (✉)
Centre National de la Recherche Scientifique (CNRS), Paris, France
e-mail: c.cottineau@tudelft.nl

Technische Universiteit Delft (TU-Delft), Faculty of Architecture and the Built Environment, Department of Urbanism, Delft, The Netherlands

D. Rybski (ed.), *Compendium of Urban Complexity*, Understanding Complex Systems,
https://doi.org/10.1007/978-3-031-82666-5_1

able to reach significant sizes[1] quite early on, between half a million and a million inhabitants in Ancient Rome [2], over a million in Baghdad in the 8th century and between 150,000 and 200,000 inhabitants in Potosí (South America) around 1560 [3]. Nowadays, depending on city definitions, the largest cities on Earth can sustain thirty to forty million inhabitants, as in Tokyo, Delhi and the Pearl River Delta (Guangzhou/Shenzhen) [4, 5].

Simultaneously, systems of cities have emerged as a higher level of organisation of human settlement, from the interactions between interdependent cities of the territorial organisation [6]. While larger cities have concentrated social, political and economic functions of command and representation, the entire system relies on the presence and interaction with cities of smaller sizes, spatially distributed across territories and serving the basic local functions of socializing, trade, spiritual gatherings, service and production. Systems of cities thus rely on a diversity of city sizes. This is the case even for city-states, which were and are parts of wider networks of cities, as exemplified by Venice and Genoa then (whose trade networks extended to Northern Africa, Central and Eastern Asia) to Dubai and Singapore now (whose airport departure boards reflect the diversity of connexions they have with other cities worldwide).

More than a mere diversity of city sizes, multidisciplinary research over the past hundred years has pointed to the fact that city sizes tend to be distributed very regularly, with a few large cities and many small cities (i.e. fat-tailed distributions). Furthermore, the main parameters of these distributions have been claimed to be quite regular and universal since Auerbach [7]'s contribution. His discovery, the subsequent work done by Lotka [8] and Zipf [9] and the reactions it generated have since fueled a plethora of literature aimed at "confronting the mystery of urban hierarchy" [10]. The Auerbach-Lotka-Zipf (ALZ) law (wrongly attributed to Zipf alone in most cases) is taught in mainstream urban studies degrees and is used in many planning and policy reports. Yet, there is still no consensus as to why most urban systems verify the ALZ law (also known as the "rank-size rule"), why some deviate from it, and what this knowledge can be useful for. In this chapter, I try to provide a critical presentation of the recent multidisciplinary debates over city size distributions, by tracing the origins of the ALZ law to earlier models of urban hierarchy (Sect. 1.2), presenting the empirical, methodological and interdisciplinary discussions it sparks (Sect. 1.3), the attempts to deviate from this model (Sect. 1.4) and the aporia of Zipf's law as a policy tool (Sect. 1.5).

---

[1] In this chapter, we use population, rather than surface or wealth, as the measuring unit of city size. Indeed, population is the most widely comparable quantity across countries and time periods, and it summarizes a significant number of other urban quantities, such as wealth produced, number of dwellings, jobs etc. [1].

1.2 Early Findings About Regularities in City Size Distribution

G.K. Zipf has become the reference name for describing the regularity of city size distributions in general, and using a power law of exponent of -1 in the distribution of city rank with city size in particular. However, these regularities had already been observed, analysed and modelled by mathematicians and physicists since 1913 at least [7]. Additionally, the contributions of Reynaud [11], Gibrat [12] and Christaller [13] on cities' geographical locations and growth had concluded that systems of cities produced regular (spatial) distributions of cities following a fat-tailed distribution of sizes. Finally, contributions to urban primacy and optimal city size contributed to the scientific understanding of urban hierarchy long before Zipf's seminal publications in the 1940s. In this section, I therefore aim to contextualize Zipf's contribution within the study of city size distribution and to recall the main theoretical conclusions on city size regularity.

1.2.1 Formalisation of a Regular Relationship Between City Sizes and Ranks

The earliest scholarly publication known to address the rank-size distribution of cities is Auerbach [7]'s article: *Das Gesetz der Bevölkerungskonzentration.*[2] This short study (translated recently from German to English [14]) starts from the observation that Germany in 1910 counted about 50 big cities, around 200 small cities, roughly 2,000 towns and nearly 100,000 villages, thus a distribution approaching a geometric progression of the number of settlements with decreasing size. Auerbach noted that if you multiplied the size s of cities by their rank r in the distribution, you obtained a quantity which was a "characteristic product" because it remained more or less constant for German cities order by decreasing size (after some variations for the largest cities)

$$p_i \cdot r_i = AK \; . \tag{1.1}$$

He called this constant AK for Absolute Concentration and used it to construct another measure of population concentration: the specific concentration index (SpK), which corresponds to AK divided by the total population P of the country. He argues with international examples that this index can complement the analysis of population density by giving an idea of the hierarchical organisation of settlements according to their size. Auerbach's formulation Eq. (1.1) was later rewritten by Singer [15][3] as

$$r_i \cdot p_i^{\beta} = A, \quad \beta = 1 \; .$$

[2] "The law of population concentration".

[3] According to Guérin-Pace [16] (I was never able to access this reference myself).

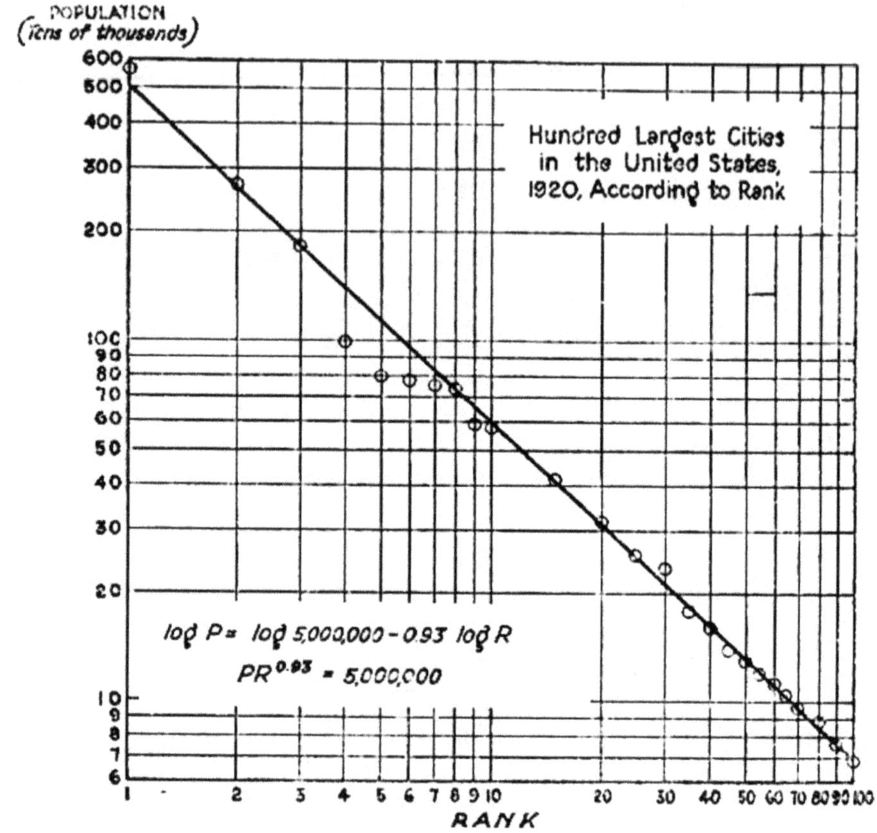

Fig. 1.1 Law of Urban Concentration. *Source*: [8], cited by Rybski and Ciccone [17]

This formalisation is equivalent (assuming $\alpha \simeq 1/\beta$) to the law which we currently know as Zipf's

$$p_i \sim r_i^{\alpha}, \quad \alpha \simeq 1\,. \tag{1.2}$$

The earliest scholarly publication known to visualise the rank-size distribution of cities on a log-log plot and to use the power law formulation of Eq. (1.2) is Lotka [8]'s article [17]. In this figure (cf. Fig. 1.1), Lotka [8] uses the visual representation of cities to estimate the value of the associated exponent α (≈ 0.9 using the 100 largest cities of the United States in 1920).

The earliest scholarly publication known to connect the power law formulation of Eq. (1.2) with the work of Pareto on income distributions is Saibante [18]. In his political economics course at the University of Lausanne, Pareto [19] famously suggested that income distributions follow a pattern best described by

$$\overline{F}(x) = \Pr(X > x) = \begin{cases} \left(\frac{x_\mathrm{m}}{x}\right)^\alpha & x \geq x_\mathrm{m} \\ 1 & x < x_\mathrm{m} \end{cases},$$

where the probability for a person picked at random to earn more than x is an inverse function of x to the power $-\alpha$, when x is over the minimum income x_m.

Building on Lotka's work, Saibante [18] "focused on the exponent [...] and thereby laid the foundation of modern empirical investigations" [17, p. 603].

The mathematical formulation chosen by Zipf relates to other disciplinary precedents such as [20] and [21]. The bibliometric works of mathematician Lotka [20] and librarian Bradford [21] had produced similar expressions to describe the frequency distribution of publishing authors, and were popularised and generalised by the linguist and philologist Zipf [22, 23]. Although he did not come up first with these results, Zipf became one of the first to establish a science of language, working extensively on empirical as well as theoretical matters and trying to explain the regularities observed systematically with quantitative methods [24]. We can explain the reason why his work is remembered today (rather than other scholars' contributions that pre-date his) by the fact that it is accessible (being written in English in a book format rather than more scattered publications) and because it spanned across fields to the social sciences.[4]

After analysing the regularity of word frequency in texts with power laws in earlier publications, G.K. Zipf turned his attention, among other topics in the social sciences, to city sizes, in particular with the books 'National Unity and Disunity' [9] and 'Human Behavior and the Principle of Least Effort' [22]. In these publications, Zipf offers both a formulation of the rank-size regularity, visual representation of empirical data, but also suggests an explanation of this relation with the idea of the "least effort" principle. In a model of economic geography [17], he shows that the optimal spatial distribution of labour in a system of settlement is affected by two opposite forces: the "force of diversification" which brings people to flock to production centers to reduce transport costs, and the "force of unification" which sees goods being transported to smaller towns where labour costs are minimised. In a situation of high transport costs, the "force of diversification" takes precedence and the populations are distributed very unevenly ($\alpha > 1$). In a situation of low transport costs, the "force of unification" takes precedence and the populations are distributed very evenly ($\alpha < 1$). In empirical situations, the two forces are counter-balanced and we find that $\alpha \simeq 1$. This explanation has clear similarities to earlier spatial models of urban settlements, although Christaller [13] is not properly referenced in [17, 22].

[4] For an analysis of the circulation of analytical models in urban studies, see [25].

1.2.2 Formalisation of a Regular Relationship Between City Sizes, Functions and Locations

The most famous spatial model for the distribution of cities is Christaller [13]'s Central Place Theory, later refined by Lösch [26]; although, as shown by Robic [27], an earlier version of the theory was written 100 years earlier in a French encyclopedia by Reynaud [11]. Central Place Theory explains the number, location, function and size of cities of a given region by the combined principles of economic rationality and spatial constraints (such as distance). It predicts that a first level of central places will service population locally by offering frequently needed goods and services. The spatial optimisation of circular catchment areas for these first level cities in an idealised isotropic plain gives way to a hexagon mesh. For goods and services of higher value and less frequent use, a smaller group of higher level cities are picked from lower level central places, whose catchment areas create an additional layer of larger nested hexagons. The resulting distribution of city sizes is therefore fat tailed in that the model creates classes of cities of identical sizes. Its rank-size graphical representation, as in Harris [28], shows a decreasing suite of steps or plateaus where Zipf's law expects a straight line.

Despite the difference in size predictions between Zipf's law and Central Place Theory, Berry and Garrison [29] find two areas of convergence. First, their predictions match in terms of classes of size. Second, the guiding principle behind Zipf's law (unification and diversification) is related to Christaller's catchment areas of goods in a Weberian localization scheme. Zipf's law represents an equilibrium between these two forces, in a similar way the hierarchy of nested hexagons in Central Place Theory represents a spatial equilibrium between servicing population with a diversity of services (in the larger cities) while keeping transport costs low (for frequently used services in smaller centres). An important difference between the models is that Zipf's law is a more generic and simple pattern to generate. Power laws are present in multiple fields of science. Indeed, there exists a wide diversity of generative models which produce a power law, from preferential attachment (e.g. [30] and multiplicative effects (e.g. [12]) to random walks and Yule [31]'s process [32, 33].

To summarize, the Auerbach-Lotka-Zipf law (subsequently ALZ) originates from a multidisciplinary background and finds precedents in the statistical work of earlier polymaths. Part of the explanation behind this regularity in urban studies stems from early models of spatial distribution of cities. Interestingly Rybski and Ciccone [17] found a disconnect between two strands of early research on city size, based on the citation patterns of early authors: the works cited by Zipf [22] which include [7, 8] and [12] and the works cited by Lösch [26] in an economics tradition (i.e. [12, 19] and [13]). Following these authors, although "Works citing Auerbach (1913) to a large extent also cite Zipf [...] typically 20% of the works citing Zipf [...] also cite Auerbach (1913)" [17, p. 606]. Additionally, in recent years, Zipf's law has become so much part of the common knowledge of urban studies that a review of 66 English-speaking journal articles reporting an empirical estimation of the exponent of the ALZ law for cities showed that only 5 of them cited [9], 37 cited [22], and a significant proportion did not cite Zipf at all [34].

1.3 Scientific Cacophony Over Zipf's Law for Cities

Despite the wide recognition of the ALZ law in geography, planning, sociophysics and urban studies, there is no consensus regarding the adequacy and opportunity of this law in describing the distribution of city sizes. Three areas of debate persist to this day. The first debate revolves around the empirical observations gathered and their contribution to corroborating or rejecting the "law". The second ongoing debate is methodological and regards the definition of the observations and the fitting techniques used to evaluate the law. Finally an interdisciplinary debate is still ongoing about the universality of the ALZ law for cities, opposing those who find more value in the model to those who find its deviations more interesting.

1.3.1 Empirical Discussions

Despite final claims of empirical validity (such as [35] and [36] based on roughly 150 US cities), seventy years of rank-size rule estimation have produced a strong disparity of results on empirical distributions of city size, and a large pool of quantitative material to compare and analyse. A first array of studies have compared results across the world, identifying countries where the city-size distribution conforms to the ALZ law and countries where it does not. A second strand of research has focused on temporal evolution, identifying when urban systems conform to the law and when they do not. A third way of analysing the diversity of empirical estimations has been through a meta-analysis.

Comparison of Countries

The first available systematic international comparison of the ALZ is Rosen and Resnick [37]'s article in the 1980s.[5] Until the development of systematic recording of urban statistics, the international comparison of rank-size distribution was restricted to a handful of countries only: the United States [15, 20, 39], the USSR [39, 40], Japan [15, 41], France and the UK [15, 42], Germany and Hungary [15] and The Netherlands [7, 43].

In their 1980 paper, economists Rosen and Resnick [37] analyse city size distributions in 44 countries (including "developing" nations at the time such as Brazil,

[5] With the exception of Allen [38]'s study of 58 countries, which is cited in the literature (e.g. in [16]) but which I have never been able to access.

Indonesia, Zaire and Mexico). Their contribution lies in the estimation of the power law exponent of the rank size distribution of cities over 50,000 inhabitants in 1970 for these urban systems. Additionally, they tested the impact of using alternative city definitions for 6 countries and provided some explanations of the deviation. They use the Pareto form of estimation, Eq. (1.3), where the Pareto exponent corresponds to $\beta \simeq 1/\alpha$ in Eq. (1.2)

$$r_i \sim p_i^{-\beta} . \tag{1.3}$$

Empirically, the parameters of this equation are estimated using its equivalent

$$\log(r_i) = B - \beta \cdot \log(p_i) . \tag{1.4}$$

They find that the estimated value of β varies between 0.809 for Morocco and 1.963 for Australia, with an average of 1.136 and 75 % of countries having urban populations distributed more evenly than predicted by the ALZ law ($\beta > 1$). According to them, this meant that "the rank-size rule should be reconsidered" [37, p. 168].

Providing updated evidence on a larger sample of countries (75), Soo [44] relies on the citypopulation.de website[6] to estimate Zipf's law on cities and agglomerations (although the precise delineation used in the website is unknown) of at least 10,000 inhabitants between 1990 and 2001. He finds that, using the Ordinal Least Square (OLS) estimation methods, such empirical data leads to rejecting the ALZ law in most countries, firstly because the distribution does not follow a Pareto/power law distribution, and secondly because "when it does, the Pareto exponent is frequently statistically different from 1, with over half the countries exhibiting values of the Pareto exponent significantly greater than 1." [44, p. 240]. An innovative aspect of his work is to compare two methods of estimation: the OLS method with the more robust Hill estimator. Using both methods as empirical evidence, Soo [44] shows that the ALZ law is rejected in most cases, in terms of significance of the exponent estimated and its value with relation to 1.

Pumain et al. [45] compared the size distribution of urban agglomerations of more than 10,000 inhabitants in BRICS countries between the 1960s and 2010s using harmonised databases built for comparative purposes. They find a large diversity of values estimated for Zipf's law in these countries around 2010, using Eq. (1.2), from $\alpha = 0.8$ in China, i.e. city sizes more evenly distributed than predicted by the ALZ law, to 1.23 in the United States, i.e. city sizes more unevenly distributed than predicted. According to these authors [45, 46], such variations are better explained when looked at dynamically (i.e. over long time periods of time).

[6] www.citypopulation.de, last accessed: 27 November 2023.

Evolution Over Time

One argument used to explain the discrepancy between the ALZ law and the kind of empirical results reported below is that the law represents an equilibrium state for a mature system of cities [35]. Until maturity is reached, a system would not exhibit a power law fit with a unity exponent. By contrast, following the predictions from Evolutionary Urban Geography [47, 48] and the New Economic Geography [49], higher economies of scale and lower transportation costs should benefit larger cities and thus increase city size unevenness (i.e. increase α or reduce β), see also Chap. 3. Given recent economic trends (drastic lowering of transportation costs and higher economies of scale in innovative goods and services due to first-mover advantages in larger cities), one would therefore expect a monotonic evolution of the ALZ exponent towards greater unevenness.[7]

Using municipality data (rather than more systematically defined morphological or functional city definitions), Parr [50] tracked the evolution of Pareto's exponent β over several decades in a dozen of countries at different stages of economic development. He identified several types of evolution: a monotonic decrease of β in Turkey, India or Egypt; a decrease followed by a "slight upturn" in Brazil, Japan, Spain, and the USSR; a clear U-shape in Sweden, France, Austria and the USA. This suggested a U-shape evolution of the Pareto exponent (or an inverse U-shape evolution of the Zipf exponent) over time within a country throughout its economic development.

Using French "urban units" with a consistent threshold over 161 years, Guérin-Pace [16] highlighted a different pattern: "when the entire system is used [threshold of 2,000 inhabitants], inequality in city size has been accentuated over time. Inversely, when only the top of the hierarchy is considered [threshold of 100,000 inhabitants], the city size inequality has been reduced." [16, p. 557]. Fifteen years later, González-Val [51]'s systematic estimation of the Pareto exponent on US "incorporated places" points to a similar conclusion. Inequality in city sizes increases between 1900 and 2000 when the 19,000+ places are taken into account, yet there is convergence in size among the 1,000 biggest places. This suggests that the ALZ model of city size distribution does not fit empirical systems' evolution over time.

Using large harmonized databases on BRICS cities over the past 50 years, Cura et al. [46] observed two remarkable patterns: 1/ the strong persistence of the peculiarities of the rank-size curve (e.g. its shape or primacy) and 2/ no single trend regarding the evolution of city size inequalities. Indeed, between 1960 and 2010, they find a monotonic increase of α for Brazil and India, a monotonic decrease for China and no clear trend for South Africa and the Former Soviet Union.

Any regular temporal evolution of Zipf's exponent is thus still undemonstrated in empirical studies. One way of resolving discrepancies between national case studies is by pulling all studies together in a meta-analysis.

[7] The same conclusion was apparently reached by Allen [38], according to Guérin-Pace [16].

Meta-Analyses

A meta-analysis is an analytical tool aimed at summarizing the quantitative effect of systematically recorded factors on a value of interest. Developed originally to analyse randomized clinical trials in medical research, meta-analyses are now used to pool results from various empirical studies also in social sciences, in order to go beyond the limits of particular studies and samples and generate more general conclusions. To this end, it contextualises the production of empirical results reported in the literature by taking into account the different choices and constraints affecting the production and reporting of the quantitative value of interest. In the case of the ALZ law for cities, context refers to three broad elements: the territorial context (in which country was the law estimated? At which date?), the statistical context (how was the regression estimated? What is the definition of cities used? Which threshold of population applies?) and the study context (was the estimation reported on its own, as part of a sensitivity analysis or a comparative study?). Theoretical authors have provided plenty of explanations as to why the exponent of the ALZ law might vary over time and across nations (cf. Sects. 1.3.1 and 1.3.1), but meta-analyses tend to show that the statistical context is more important in determining the values of the exponent reported in the literature.

Nitsch [52] conducted the first meta-analysis of the ALZ law for cities. Selecting 29 references from the EconLit database, he was able to quantify the effect of estimation specifications on reported 515 values of Pareto exponents. He concluded that "the estimated exponent in a Zipf regression is on average not 1.0." [52, 97]. Furthermore, he noted that the value of the estimated exponent tended to vary with the type of city definition used in the study (i.e. metropolitan areas vs. built-up areas or municipalities), the year of the estimation and the number of other ALZ exponents estimated in the study. Rather than a picture of universal law, the work of Nitsch [52] pointed to a broad regularity of size distribution whose exponent is very sensitive to the technical specifications of its estimation.

More recently, Cottineau [53] extended the meta-analysis to include 1962 estimations from 86 studies and added more variables to test the effect of the territorial context and to control for the regression method (OLS, Maximum Likelihood, etc.). She finds that "no iron Zipf's law exists for cities" [53, p. 20] and estimates the role of technical specifications in the variation of reported exponent values to about 40 %, i.e. much more than the variation attributable to territorial, temporal and urban differences between countries. Technical specifications pertains to methodological choices made during data collection and statistical implementation of the regression. They represent the bulk of the methodological discussions on-going around the ALZ law for cities.

1.3.2 Methodological Discussions

We highlight four methodological elements to which recent debates attribute an effect on the estimation of the ALZ exponent: the scale of the system of cities analysed, the city definition chosen, the population cutoffs for including cities and the estimation methods selected.

What Scale for the System?

Identifying systems of cities is not straightforward. It means following multi-faceted inter-city interactions and tracing a limit where reciprocal links become stronger within the limit than with cities outside the limit [47]. Because national boundaries facilitate interurban circulations and data collection, most authors default to that scale to estimate the ALZ law. In the MetaZipf database [53],[8] estimations of the rank-size curve with cities selected at the country level represent 80 % of all 2188 estimations. The remaining estimations represent city systems at the sub-national (18 %) or continental (2 %) level. The sub-national scale can be justified in continental-sized countries, for example in Brazil [54]. It seems less appropriate when it results from data collection constraints, such as England and Wales–that is, without Scotland [15]. On the other hand, the continental scale can be justified when interurban interactions are found to exceed national borders, as in Europe or the former Soviet Union [46]. In any case, an appropriate delineation of the city system is important because only an internally consistent system should lead to a strict adherence to the ALZ law for large elements, as demonstrated by Cristelli, Batty, and Pietronero [55]. A correlated issue to that of the system scale is the issue of city delineation.

What is a City?

Delineating cities is not straightforward either. It means favouring one or more features of cities (political power, morphology, functional integration–depending on whether one focuses on people or the built environment) and tracing a limit where the concentration of urbanity seems to drop according to the criterion used.

Along with Rosen and Resnick [37], I would argue that the "entire metropolitan area is the most desirable choice for an urban unit as it represents an integrated economic unit." [37, p. 170]. However, many national statistical bodies do not provide such delineations and offer only municipal level data (the improperly named 'city proper'). Therefore many authors default to that delineation to estimate the ALZ law. In the previously cited MetaZipf database, estimations of the rank-size curve with cities delineated as municipalities rather than morphological or functional agglomerations represent 58 % of all estimations. The remaining estimations represent cities

[8] https://github.com/ClementineCttn/MetaZipf.

defined as morphological agglomerations (24 %), as functionally integrated units–metropolitan areas for instance–(6 %) or a mix of those (12 %). The municipal delineation could be justified for historical periods when municipalities still contained most of the urban footprint and activity. It seems less appropriate nowadays, especially when other definitions are available (or possible to construct, by aggregating municipal data based on morphological and/or functional criteria). In any case, an appropriate delineation of the city system is important because it affects the value of the power law exponent measured [44, 53, 56].

Population Cutoff and Sample Size

Another aspect of city definition pertains to the minimum population size for an urban aggregate to be considered a city. This minimum population (i.e. cutoff point of the distribution) has a mechanistic impact on the number of cities considered to examine the ALZ law: with low population cutoffs (from a few hundred to a few thousand people), many small settlements are considered as cities and the number of observations is large; when high population cutoffs are applied (hundred of thousands or million people), only a handful of cities are considered in the regression for most countries. The issue with this cutoff point was already mentioned by Auerbach [7] but remains relevant because there is no absolute justification for a given cutoff (no magic number distinguishes cities from non-urban settlements everywhere and at all times), yet the value of the cutoff chosen has a strong impact on regressed value of the ALZ exponent [37, 51, 53]. Using Monte Carlo simulations of US cities with a random rolling sample method, Nota and Song [57] found that "in 1980, 1990, and 2000, an increase of approximately one percent in the number of urban areas used in regression would cause a 0.16 percent reduction in the value of the estimated exponent. [...] In other words, the rank-size rule is not necessarily an economic regularity and may in fact be a statistical phenomenon" [57, p. 30].

As noted by Rosen and Resnick [37], the population cutoff and sample size used to estimate the ALZ law would not be important if the power law fit was perfect. "But since deviations from the Pareto curve occur, our choice of sample size becomes important" [37, p. 171]. The question of deviations and their meaning is analysed in Sect. 1.4.

Fitting Technique and Estimation Methods

Finally, the fitting technique used to examine the ALZ law empirically has been found to play a role in the value of the exponent estimated. Soo [44] compares the results of the regression of the ALZ law for a variety of countries with the Ordinary Least Squares (OLS) and the Hill estimator. In the first case, the best fitting line is found heuristically, whereas in the second case, a Pareto distribution is assumed in the first place. The author concludes that "the results we obtain depend on the estimation method used, and in turn, the preferred estimation method would depend on our sample size and on our theoretical priors–whether or not we believe that Zipf's Law holds" [44, p. 261]. In a meta-analysis of empirical estimates of the ALZ law for cities, Cottineau [53] considers the four dominant estimation methods present in the literature: OLS, the likelihood method (which includes the Hill estimator), Gabaix and Ibragimov [58]'s modification of the traditional OLS regression (which simply replaces R_i by $R_i - 0.5$) and Markov chains. She finds that the likelihood method is indeed significantly associated with a lower Pareto estimate[9] (OLS meta-analysis) and that Gabaix and Ibragimov [58]'s estimation method leads to estimate systematically higher Pareto exponents on average (fixed-effect meta-analysis). These effects add significant, unnecessary (and usually overlooked) noise to the interpretation of the ALZ law for cities and its deviation.

1.3.3 Disciplinary Angles

An important aspect of the multidisciplinary interest in the ALZ law (including biology, sociology, physics and economics [59]) is that it reflects different attitudes towards modelling, residuals and deviations (cf. Sect. 1.4). In fields such as economics and physics, the model takes the centre stage, and its status is either validated or rejected. By contrast, geographers will use the model as a "generality" filter and will focus on understanding the local deviations to the model. In a "meta-"meta-analysis, Cottineau [34] tested the influence of this potential disciplinary bias on the reporting of estimation results in empirical studies of Zipf's law by authors from various disciplines. She hypothesised that the usual publication bias, whereby "positive" results (i.e. validating an hypothesis) get more easily published than "negative" results (i.e. the absence of significant evidence), play differently for different disciplines, since positive results in physics and economics consist in validating the ALZ law whereas positive results in geography can mean rejecting the model in favour of its deviations. She computed similarity matrices between 66 empirical articles estimating a coefficient for Zipf's law, in terms of their wording, the references cited, and the disciplines of the journals cited (cf. Fig. 1.2). She shows 1/ that similar wording between pairs of articles is significantly associated with similar values of estimated coefficient (average and dispersion) and 2/ that similar citation patterns between pairs of articles is significantly associated with similar values of estimated coefficient (average only). "These results point towards the existence of a combined publication and reporting bias [...] Indeed, despite the fact that city sizes could be observed identically by all researchers using the same specifications, I find that different authors tend to delineate cities and report Zipf estimations differently depending on the field they come from, which translates into the words they use and the citations they mobilise" [34, p. 1459].

[9] In the original article, the coefficient of this regression is positive since the regression of Zipf's law uses the Lotka equation form, with the population expressed as a function of the rank, unlike Eq. (1.4).

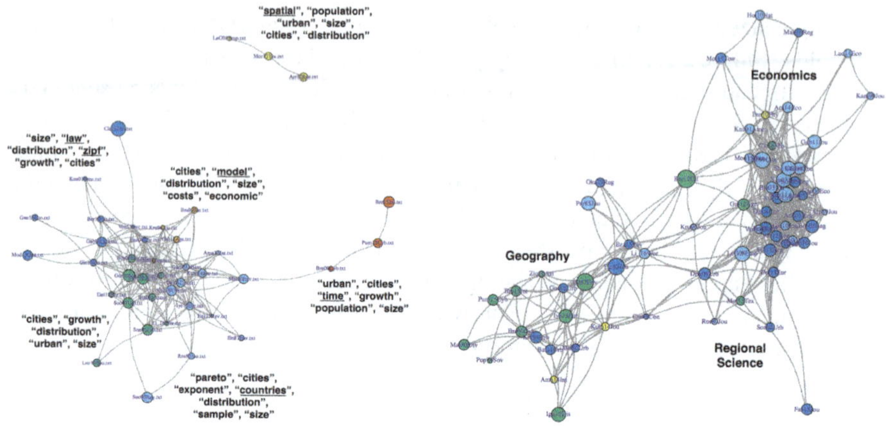

Fig. 1.2 Wording and citation similarity between articles reporting an empirical estimation of the ALZ law for cities. *Source*: [34, Figs. 4–6] distributed under the terms of the Creative Commons CC BY license

1.4 Deviating from the ALZ Law

Beyond the variation of value of the β coefficient discussed above, some analyse the poor fit to a power law model in terms of meaningful curve deviations. This section reviews the theoretical propositions to explain both the deviation from the value of 1 for the β coefficient and the deviation from a power law distribution.

There are two main sets of explanation of empirical deviation to the ALZ law for cities. The first one relates to theories of socioeconomic development and their consequence in terms of population settlement (Sect. 1.4.1). This type of top-down explanations is in line with Zipf's original take on why city sizes are so regularly distributed. The second set of explanations (Sect. 1.4.2) corresponds more directly to urbanisation processes. This type of bottom-up explanation emanates from the study of cities themselves and their evolution.

1.4.1 Socioeconomic Developments

A first strand of explanations for the deviation from the ALZ law suggests that if a system of cities does not fit a power law distribution of their sizes, or if the exponent of this power law differs from 1, it means that the system has not reached maturity and is therefore on its way there. This type of explanation encompasses both theories of economic development (whereby "wealthier countries […] have more evenly distributed populations [… because] high income allows a country to support a network of intermediate-sized cities as demand thresholds are passed" [37, p. 182]) as well as theories of territorial integration (whereby newly integrated federal states tend to duplicate small to medium-sized cities which were the heads of previously independent systems of cities, as 19th century India and China [60] or the Soviet Union [28]). The lack of integration is supposed to produce rank-size curves which are convex instead of linear in a log-log plot, whereas the lag in socioeconomic development can lead to convex curves (when small cities merge with one another through spatial growth [61]), to concave curves (when urban growth rates are autocorrelated over time [62]) when the primate cities are larger than expected), or simply to different exponent values.

Finally, Pumain [47] suggests that systems of cities tend to be more uneven over time because larger cities grow faster on average, due to first-mover advantages in innovations. This is another illustration of the autocorrelation of growth rates which would lead to a deviation from the ALZ law in terms of the value of its coefficient.

1.4.2 Urbanisation Patterns

The explanations related to urbanisation processes are the most powerful in predicting the deviation of the empirical exponent to its theoretical value of 1 [53]. One of them relates to the coevolution of urban systems and transportation systems. Therefore, recently urbanised systems are expected to be more unequal than ancient ones because the transportation networks available at the time of urbanisation were slower in old systems, a larger amount of small cities were necessary. In areas urbanised with railroads and highways, these small cities tend to have been short-circuited [47, 63]. According to Morrill [64], small territories have more uneven distribution of city sizes than large countries because the concentration of power in the largest cities is not balanced by a sufficiently large set of secondary cities, which is seconded by Rosen and Resnick [37]'s empirical inquiry. Finally, the spatial dependence of urban growth (or the positive autocorrelation of growth rates), which contrary to Gibrat's model assumptions,[10] can produce empirical deviations from the ALZ law [65, 66].

1.5 Beyond Zipf's Law

The doubt formulated around the ALZ law and the reduction of city systems to a single value lead many authors to consider new paths to better describe city size distributions and the regularities of urban systems more generally. The first approach is to consider other distributions than power laws. The second is to reintegrate geographical space in the analysis of city sizes.

[10] Gibrat's model being the model most commonly accepted to generate Zipf's regularity [35].

1.5.1 From Power Laws to Other Distributions

The most frequent distribution preferred to the power law is the lognormal distribution [67–69]. There are two main reasons for this. Firstly, these distributions belong to the same type of mathematical distributions, i.e. skewed distributions [70]. Empirical data such as city populations are usually not sufficient to detect and distinguish significantly between these distributions [71]. Secondly, the lognormal distributions is usually preferred to a power law when analysing city sizes because the generative mechanisms suggested to explain them tend to produce lognormal distributions rather than power laws. This is the case of [30]'s preferential attachment model or Gibrat's stochastic model [35].

Other distributions have been suggested, like the Beta or Lavalette distributions [72], or hybrid forms of distributions [73–75], for instance a Pareto upper tail combined with a lognormal lower tail [76]. This strand of research trades the simplicity of the ALZ model for a more accurate representation of city size distribution. It must be noted that this quest can only achieve significant results if it relies of carefully delineated and harmonised urban data, since the quality of any statistical fit depends heavily on the quality of the underlying data [46].

1.5.2 From Power Laws to Spatial Models

A more radical improvement of the model involves the consideration of geographical space and spatial relations in the study of city size distributions. For instance, Gallo and Chasco [65] consider spatial dependence in their estimation of Zipf's law, using a spatial autoregressive model with spatially autocorrelated errors. They find statistical evidence that considering spatial dependence in urban growth impacts the estimated value of the ALZ law for cities. Bergs [75] demonstrates the necessity for economic approaches to analyse the ALZ law spatially, both with theoretical and empirical datasets.

The consideration of spatial dependence in empirical estimations goes hand in hand with the focus on spatial models to generate and potentially explain the emergence and stability of city size distributions. Such models range from social physics models (for instance [77]) with their call for simplicity, to more complex agent-based models involving non-deterministic interactions (for instance [78]).

1.6 Conclusion

> Despite the early discovery, the quest for a robust theoretical model to explain such an empirical distribution of cities remains elusive; starting with Christaller (1933) who described Auerbach's finding as 'a most incredible law,' yet essentially was 'nothing more than just

> playing with numbers'. The view that the Zipf's Law is 'atheoretical' carries on to the present date. [79, p. 2439]

More than a century since the first attempts at summarizing city size distributions with an elegant and simple mathematical expression, urban studies scholars can be left with a demoralising impression regarding the rank-size regularity and the ALZ law. Past the initial sense of mystery and pleasing simplicity, many question the future of this model. Is the law pointless? What should we do with this statistical relationship? Since much empirical evidence rejects the law and numerous self-reinforcing process can generate a power law distribution, can we really use the ALZ law to validate our theories on how cities grow and why urbanisation differ from country to country? Does it have any value as a policy tool?

This short review of the recent literature on the ALZ law suggests that the rank-size curve is still a popular model to analyse urbanisation patterns and to explain national differences. However, the cacophony of interpretations derives from the wide variations in urban definitions and estimation techniques. Until these differences are accounted for across disciplines and academic traditions, theoretical models will be challenging to validate empirically. Until theoretical models are validated empirically, it seems adventurous to try to use the ALZ law as a policy tool, let alone as a policy objective.

Acknowledgements The author thanks Denise Pumain, Marie-Claire Robic, Michael Batty, Robin Morphet and John Parr for their insight and many discussions about Zipf, the ALZ law and the urban hierarchy over the past decade.

References

1. D. Pumain, Une théorie géographique des villes. Bulletin de la Société Géographique de Liége **55**(2), 5–15 (2010)
2. G.R. Storey, The population of ancient Rome. Antiquity **71**(274), 966–978 (1997). https://doi.org/10.1017/S0003598X00085859
3. A. Aruska de Souza Santos, S. AS Iamamoto, The Difficult Legacy of Mining in Past and Contemporary Potosí and Ouro Preto, in *Journal of Latin American Geography* 18.3 (2019), pp. 110–131. https://www.jstor.org/stable/48618853
4. United Nations, *The world's cities in 2018* (Tech. rep. New York, United, 2018)
5. O.E.C.D. European Commission, *A new perspective on urbanisation* (OECD, Tech. rep, 2020)
6. D. Pumain, *Hierarchy in natural and social sciences*. (Springer, 2006)
7. F. Auerbach, Das Gesetz der Bevölkerungskonzentration, in *Petermanns Geogr. Mitteilungen* 59.74 (1913), pp. 74–76. https://zs.thulb.uni-jena.de/receive/jportal_jparticle_00586390
8. A.J. Lotka, *Elements of Physical Biology*. (Williams and Wilkins, 1925)
9. G. Kingsley Zipf, *National unity and disunity; the nation as a bio-social organism.* (Principia Press, 1941)
10. P. Krugman, Confronting the mystery of urban hierarchy. Journal of the Japanese and International economics **10**(4), 399–418 (1996). https://doi.org/10.1006/jjie.1996.0023
11. J.R. Villes, Encyclop é die nouvelle **8**, 670–687 (1841)

12. R. Gibrat, *Les inégalités économiques: applications aux inégalités des richesses, à la concentration des entreprises... d'une loi nouvelle, la loi de l'effet proportionnel*. Libr. du Recueil Sirey, 1931
13. W. Christaller, *Die Zentrale Orte in S ü ddeutschland* (Fischer, Jena, 1933)
14. F. Auerbach, A. Ciccone, The law of population concentration. Environment and Planning B: Urban Analytics and City Science **50**(2), 290–298 (2023). https://doi.org/10.1177/23998083221147139
15. H.W. Singer, The "courbe des populations.âŁž A parallel to Pareto's Law, in The Economic Journal **46**(182), 254–263 (1936). https://doi.org/10.2307/2225228
16. F. Guérin-Pace, Rank-size distribution and the process of urban growth. Urban studies **32**(3), 551–562 (1995). https://doi.org/10.1080/00420989550012960
17. D. Rybski, A. Ciccone, Auerbach, Lotka, and Zipf: pioneers of power-law city-size distributions. Archive for History of Exact Sciences **77**(6), 601–613 (2023). https://doi.org/10.1007/s00407-023-00314-0
18. M. Saibante, La concentrazione della popolazione, in *Metron–Rivista Internszionale di Statistica* 7.2 (1928), pp. 53–99. https://ebiblio.istat.it/digibib/Metron/MetronV7N2_1928.pdf
19. V. Pareto, *Cours d' é conomie politique*. (Rouge, 1896)
20. A.J. Lotka, The frequency distribution of scientific productivity. Journal of the Washington academy of sciences **16**(12), 317–323 (1926)
21. S.C. Bradford, Sources of information on specific subjects. Engineering **137**, 85–86 (1934)
22. G. Kingsley Zipf, *Human behavior and the principle of least effort: an introd. to human ecology*. (Addison-Wesley Press, 1949)
23. Y.-S. Chen, F.F. Leimkuhler, A relationship between Lotka's law, Bradford's law, and Zipf's law. Journal of the American Society for Information Science **37**(5), 307–314 (1986). https://doi.org/10.1002/(SICI)1097-4571(198609)37:5<307::AID-ASI5>3.0.CO;2-8
24. C. Prün, R. Zipf, Biographical notes on G. K. Zipf, in *Glottometrics* 3 (2002), pp. 1–10. https://www.ram-verlag.eu/wp-content/uploads/2018/08/g3zeit.pdf#page=5
25. C. Cottineau et al., The role of analytical models and their circulation in urban studies and policy, in *Urban Studies* (2024), p. 00420980241237410. https://doi.org/10.1177/00420980241237410
26. A. Losch, *Economics of location*. (Yale University Press, 1954)
27. M.-C. Robic, Cent ans avant Christaller... une théorie des lieux centraux, in *L'Espace g é ographique* (1982), pp. 5–12
28. C. Dennison Harris, *Cities of the Soviet Union: studies in their functions, size, density, and growth* (Published for Association of American Geographers by Rand McNally, Chicago, 1970)
29. B.J.L. Berry, W.L. Garrison, Alternate explanations of urban rank-size relationships. Annals of the Association of American Geographers **48**(1), 83–90 (1958). https://doi.org/10.1111/j.1467-8306.1958.tb01559.x
30. H.A. Simon, On a class of skew distribution functions. Biometrika **42**(3/4), 425–440 (1955). https://doi.org/10.2307/2333389
31. U.G. Yule, A mathematical theory of evolution, based on the conclusions of Dr. JC Willis, FRS, in Philosophical Transactions of the Royal Society of London Series B **213**, 21–87 (1925). https://doi.org/10.1098/rstb.1925.0002
32. M. Mitzenmacher, A brief history of generative models for power law and lognormal distributions. Internet mathematics **1**(2), 226–251 (2004). https://doi.org/10.1080/15427951.2004.10129088
33. M.E.J. Newman, Power laws. Pareto distributions and Zipf's law, in Contemporary physics **46**(5), 323–351 (2005). https://doi.org/10.1080/00107510500052444
34. C. Cottineau, What do analyses of city size distributions have in common?" In: *Scientometrics* (2022), pp. 1–25. https://doi.org/10.1007/s11192-021-04256-8
35. X. Gabaix, Zipf's law for cities: an explanation. The Quarterly journal of economics **114**(3), 739–767 (1999). https://doi.org/10.1162/003355399556133
36. B.J.L. Berry, A. Okulicz-Kozaryn, The city size distribution debate: Resolution for US urban regions and megalopolitan areas. Cities **29**, S17–S23 (2012). https://doi.org/10.1016/j.cities.2011.11.007

37. K.T. Rosen, M. Resnick, The size distribution of cities: an examination of the Pareto law and primacy. Journal of urban economics **8**(2), 165–186 (1980). https://doi.org/10.1016/0094-1190(80)90043-1
38. G.R. Allen, The 'courbe des populations', a Further Analysis. Bulletin of the Oxford University Institute of Statistics **16**, 179–189 (1954)
39. F.T. Moore, A note on city size distributions. Economic Development and Cultural Change **7**(4), 465–466 (1959). https://doi.org/10.1086/449818
40. V.R. Popov, Investigation of the System of Urban Places of Crimea Oblast. Soviet Geography **15**(1), 19–23 (1974). https://doi.org/10.1080/00385417.1974.10770642
41. A. Okabe, An expected rank-size rule: A theoretical relationship between the rank size rule and city size distributions. Regional Science and Urban Economics **9**(1), 21–40 (1979). https://doi.org/10.1016/0166-0462(79)90019-X
42. B. Ward, City structure and interdependence. Papers of the Regional Science Association **10**(1), 207–221 (1962). https://doi.org/10.1007/BF01934688
43. M.C. Deurloo, De wet der urbane concentratie. Tijdschrift voor economische en sociale geografie **63**(5), 306–314 (1972). https://doi.org/10.1111/j.1467-9663.1972.tb00557.x
44. K. Tong Soo, Zipf's Law for cities: a cross-country investigation, in *Regional science and urban Economics* 35.3 (2005), pp. 239–263. https://doi.org/10.1016/j.regsciurbeco.2004.04.004
45. D. Pumain et al., Multilevel comparison of large urban systems, in *Cybergeo: European Journal of Geography* 706 (2015). https://doi.org/10.4000/cybergeo.26730
46. R. Cura et al., The old and the new: qualifying city systems in the world with classical models and new data. Geographical Analysis **49**(4), 363–386 (2017). https://doi.org/10.1111/gean.12129
47. D. Pumain, Pour une théorie évolutive des villes, in *L'Espace géographique* (1997), pp. 119–134
48. D. Pumain, Une théorie géographique pour la loi de Zipf. R é gion et d é veloppement **36**(31–54), 125–150 (2012)
49. M. Fujita, P.R. Krugman, A. Venables, *The Spatial Economy: Cities, Regions, and International trade*. (MIT press, 1999)
50. J.B. Parr, A note on the size distribution of cities over time. Journal of Urban Economics **18**(2), 199–212 (1985). https://doi.org/10.1016/0094-1190(85)90017-8
51. R. González-Val, The Evolution of US City Size Distribution from a Long-Term Perspective (1900–2000). Journal of Regional Science **50**(5), 952–972 (2010). https://doi.org/10.1111/j.1467-9787.2010.00685.x
52. V. Nitsch, Zipf zipped. Journal of Urban Economics **57**(1), 86–100 (2005). https://doi.org/10.1016/j.jue.2004.09.002
53. C.C. MetaZipf, A dynamic meta-analysis of city size distributions, in *PloS one* 12.8 (2017), e0183919. https://doi.org/10.1371/journal.pone.0183919
54. L. Suarez-Villa, Rank size distribution, city size hierarchies and the beckmann model: some empirical results. Journal of Regional Science **20**(1), 91–95 (1980). https://doi.org/10.1111/j.1467-9787.1980.tb00628.x
55. M. Cristelli, M. Batty, L. Pietronero, There is more than a power law in Zipf, in *Scientific reports* 2 (2012), p. 812. https://doi.org/10.1038/srep00812
56. P. Veneri, City size distribution across the OECD: Does the definition of cities matter? Computers, Environment and Urban Systems **59**, 86–94 (2016). https://doi.org/10.1016/j.compenvurbsys.2016.05.007
57. F. Nota, S. Song, Further analysis of the Zipf's law: Does the rank-size rule really exist?" In: *Journal of Urban Management* 1.2 (), pp. 19–31. https://doi.org/10.1016/S2226-5856(18)30058-X
58. X. Gabaix, R. Ibragimov, Rank-1/2: a simple way to improve the OLS estimation of tail exponents. Journal of Business & Economic Statistics **29**(1), 24–39 (2011). https://doi.org/10.1198/jbes.2009.06157
59. M.J. Beckmann, City hierarchies and the distribution of city size. Economic Development and Cultural Change **6**(3), 243–248 (1958). https://doi.org/10.1086/449769

60. G.A. Johnson, Rank-size convexity and system integration: a view from archaeology. Economic geography **56**(3), 234–247 (1980). https://doi.org/10.2307/142715
61. Y. Ijiri, H.A. Simon, Interpretations of departures from the Pareto curve firm-size distributions. Journal of Political Economy **82**(2), 315–331 (1974). https://doi.org/10.1086/260194
62. D.R. Vining Jr., Autocorrelated growth rates and the Pareto law: a further analysis. Journal of Political Economy **84**(2), 369–380 (1976). https://doi.org/10.1086/260440
63. F. Moriconi-Ebrard, *L'urbanisation du monde depuis 1950* (Anthropos, Economica, 1993)
64. R.L. Morrill, *The spatial organization of society*. (Duxbury Press, 1970)
65. J. Le Gallo, C. Chasco, Spatial analysis of urban growth in Spain, 1900–2001. Empirical economics **34**(1), 59–80 (2008). https://doi.org/10.1007/s00181-007-0150-5
66. K.-M. Cheng, Y. Zhuang, Spatial Econometric Analysis of the Rank-size Rule for the Urban System: A Case of Prefectural-level Cities in China's Middle Area. Scientia Geographica Sinica **32**(8), 905–912 (2012)
67. J.B. Parr, K. Suzuki, Settlement populations and the lognormal distribution. Urban Studies **10**(3), 335–352 (1973). https://doi.org/10.1080/00420987320080471
68. B.T. Robson, *Urban growth: an approach*. (Methuen, 1979)
69. J. Eeckhout, Gibrat's law for (all) cities. American Economic Review **94**(5), 1429–1451 (2004). https://doi.org/10.1257/0002828043052303
70. M. Barbut, A family of distributions: from Paretian to contra-Paretian. Applications to the study of urban concentration and its evolution, in *Cybergeo, European Journal of Geography* 266 (2004)
71. A. Clauset, C. Rohilla Shalizi, M.E.J. Newman, Power-law distributions in empirical data, in *SIAM review* 51.4 (2009), pp. 661–703. https://doi.org/10.1137/070710111
72. O. Fontanelli et al., Beyond Zipf's law: the Lavalette rank function and its properties, in *PloS one* 11.9 (2016), e0163241. https://doi.org/10.1371/journal.pone.0163241
73. M. Levy, Gibrat's law for (all) cities: Comment. American Economic Review **99**(4), 1672–75 (2009). https://doi.org/10.1257/aer.99.4.1672
74. Y. Malevergne, V. Pisarenko, D. Sornette, Testing the Pareto against the lognormal distributions with the uniformly most powerful unbiased test applied to the distribution of cities, in *Physical Review E* 83.3 (2011), p. 036111. https://doi.org/10.1103/PhysRevE.83.036111
75. R. Bergs, Spatial dependence in the rank-size distribution of cities–weak but not negligible, in *PloS one* 16.2 (2021), e0246796. https://doi.org/10.1371/journal.pone.0246796
76. Y. Ioannides, S. Skouras, US city size distribution: Robustly Pareto, but only in the tail. Journal of Urban Economics **73**(1), 18–29 (2013). https://doi.org/10.1016/j.jue.2012.06.005
77. D. Rybski, A. García Cantú Ros, J.P. Kropp, Distance-weighted city growth, in *Physical Review E* 87.4 (2013), p. 042114. https://doi.org/10.1103/PhysRevE.87.042114
78. C. Schmitt et al., Half a billion simulations: Evolutionary algorithms and distributed computing for calibrating the SimpopLocal geographical model. Environment and Planning B: Planning and Design **42**(2), 300–315 (2015). https://doi.org/10.1068/b130064p
79. Y. Mansury, L. Gulyas, The emergence of Zipf's Law in a system of cities: An agent-based simulation approach. Journal of Economic Dynamics and Control **31**(7), 2438–2460 (2007). https://doi.org/10.1016/j.jedc.2006.08.002

Chapter 2
Urban Scaling Laws

Fabiano L. Ribeiro and Vinicius M. Netto

Abstract Understanding how size influences the internal characteristics of a system is a crucial concern across various fields. Concepts like scale invariance, universalities, and fractals are fundamental to this inquiry and find application in biology, physics, and particularly urbanism. Size profoundly impacts how cities develop and function economically and socially. For example, what are the pros and cons of residing in larger cities? Is life really more expensive or less safe in larger cities? Or do they really offer more opportunities and generally higher incomes than smaller ones? To address such inquiries, we utilize theoretical tools from scaling theory, enabling a quantitative description of how a system's behavior changes across different scales, from micro to macro. Drawing parallels with research in biology and spatial economics, this chapter explores recent discoveries, ongoing progress, and unanswered questions regarding urban scaling.

2.1 Introduction

Understanding how size affects the internal properties of a system is a critical issue in many areas of knowledge. For instance, when a new drug is tested in mice, how is it possible to extrapolate the same properties to the human body, which is three orders of magnitude larger? Or, in the context that we are interested in here, what are the advantages or disadvantages of living in a city larger than another? That is, is living in a bigger city more expensive or dangerous, or does it bring more opportunities for interaction and materializing ideas or even achieving a higher income than living in a smaller one? How can we know if a city of a given population needs one more hospital or a petrol station? To answer these questions, we need to use a set

F. L. Ribeiro (✉)
Department of Physics, Universidade Federal de Lavras (UFLA), Lavras, Brazil
e-mail: fribeiro@ufla.br

V. M. Netto
Research Centre for Territory, Transports and Environment (CITTA), Faculty of Engineering (FEUP), University of Porto, Porto, Portugal
e-mail: vmnetto@fe.up.pt

D. Rybski (ed.), *Compendium of Urban Complexity*, Understanding Complex Systems,
https://doi.org/10.1007/978-3-031-82666-5_2

of theoretical tools known as *scaling theory* [1–3], which allows for a quantitative description of how a system changes its behavior from one scale to another, or, more precisely, from a micro-scale to a macro-scale. Scaling invariance, normalization group, critical phenomena, universalities, and fractals are examples (not exhaustive) of concepts that form the core of this theory and they find applicability in fields such as biology, physics, engineering, and, for this chapter, urbanism.

A naive approach to scaling from one system to another is performing a simple linear data extrapolation. For example, if we effectively use 1 ml of a drug in a mouse weighing 40 g, then we might assume that we need to use 1 L of the same drug for a human weighing 40 kg (1,000 times bigger). Similarly, if a city of 100,000 inhabitants works well with 10 petrol stations, then we might assume that another city with 1,000,000 inhabitants will work well with 100 petrol stations. However, this simplistic linear extrapolation of the data does not work in these contexts. Biological and urban systems do not scale in a purely multiplicative way, meaning that a simple multiplication of the results obtained by the prototype (a mouse or a small city) will not allow us to obtain the desired effect in another scale of interest (a human body or a large city). In other words, both biological entities and cities are nonlinear scaling systems, and this is what we aim to explore in this chapter.

Many systems that present such non-linearity are (usually) governed by a power-law of the form

$$Y = Y_0 N^{\beta}\,, \tag{2.1}$$

where N is the metric that represents the size of the system, Y is a dependent variable, Y_0 is a constant, namely the intercept, and β is the *scaling exponent*. In the case of biology, this equation represents well how the metabolic rate, which is the amount of energy expended by an organism per time, relates to its body mass [6, 7], as shown in Fig. 2.1. The straight line in a log-log plot is the signature of a system well described by a power-law equation like Eq. (2.1), and this is the case for mass and metabolic rate in biology, as evidenced in this figure. For vascular beings larger than $\sim 10^{-4}$ g, as poikilotherms (animals that vary their internal body temperature) and homeotherms (animals that maintain a constant internal body temperature), we have the so-called *Kleiber's law*, that is the particular case of Eq. (2.1) with $\beta = 3/4$ [8]. The fact that β is different from 1 (less than 1 in this case) means that the metabolic rate scales non-linearly (sub-linearly in this case) with the organism's size. In addition to the metabolic rate, other quantities also scale non-linearly with the size of the organism, such as heart rate and brain size.

In the case of a city, we have a similar situation when we analyze many urban metrics as a function of the city population size. For instance, Fig. 2.2 shows that Eq. (2.1) describes the empirical data well (straight line in the log-log plot) when Y represents the urban gross domestic product (GDP) and N represents the city population size of cities in Brazil and the United States [9–11]. Despite the fact that the two sets of data are vertically different, characterized by a different intercept due to economic differences between these two countries, the slopes of the straight lines that characterize the data are approximately the same.

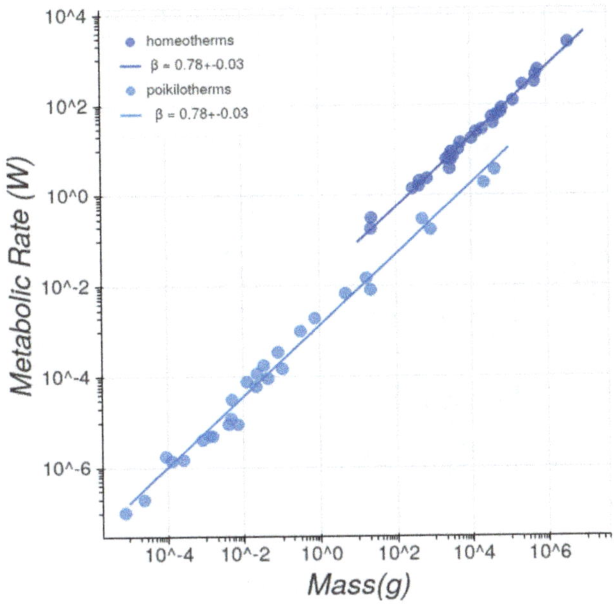

Fig. 2.1 Metabolic rate as a function of body mass for two vascular taxonomic groups: homeotherms (organisms that maintain a constant internal body temperature) and poikilotherms (organisms that vary their internal body temperature). A power-law of the type given by Eq. (2.1) is observed for all taxonomic groups (the straight lines capture the trend of points). The parameter Y_0 (the intercept) varies from group to group, but β is approximately constant and sub-linear ($\beta < 1$). The data were extracted directly from the references [4] (homeotherms) and [5] (homeotherms and poikilotherms). Source: modified after [6, Fig.1]

These empirical facts suggest that, despite the remarkable complexity underlying biological and urban phenomena, the relationships between metabolic rate and body mass, and between GDP and city population size seem to obey a relatively simple relationship, given by the power-law Eq. (2.1).

A power-law equation has some interesting properties, but maybe the most interesting for our purpose here is that it is scale-invariant if β is fixed. Let us explain this in more detail. Suppose the size metric changes from N to λN, where λ is an amplification factor. Following Eq. (2.1), one gets

$$Y(\lambda N) = Y_0(\lambda N)^\beta = Y_0' N^\beta \,, \tag{2.2}$$

where $Y_0' \equiv Y_0 \lambda^\beta$ is another constant. Consequently, $Y(\lambda N) \propto Y(N)$ (from Eqs. (2.1) and (2.2)), which means that the behavior of the system does not change when the scale changes. This is the *scale invariance*. In other words, one can say that systems that are governed by a power-law, such as biological entities or cities, appear to behave the same way regardless of the scale we use to look at it [13]. In short,

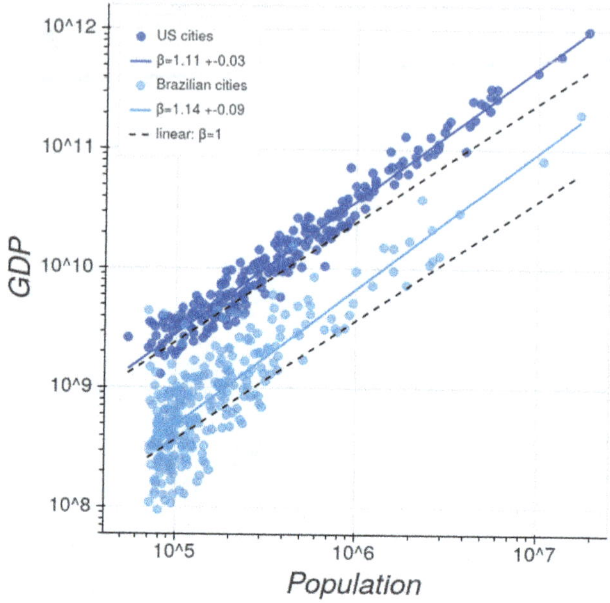

Fig. 2.2 Example of urban scaling. The urban gross domestic product (GDP) is plotted as a function of population N for cities in Brazil and the United States (on a double logarithmic scale). The solid lines (dark and light blue) are the best fits that capture the trend of the points and reveal the power-law property. Both countries exhibit scaling exponents $\beta > 1$ (super-linear), with very similar values despite the socio-economic differences between the two countries, which justifies the difference in the vertical position of the straight lines (different intercepts Y_0). The dashed lines indicate linearity ($\beta = 1$) and are included only to highlight the super-linear behavior of the data trend. Source: modified after [12, Fig.1]

these systems do not have a natural scale and, because of this, they are also called scale-invariant systems [14, Chap. 4].

To illustrate the concept of scale-invariance, let us present one counterexample and one example where scale-invariance hold true. A person who is 2.5 m tall (the height of the largest living human being) is undoubtedly an exceptional case, as the distribution of people's heights follows a normal distribution, which has a typical size or scale [13]. Therefore, the distribution of people's heights is not scale-invariant. However, the distribution of city sizes is governed by a power-law equation (the so-called *Zipf's law*). Consequently, the city-size distribution is scale-invariant, implying that cities like New York or São Paulo are not exceptional in size. Despite being the largest cities in their respective countries, these cities have the expected sizes given the system of cities to which they belong [15].

2.2 History

The attempt to explain how the properties of things change when their size increases is not new. For instance, Galilei studied this question and discussed how the geometry of things limits the existence of arbitrarily large living beings [3]. However, it was only in the 19th century that a systematic way to quantify scaling was developed, particularly in biology. In fact, a power-law of the form Eq. (2.1) was first observed in biology in 1839 by Saurus and Rameaux [16]. They noticed that the metabolic rate per unit weight decreases with increasing animal size. At the end of the 19th century, Max Rubner [17, 18] studied dogs and found that the energy production per square meter of the body surface is constant with the animal's size, leading to a power-law relationship between energy and body mass with $\beta = 2/3$. Based on this observation, he postulated that living organisms evolved, by natural selection, to a state in which body mass should follow a surface scaling law and thus be able to radiate excess heat. He proposed that the numerical value of the *allometric (scaling) exponent* would be a natural response to the release of heat by the organism, establishing the relationship between the surface area and volume of the organism. This idea led to a theoretical exponent $\beta = 2/3$. Rubner's theory, known as the *surface hypothesis*, seemed to be reasonably coherent and was accepted for 50 years. However, more careful experiments were performed at the beginning of the 20th century. Among these studies, we highlight the works of Krogh [19] and Kleiber in 1932 (the best-known study) [20]. From the analyzed data set, an experimental value of $\beta \approx 3/4$ was observed, which differed from the theoretical result that was accepted until then. This result is now known as the $3/4$ *Kleiber's law*. The sub-linear behavior of the scaling exponent ($\beta < 1$) implies that larger animals demand less energy per cell, indicating a "scaling economy". Nowadays, we know that the 3/4 Kleiber's law applies only to vascular beings and not to prokaryotes and protists organisms, which present super-linear and linear scaling properties, respectively [8].

After an extensive debate on the precise value of the scaling exponent in biology, fundamental work was done in the late 1990s by the theoretical physicist Geoffrey West, together with the biologists James Brown and Brian Enquist. They proposed a model based on the efficiency of nutrient distribution within organisms to explain the scaling law in biology. The theory is based on three premises: (i) a fractal distribution network, (ii) terminal units that do not vary with organism size, and (iii) natural selection, that favors a distribution network that minimizes energy waste. This model yields a scaling exponent $\beta = 3/4$ [6, 7, 21–23]. However, given its own premises, this theory only applies to organisms with closed circulatory systems. What is curious about all these ideas is that while Galileo explained the impossibility of arbitrarily large animals using simple geometry, a broader understanding of scaling laws involves more complex concepts, such as networks and fractals.

Continuing with the historical development of scaling understanding, in 1932 the linguist George Zipf realized when analyzing the frequency of word use in texts that a relatively small set of words are used many times, while the vast majority of words are rarely used [24]. By quantifying the frequency of word use, he noticed that the most used word in a text, book, or newspaper was, on average, used twice as much as the second most used word and three times as much as the third most used word, and so on. Zipf called this pattern the *Rank versus Frequency rule*, but today it is simply known as *Zipf's law*. Zipf (1949) [25] and (previously) Auerbach (1913) [26, 27] also realized that the same properties are observed in city size distributions. The most populous city in a country is typically twice as large as the second most populous city, and so on, analogously to the frequency of words. Today, we know that these ranking rules apply to many other contexts, and because of this, power-law distribution with exponent -1 are often called *Zipf's law* [13, 28] (for more on city size distributions, see Chap. 1).

Scaling relationships do not only apply to the distribution of a single variable, the population size, but also encompass the way other urban and socio-economic variables respond to changes in size [29]. An example of this can be seen in the connection between the concepts of *increasing returns to scale* and *agglomeration effects* in spatial economics and urban theory (more details in Chap. 3). Increasing returns are characterized by the super-linear scaling of specific socio-economic variables. Their connection to agglomeration effects can be traced back to the work of economist Alfred Marshall at the end of the 19th century [30]. Facilitated by larger size and greater population density, the clustering of economic activity in cities has been associated with the advantages of close proximity between economic agents in the form of access to a larger pool of labor, wider ranges of suppliers in intermediary exchanges, and proximity to local markets, leading to shared infrastructure, and lower transport and communication costs [31]. The advantages of agglomeration can intensify with the concentration of companies of the same industry, leading to regional or urban specialization, including labor. This specialization can also foster knowledge spillovers *within* an industry, known as Marshall-Arrow-Romer externalities or simply *Marshall's scale economies* [30], promoting innovation and growth in productivity and employment.

In addition to specialization, scale in urban agglomeration can also lead to industrial *diversity* (*Jacobs economies* [32, 33]). In larger cities, the concentration of economic activity can create a diverse mix of industries. Communication becomes more extensive with decreasing distances [34]. Combined, economic diversity and proximity encourage cross-fertilization of ideas and technologies through knowledge spillovers *between* industries [33, 35]. This setting can lead to a virtuous process where innovation and productivity increase, attracting more businesses to the city, further diversifying the economy, and stimulating economic resilience, like the ability to recover from economic shocks [36]. For example, doubling the size of a city by grouping different industries could increase productivity by 3 % to 8 % [37–39]. Urban areas with greater industrial diversity experience higher income growth. More broadly, evidence of the relationship between urban scaling and increasing returns can be found in productivity [40, 41], employment, and income [42]. In turn, cities can benefit from agglomeration economies in the provision of public goods and services, such as transportation, education, and healthcare. Due to their greater population density and scale, larger cities are often able to provide these services more efficiently and at a lower cost per capita than smaller cities or rural areas.

In recent years, we have seen a larger trans-disciplinary effort from the scientific community to understand urban scaling properties. Apart from work on increasing returns such as agglomeration and diversity effects, particularly since the 1960s in urban theory [32] and spatial economics [33] and strongly supported by data since the 1990s [35, 39], more recent contributions have emerged from geography and physics. This field has gained ground since the turn of the 21st century, thanks mainly to the increase in computational capacity, which allowed an unprecedented acquisition of urban data. It is in this context that more robust interpretations of these laws emerge, mainly driven by the work of Denise Pumain [43], Luis Bettencourt, Geoffrey West [9, 10] and many others [12, e.g.] who, based on extensive empirical evidence, were able to formulate theories to explain how urban metrics scale with population size. Today, urban scaling is one of the key areas in the so-called new science of cities [29, 44]. The following sections will address more details about these theories and associated empirical evidence.

2.3 Empirics

Empirical evidence suggests that the scaling exponent β depends strongly on the type/category of the urban variable. The seminal work of Bettencourt and colleagues [9] proposed three categories/families of urban variables related to scaling: (i) socio-economic; (ii) structure and infrastructure; and (iii) individual needs. Figure 2.3 shows the scaling exponent distribution for these three categories when a sample coming from 70 papers is considered [45].

The first category of variables is associated with socio-economic variables, such as communicable diseases and economic entities or activities (including GDP and wages), in which the scaling exponent is typically distributed around $\beta \approx 1.15$. The fact that this exponent is greater than 1 (the super-linear regime) means that when city size increases, the per-capita number of such variables also increases, as shown in Fig. 2.4 left. In simple words, *larger cities are richer*. The second category of variables is associated with urban structures, such as the total length of streets and the total area covered by streets (fundamental allometry), and with infrastructures such as the total length of electrical cables or the number of gas stations, in which the scaling exponent is distributed typically around $\beta \approx 0.85$. The sub-linear scaling exponent means that when city size increases, the per-capita number of these structural and infrastructural quantities decreases, as shown in Fig. 2.4 right. This implies a scaling economy, which means that *larger cities do more with less*. A third category of variables is associated with individual needs, such as water and electricity consumption, the number of jobs, and the number of rented houses, in which the scaling exponent is distributed typically around $\beta \approx 1$, indicating a linear relationship between these variables and the city population size. In the following subsections, we present more details about the scaling behavior of each of these categories.

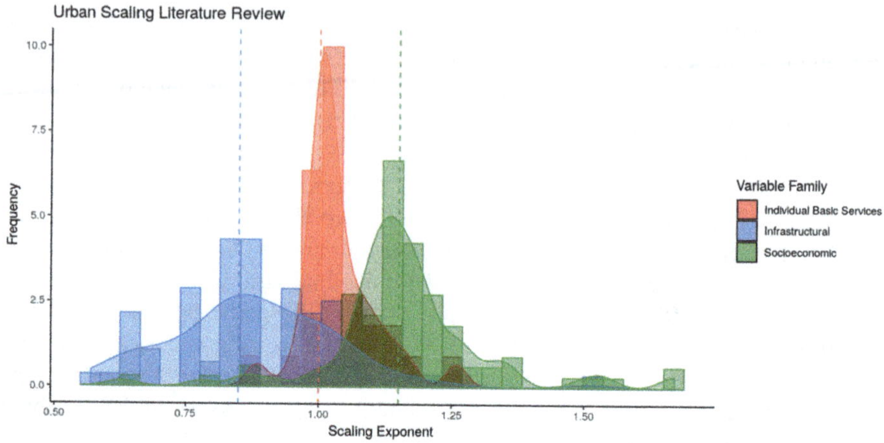

Fig. 2.3 Distribution of values of the scaling exponent β for three different categories/families of urban variables. The samples are composed of data extracted from 70 papers. In total, more than 550 data points were extracted for a plethora of variables. From these data, it is possible to see how variables of each type are distributed in the sub-linear, linear, and super-linear regimes. The vertical dashed lines represent $\beta = 0.85$ (blue), $= 1$ (red), and $= 1.15$ (green). Source: [45]

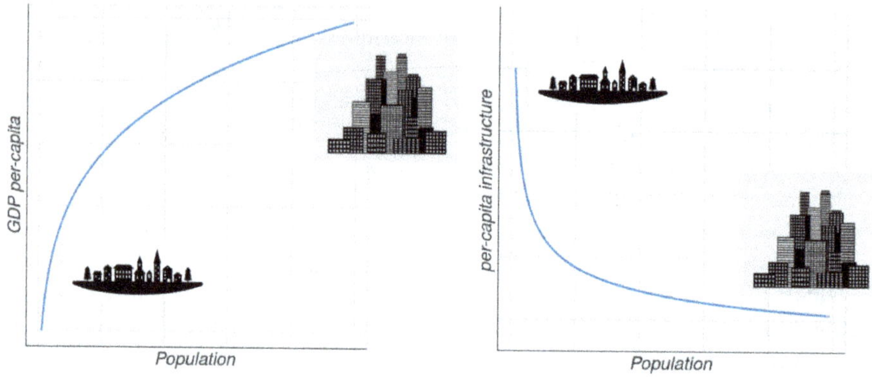

Fig. 2.4 Left: typical GDP per capita (Y/N) as a function of population size (N). The monotonic increasing curve (due to the super-linear scaling of Y with N) implies that larger cities have more GDP per inhabitant. In simple words, larger cities are richer. Right: typical infrastructure per capita (Y/N) as a function of population size (N). The monotonic decreasing curve (due to sub-linear scaling of Y with N) implies that larger cities require less infrastructure per inhabitant. In simple words, larger cities do more with less

2.3.1 *Socio-Economic Variables*

The finding $\beta > 1$ for socio-economic variables means that when we increase the size of a city, the GDP or number of interactions increases more than proportionally. That is the so-called *increasing returns to scale* in larger cities [10, 11, 46]. To understand this, consider the following example. Suppose that a specific city increases its population by 1 %, then if $\beta = 1.15$, the GDP of the same city will increase by 1.15 %. That is because the derivative of Eq. (2.1) yields $dY = Y_0 \beta N^{\beta-1} dN = Y_0 N^{\beta} \beta dN/N$, and consequently $dY/Y = \beta dN/N$. That means if $\beta \approx 1.15$, the increment in the urban metric (dY/Y) is approximately 1.15 times bigger than the population increment (dN/N). But please note that this numeric example only works for small population increments.

The increasing returns to scale observed can be attributed to the interplay between properties of the urban structure and human interactions–the key variable driving socio-economic development. Empirical evidence suggests that social interactions grow super-linearly with population size, as evidenced by analyses of phone call data and measures of interpersonal contacts [11, 13, 47]. This increased interactivity is thought to be a catalyst for creativity and productivity, resulting in social and economic benefits [33]. Therefore, larger cities with larger populations are likely to provide more opportunities for interactions and contacts between people, leading to wealthier and more creative cities [10, 12, 48, 49]. Section 2.4 presents more details about this discussion.

However, the increased connectivity between people in larger cities might have a downside. Large cities are more susceptible to epidemics and communicable diseases, such as AIDS [9] and COVID-19[50], which increase super-linearly with the size of a city. Alternatively, to the increasing return to scale, the super-linearity of socio-economic variables can also be a consequence of the imbalance between larger and smaller cities within a specific system of cities (e.g., a country). In other words, super-linearity can also represent a situation in which wealth is disproportionately concentrated in larger cities, as observed in some countries [51, 52].

2.3.2 Structure and Infrastructure Variables

Regarding structure and infrastructure variables, the scaling exponent is approximately $\beta = 0.85$, as depicted in Fig. 2.3. Additionally, the scaling relationship between area and population, known as *fundamental allometry*, varies across different countries but consistently remains below 1 [53], indicating a sub-linear regime. In analogy to the increasing returns, $\beta = 0.85$ implies that a small increment in city size results in a per-capita infra/structure reduction of around 15 %. For example, when the size of a city increases, the number of gas stations per inhabitant tends to decrease (see Fig. 2.4 right). The sub-linear scaling of this kind of variable is usually explained as a consequence of social network properties and spatial constraints, acting in a bottom-up generation process [11, 12].

Interestingly, the scaling of urban infra/structure has a qualitative analogy with scaling in biology. While there is a lower per-capita demand for amenities and services in larger cities, there is also a lower energy-per-cell demand in large animals. West and his collaborators [7, 21, 22] justify this analogy by arguing that both living organisms and cities are composed of self-similar and fractal resource distribution networks. In both systems, there is an underlying network structure (for transport, supply, etc.) that mediates the interactions among cells or among individuals [10].

However, the quantitative analysis differs between cities and biology. While in cities, the scaling exponent is around $\beta = 0.85$–between infrastructure and population size -, in the case of biology, the exponent is around 0.75–between energy expended/demanded and animal mass. This means biology is more efficient than cities: while a slight increase in city size leads to a 15 % reduction of infra/structure per capita, a slight increase in a living organism leads to a reduction of 25 % of energy demanded per cell. We could think that biology is more efficient because it has a much longer evolutionary history than cities. It would follow that cities would evolve towards a better economy of scale in the coming centuries or millennia. However, some empirical studies show that the urban scale exponent might remain constant over decades [54, 55], and would not be actually evolving. In this sense, perhaps the explanation for this quantitative difference between biology and cities is that they are fundamentally different systems despite presenting some qualitative similarities. For instance, while in biology individuals are constantly eliminated by natural selection, cities are comprised of highly durable structures that, once built, tend to last for decades or centuries, being changed at significant human and material cost [56, 57].

2.3.3 *Individual Needs Variables*

The third type of variable scales linearly with population size ($\beta \approx 1$). These include urban variables related to individual needs, such as water consumption, number of jobs, and number of rented houses. The linearity of these variables means that the size of the city does not affect the per-capita values of these quantities. In other words, it does not matter if a person lives in a small town or in a big city; the amount of water each individual consumes, on average, is the same. Consequently, the total water consumption in the city, as a whole, grows proportionally to the increase in the number of inhabitants. The same situation holds for other individual needs variables, as evidenced by many empirical pieces [9].

2.3.4 *Deviations from this Categorization*

In addition to the success of this categorization, which has opened doors for a new understanding of urban mechanisms and the application of quantitative methods, it is interesting to mention some works that have questioned the limits of this categorization or even proposed alternative classes.

For example, Strano et al. [51] demonstrated that low-income European cities exhibit superlinear scaling between *gross metropolitan product* and population, while high-income cities show linear scaling. Additionally, Meirelles et al. [46] identified a fourth category of variables associated with structures and infrastructures that depend strongly on top-down mechanisms under planning, such as public investments or political and economic incentives. In this case, structure and infrastructure variables that would be expected to have a sub-linear scaling exponent, according to Bettencourt et al.'s categorization, are found to have a super-linear scaling exponent due to the action of external factors. This is the case, for instance, of sewage treatment systems and health facilities in Brazil.

The lack of a widely adopted definition for city boundaries is another factor that interferes not only with the predicted values of the scaling exponent but also can lead to transitions between different scaling regimes, such as from a super-linear to a sub-linear regime or vice versa [58, 59]. More detailed discussion on this topic can be found in Sect. 2.5, which delves into criticisms of urban scaling.

2.3.5 Universalities

Recent works have supported the robustness of the quantitative urban scaling exponent estimates (e.g. [54]). It has been observed not only in different countries, but also for different historical periods [60]. These pieces of evidence suggest that the robustness of the urban scaling value could be, in fact, the manifestation of some kind of universality. That is, there may be fundamental principles that govern the growth and organization of cities that are independent of specific cultural, historical, and geographic factors. Optimistically thinking, if this universal law is true, then more comprehensive theories can be identified and proposed to explain urban phenomena. Understanding these principles and the underlying processes could have relevant implications for urban planning, sustainability, and the development of cities around the world.

However, there is also evidence against the idea of universality. For example, the ones cited in the last subsection, among other evidence [61, 62]. After all, the identification and validation of such dynamics, whether universal or not, should help decisively in the formulation of urban policies in order to identify opportunities and improve the quality of life of citizens. Therefore, further research is needed to clarify the possible universality of urban scaling and its implications for urban planning and policy-making.

2.4 Possible Explanations

Why does urban scaling emerge at all? This question is at the core of an important debate in the recent literature. While several models attempt to explain this emergence, a recent review [12] highlights the lack of consensus on the fundamental process behind the phenomenon of urban scaling. While some models have succeeded in explaining specific aspects of cities, a more fundamental and universal explanation, that is, a *urban scaling theory*, has yet to be developed. So far, we have a collection of ideas and insights that, at best, explain the scaling of specific urban sectors.

Regarding the fundamental allometry, which defines the sub-linear scaling relationship between cities' area and population, the Bettencourt model [10] offers an explanation. It posits that individual socioeconomic production should be sufficient to cover transportation costs, which guarantees accessibility and ensures full integration of the city. This idea conducts that the fundamental allometry scaling exponent, say β_{FA}, scales sublinearly as $\beta_{FA} = D_f/(D_f + 1)$, where D_f is the fractal dimension of the city's built-up area. Another model that addresses fundamental allometry is the one proposed by Louf and Barthelemy [63, 64]. They propose that congestion resilience might contribute to the sublinearity between area and population size. This idea yields results similar to the Bettencourt model despite having different underlying premises. Further details on this discussion can be found in [12].

The underlying principle of urban structures and infrastructures scaling usually stems from network properties. For instance, Bettencourt explains its sublinearity by invoking hierarchical network properties, akin to West, Brown, and Enquist's scaling theory for biological systems (see Sect. 2.2). Molinero & Thurner [65] add that geometrical considerations are a primary factor influencing urban scaling. They argue that the infra/structure scaling exponent, namely β_{infra}, arises from the interplay between two fractal structures: one formed by the spatial distribution of people, characterized by fractal dimension D_P, and the other formed by the street network, characterized by fractal dimension D_{infra}. They show this exponent is determined by the ratio between the fractal dimensions of these two structures, that is $\beta_{\text{infra}} = D_{\text{infra}}/D_P$. Since $D_P \geq D_{\text{infra}}$ (due to people being embedded in 3D space while streets are in 2D space), the emergence of sub-linearity occurs due to the difference in fractal dimensions of these structures. Alternatively, Ribeiro et al. [11] concentrate on the number of amenities required in a city to fulfill the needs of its inhabitants. They propose a gravity interaction model between consumers and firms based on distance, characterized by a decay exponent γ. This approach was motivated by the tendency of citizens to choose nearby locations for purchasing goods, resulting in $\beta_{\text{infra}} = \gamma/D_P$. This means that urban scaling could be explained by the relation between the accessibility (given by the decay exponent γ) and the geometric/spatial population distribution properties (characterized by D_P). As a smaller value of γ indicates a greater range that people are willing to travel, this result suggests that cities with better mobility promote larger infrastructure scale economies.

The diversity of ideas presented in the last paragraph makes it clear that there is no consensus about the primary mechanisms that lead to fundamental allometry and infrastructure scaling. However, within the context of socioeconomic variables, it is nearly unanimous in the scientific community that the driving force behind the emergence of super-linear scaling is the *intensity and frequency of human interactions*. According to empirical findings, the number of contacts people have–for instance, measured by the number of mobile phone contacts, which grows super-linearly with the city population size [47]. Then, hypothetically, a series of other urban metrics, such as GDP, number of patents, or cases of socially transmissible diseases, also scale super-linearly. The larger the city, the greater the density and, consequently, the more opportunities for people to interact. With a larger number of interactions, more ideas and wealth are created, in the sense that connectivity is the motor that generates increasing returns to scale [33, 48]. Within this context, it is proposed in [12] a general framework to connect the number of interactions and socio-economic urban scaling, as summarized in the scheme presented in Fig. 2.5. This scheme yields a general equation

$$Y \sim N^2 n_c , \tag{2.3}$$

when it is assumed that the socioeconomic output from one interaction, namely g, is scale-independent. Here, n_c is the density of contacts, that is, the ratio between the average number of contacts (friends, clients, etc..) of a single person, namely $\langle k_i \rangle$, and the city population size; that is $n_c = \langle k_i \rangle / N$. In [12], it is shown that the general Eq. (2.3) embraces many mathematical models in the literature, and these models differentiate each other only in the way they propose to estimate or determine the density of contacts n_c.

For instance, the Bettencourt model [10] employs cross-sectional ideas to infer human interaction. This model posits that the density of contacts is given by the ratio between the area that a single person effectively uses, say a, and the total

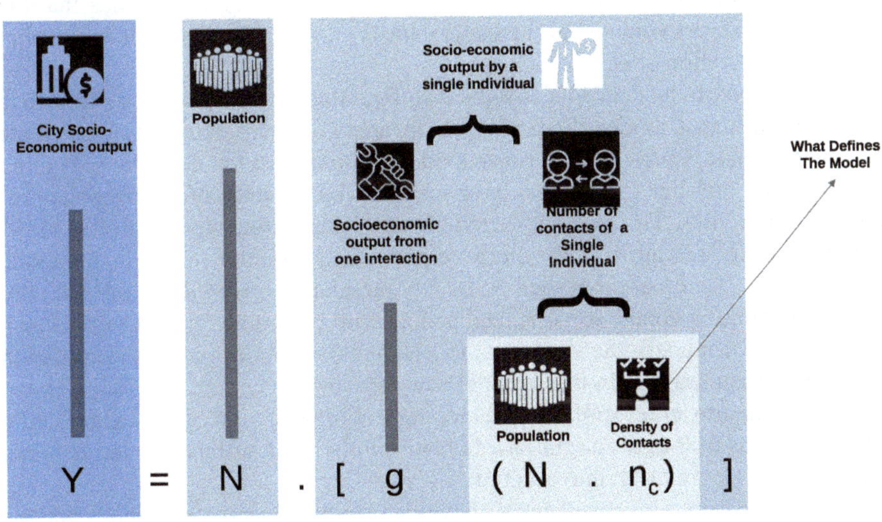

Fig. 2.5 Diagram illustrating the general framework for understanding how the number of contacts relates to the socio-economic output Y (e.g., GDP). The idea is that Y can be understood as the sum of all individual socio-economic outputs of the city, that is, $Y = Ny$, where y is the (average) socioeconomic output by a single individual. In turn, y is the result of the socioeconomic output generated by the interaction between a typical individual and his/her contacts. That is, y is given by the average number of contacts multiplied by the socio-economic output of the interaction with one single contact; that is, $y = \langle k_i \rangle g$. This idea yields the general relation $Y = gN^2 n_c$, where $n_c = \langle k_i \rangle / N$, is the density of contacts. In fact, the density of contacts n_c is what defines some particular model in the literature. Source: modified after [12, Fig.4]

infrastructured area of the city, say A_n. More especifically, $n_c = a/A_n$. Then if a is scale-invariant and $A_n \sim N^{\beta_{\text{infra}}}$ (scaling of the infrastructure, as explained before), then $Y \sim N^{2-\beta_{\text{infra}}}$, and consequently the socio-economic scaling exponent, namely β_{se}, is determined by $\beta_{\text{se}} = 2 - \beta_{\text{infra}}$.

Another form to infer the density of contacts is to use the gravity idea, that is, considering that the probability of two-person interactions decays with the distance as $p(r) \sim 1/r^{\gamma}$, as suggested in [11]. The decay exponent γ controls the interaction range (as mentioned before). What emerges from this model is that the density of contacts scales as $n_c \sim N^{-\gamma/D_P}$ and consequently, from the general Eq. (2.3), one has $Y \sim N^{2-\gamma/D_P}$. That is, this idea leads to $\beta_{\text{se}} = 2 - \gamma/D_P$, which means that, once again, the gravity model explains urban scaling by the relation between accessibility and population spatial distribution. Note that the superlinearity ($\beta_{\text{se}} > 1$) only happens if γ is sufficiently small (in fact, smaller than D_P), which implies that the superlinearity is a consequence of the range of interaction of the people being sufficiently large. That is, increasing return to scaling is a consequence of a fully integrated city.

What is remarkable is that, despite the Bettencourt and gravity models being fundamentally different, both suggest that the integrity of the city fosters increasing returns to scaling and infrastructure economies. The convergence of different models to the same conclusion, namely the importance of city integrity, implies that the integration of diverse individuals across distances may be a fundamental component of the agglomeration effect.

Indeed, beyond the concepts discussed so far, alternative explanations for urban scaling exist. In fact, as identified in [12], there are two groups of models to elucidate this phenomenon: intra- and inter-city models. In intra-city models, the prevailing notion is that non-linearity primarily arises as a consequence of endogenous processes within cities. The examples cited above illustrate this type of model. However, it seems odd to attribute the entire responsibility for inferring socio-economic output solely to internal properties of the city population, given that cities are constantly exchanging goods, services, and people with each other. In this sense, some authors propose models–the inter-city models–that consider exogenous information to explain urban scaling. In these models, factors such as commuting [66, 67], the influence of nearby cities [68], and the hierarchical organization expressed by Zipf's law [52, 69], as well as the migration of creative individuals attracted to larger urban centers [70, 71], are taken into account.

Of course, works in urban theory have dealt with aspects of people's actions and performance in connection with urban structures since Alexander's pioneering works in the 1960s [56, 72], a tradition unfolded into approaches to street networks in Hillier's [73] space syntax, paralleled by efforts in urban morphology stemming from Jacobs' urban vitality theory [32] and Martin and March's [74] work on built form, and by spatial interaction modeling by Batty [75] and others. Indeed, the spatial dimension of cities as a factor with explanatory power over social processes lies at the heart of urban theory. However, the explicit connection between those spatially rich structural views of cities with scaling effects is yet to be fully pursued.

In summary, urban scaling is a complex phenomenon resulting from the interplay of factors such as population, density, hierarchical organization, and geometry (expressed by network structure, spatial distributions, topography, physical barriers, and fractality) that can either enhance or hinder interactions between people. From a more simplistic perspective, we can say that the spatial structure of a city plays a crucial role in determining the nature and extent of human interactions, which in turn shapes the scaling of various urban metrics. However, there is still much more to be understood.

2.5 Criticism

While the empirical findings described in the preceding sections support the notion that urban metrics are related non-linearly to city population size, there are still counterexamples or contradictions that require further investigation and resolution.

An important issue of urban scaling, as expressed by the power-law given in Eq. (2.1), is that it implicitly assumes that urban metrics are dominated by city population size and does not account for the relationships between cities. Cities are complex open systems governed by a virtually infinite number of economic, social, and physical processes. From an optimistic point of view, one could argue that the power-law Eq. (2.1) is a successful first-order approximation and that the interference of exogenous processes is a second-order negligible perturbation. Under this perspective, one could contend that all city complexities vanishes, leaving only one piece of information to infer the urban metric measure, i.e. the city population size N. However, from a critical point of view, it is challenging to accept that a city's performance and economy would depend solely on a single variable in an increasingly interconnected contemporary world.

Firstly, the very nature of this variable, i.e. the idea of using population size as a proxy for city size, can be seen as a problem (see [59, 65, 76, 77]). The representation of cities as zero-dimensional objects–the absence of the spatial dimension of cities–in data-driven research ignores two facts: that there are infinite shapes and configurations to accommodate the same population; and that such configurations or structures matter. The performance of cities can vary significantly based on their internal structures, even if they share the same population and geographic size (see Chap. 12).

A step further, cities are social products. Like other complex social phenomena, they resist calls to universality and other possible temptations emanating from the field of social physics. Cities are highly subject to cultural and regional variations that enormously enrich them as artifacts and urban life as collective experiences. So far, the urban scaling approach has been mostly indifferent to such diversity. Of course, one could argue that this is precisely the point. However, social and spatial differences may have causal effects. For instance, different urban structures and the cultural spatial information encoded in them [78] might help explain why populations in certain world regions achieve larger economic outputs, illustrated in Fig. 2.2, or are better at spatial navigation [79].

A recurrent and more pragmatic critical point is the dependence of the estimated scaling exponent value on the definition of a city [76, 80]. Although we can easily identify a city by sight, it is not clear where its boundaries begin and end. Recent works have demonstrated that the definition of the urban boundary affects scaling exponents, and in some cases, their values can change the regime/classification that a given metric below [58, 59]. As Cottineau et al. [58] reported, "different scaling regimes can be encountered for the same territory, time, and attribute, depending on the criteria used to delineate cities". A striking example of this inconsistency is the analysis of carbon emissions [81]. When *urban areas* are used, a super-linear regime is obtained, indicating that *small cities are greener*. However, for *combined statistical areas*, a sub-linear regime is obtained, indicating the opposite, that *larger cities are greener*. Besides the city boundaries definition, the selection of statistical methods used in the analysis can also impact the estimate of the scaling exponent, as highlighted in a study by Leit a o et al. [82]. Therefore, further research is needed to shed light on this aspect. For a more complete discussion of the deficiencies and criticisms of urban scaling laws, see [80].

In short, it is fair to say that there are many factors beyond population size that can promote gains in socio-economic performance or optimization in urban structures and infrastructures, as seen in decades of research in urban theory (e.g. [44, 73, 74]), and in the increasing returns literature in spatial economics (e.g. [30, 33, 35], to mention a few).

2.6 Open Issues/Questions

The apparent universality in some aspects of urban phenomena, particularly with regards to the scaling hypothesis discussed here, paves the way for the creation of a *Unified Urban Theory* (UUT) [83–85]. The UUT would be a quantitative theory that combines a few premises and derives many different observed urban patterns as particular cases [86]. The possibility of achieving such a feat indicates that, with the advent of big data and high computational performance, urban science is gaining the status of a "quantitative science". If UUT can indeed be developed, it would provide a set of systematic approaches to explaining and predicting urban features, benefiting planners and public administrators. As we argued above, the crucial challenge here lies in the very nature of cities. Like other social phenomena, cities have substantial cultural differences, both physical and symbolic, full of causal potential, which easily evades general theories.

Another issue is whether scaling laws in urban phenomena are analogous to those in biological systems. For instance, are the scaling laws in biology and urban scaling laws manifestations of the same physical principle? If so, what principle is this? If not, what differentiates these two systems? And how could we demonstrate it (or not) in a technical and quantified manner without relying on mere speculation? Questions like these may well guide future investigations. For instance, scaling in biology appears to be analogous to urban structure and infrastructure scaling, as both present sub-linear scaling. However, their scaling exponent values differ quantitatively. While data and theory on vascular organisms suggest that $\beta \approx 0.75$ (see Secs. 2.1 and 2.2), in the case of urban infrastructure, we have $\beta \approx 0.85$. This quantitative difference implies that biology is more efficient in terms of resource economy. But why is there such a quantitative difference? One naive answer could be that biology had billions of years to evolve while cities only had a few thousand years. Based on this statement, one could argue that cities still need to evolve in order to be as scaling-efficient as biology. However, the data also suggest some kind of stability of the urban scaling exponent when the time evolution of urban systems is analyzed, suggesting that urban systems are already at the equilibrium scaling stage. This shows that the analogy between biology and cities is still a very open point that must be elucidated.

Finally, in response to the critical issues raised above, we also think that the question of scaling properties and effects must be connected with the fundamental problem of how spatial structures become part of the socio-economic dynamics cities express and support. A key question here is how such structures might interfere with the role of population size in harnessing or hindering its power for super-linear or sub-linear behavior as positive effects on socio-economic performance and sustainability. Urban structures characterized by poor distributions of access and opportunities make people's movement and efforts more difficult, fragile, and harmful to society and its environment, e.g. being more subject to fluctuations like traffic congestion and unpredictability such as natural or man-made disasters. Cities are vital in that sense: sustainable societies need well-structured cities.

2.7 Summary

In this chapter, we discussed recent findings, ongoing advancements, and open questions on urban scaling. Exploring parallels to other research areas in urban studies and spatial economics, our overview covered the following key points.

- Following the development of scaling law concepts in biology, there has been a recent trans-disciplinary effort to understand urban scaling properties. Contributions from geography, physics, and computational analysis have led to the formulation of theories explaining how urban metrics scale with population size. However, scaling relationships extend beyond population size and encompass other urban and socio-economic variables. Concepts like increasing returns to scale and agglomeration effects in spatial economics (see Chap. 3) and urban theory have captured how scale, along with proximity and diversity, contribute to productivity, innovation, and economic growth.
- Different types of urban variables have consistently shown specific behaviors regarding scaling and can be grouped into three categories: (i) socio-economic

variables (super-linear); (ii) urban structures and infrastructures (sub-linear); and (iii) individual needs (linear).

- The consistency of urban scaling exponent estimates across different countries and historical periods suggests the possibility of a universal phenomenon governing the growth and organization of cities, independent of cultural, historical, and geographic factors. However, some evidence also suggests that cultural and political conditions may challenge universality in urban scaling. More research is necessary to determine whether urban scaling is influenced by specific contextual factors.
- The number of human interactions is believed to be the driving force behind super-linear scaling of socio-economic variables. However, there is little consensus on the fundamental processes behind the details. Natural or artificial features, such as efficient transport infrastructure or geographic factors, might interfere with human interaction and impact scaling effects. The spatial structure of a city may also play a crucial role in shaping human interactions and, consequently, the scaling of urban metrics. As a complex phenomenon, urban scaling can be influenced by population, densities, urban structures, hierarchical organization–and possibly more.
- Critical observations pointed out that the definition of a city's boundaries has a significant impact on scaling exponent estimates. Furthermore, relying solely on a single variable, i.e. population size, to explain a city's performance and economy is challenging. Population size as a proxy for city size overlooks the spatial dimension of cities and the diverse internal structures they can have, which can significantly impact interactions.
- Open questions and future directions include the possibility of creating a unified theory, which is particularly challenging when context-dependent features that help define the unique identities of cities as social products are considered. That said, the similarities and differences between scaling laws in biology and urban phenomena continue to appeal as key areas of investigation, including the time evolution of urban systems and the stability of the scaling exponent. And a major research avenue begs the question of how durable urban structures may harness or hinder the power of population size in super-linear or sub-linear effects on socio-economic and environmental sustainability.

Acknowledgements F. L. Ribeiro thanks CNPq (Grants 403139/2021-0 and 424686/2021-0) and Fapemig (Grant APQ-00829-21) and APQ-06541-24 (FAPEM16) for financial support. V. M. Netto thanks CITTA UPorto and CEEC FCT (grant 2023.07510.CEECIND), FAPERJ (grant E-26/211.381/2021) and CNPq (grant 315086/2020-3).

References

1. L.P. Kadanoff, *Statistical Physics: Statics, Dynamics and Remormalization* (World Scientific, Singapore, 2000)
2. H. Eugene Stanley, *Introduction to Phase Transitions and Critical Phenomena*. (Oxford: Oxford University Press, 1987)

3. Geoffrey West, *Scale: The Universal Laws of Growth, Innovation, Sustainability, and the Pace of Life in Organisms, Cities, Economies, and Companies* (Penguin Press, 2017)
4. J. Allman, *Evolving Brains*. Ed. by Scientific American Library. New York, 1999
5. A.M. Hemmingsen, Energy metabolism as related to body size and respiratory surfaces, and its evolution, in *Rep Steno Mem Hosp Nord Insulinlab* 9.1-110 (1960)
6. F.L. Ribeiro, W. RLS Pereira, A gentle introduction to scaling relations in biological systems, in *Revista Brasileira de Ensino de F í sica* 44 (2022), e20210291. https://doi.org/10.1590/1806-9126-RBEF-2021-0291
7. G.B. West, J.H. Brown, B.J. Enquist, A general model for the origin of allometric scaling laws in biology. Science **276**(5309), 122–126 (1997). https://doi.org/10.1126/science.276.5309.122
8. J.P. DeLong et al., Shifts in metabolic scaling, production, and efficiency across major evolutionary transitions of life. Proceedings of the National Academy of Sciences **107**(29), 12941–12945 (2010). https://doi.org/10.1073/pnas.1007783107
9. L.M.A. Bettencourt, J. Lobo, D. Strumsky, Invention in the city: Increasing returns to patenting as a scaling function of metropolitan size. Research Policy **36**(1), 107–120 (2007). https://doi.org/10.1016/j.respol.2006.09.026
10. L.M.A. Bettencourt, The origins of scaling in cities. Science **340**(6139), 1438–41 (2013). https://doi.org/10.1126/science.1235823
11. F.L. Ribeiro et al., A model of urban scaling laws based on distance-dependent interactions, in *Royal Society Open Science* 4.160926 (2017). https://doi.org/10.1098/rsos.160926
12. F.L. Ribeiro, D. Rybski, Mathematical models to explain the origin of urban scaling laws. Phys. Rep. **1012**, 1–39 (2023). https://doi.org/10.1016/j.physrep.2023.02.002
13. M.E.J. Newman, Power laws, Pareto distributions and Zipf's law. Contemp. Phys. **46**(5), 323–351 (2005). https://doi.org/10.1080/00107510500052444
14. A. Barabási, *Network Science*. Cambridge University Press; 2016. URL: https://barabasi.com/f/623.pdf
15. G. West, *Why New York Is Just an Average City*. online. Nautilus: science connected. 2017. URL: https://nautil.us/why-new-york-is-just-an-average-city-236671/
16. T. Robiquet, Rapport sur un mémoire addressé à l'Académie Royale de Médecin par MM Sarrus et Rameaux. Bull Acad R Med Belg **3**, 1094–1100 (1839)
17. M. Rubner, Über den Einfluss der Körpergrösse auf Stoff- und Kraftwechsel. Z. Biol. **19**, 536–562 (1883)
18. L. von Bertalanffy, Quantitative Laws in Metabolism and Growth. The Quarterly Review of Biology **32**(3), 217–231 (1949). https://doi.org/10.1086/401873
19. A. Te Krogh, *Respiratory Exchange of Animals and Man* (London, UK, Longmans, 1916)
20. M. Kleiber, Body size and metabolism. Hilgardia **6**(11), 315–353 (1932). https://doi.org/10.3733/hilg.v06n11p315
21. G.B. West, The Fourth Dimension of Life: Fractal Geometry and Allometric Scaling of Organisms. Science **284**(5420), 1677–1679 (1999). https://doi.org/10.1126/science.284.5420.1677
22. G.B. West, J.H. Brown, Life's Universal Scaling Laws. Physics Today **57**(9), 36–42 (2004). https://doi.org/10.1063/1.1809090
23. V.M. Savage, E.J. Deeds, W. Fontana, Sizing up allometric scaling theory, in *PLoS Computational Biology* 4.9 (2008). https://doi.org/10.1371/journal.pcbi.1000171
24. G. Kingsley Zipf, *Selected studies of the principle of relative frequency in language*. Harvard university press, 1932
25. G. Kingsley Zipf, *Human Behavior and the Principle of Least Effort: An Introduction to Human Ecology*. 1949. ISBN: 161427312X
26. F. Auerbach, A. Ciccone, The Law of Population Concentration, Environment and Planning B: Urban Analytics and City. Science **50**(2), 290–298 (2023). https://doi.org/10.1177/23998083221147139
27. D. Rybski, A. Ciccone, A. Lotka, Zipf-pioneers of power-law city-size distributions. Arch. Hist. Exact Sci. (2023). https://doi.org/10.1007/s00407-023-00314-0
28. A. Akira Toda, Zipf's Law: A Microfoundation, in *SSRN, formerly known as Social Science Research Network* (2017). https://doi.org/10.2139/ssrn.2808237

29. M. Batty, Scaling in city size distributions, Environment and Planning B: Urban Analytics and City. Science **50**(2), 287–289 (2023). https://doi.org/10.1177/23998083231155725
30. A. Marshall, *The Growth of Economic Science, Appendix B of the Principles of Economics*. 1890
31. P. Krugman, Urban concentration: the role of increasing returns and transport costs. International Regional Science Review **19**(1–2), 5–30 (1996). https://doi.org/10.1177/016001769601900202
32. J. Jacobs, *The life and death of great American cities* (Random House, New York, 1961)
33. J. Jacobs, *The Economy of Cities* (Vintage, 1969)
34. T.J. Allen et al., Managing the flow of technology: Technology transfer and the dissemination of technological information within the R&D organization, in *MIT Press Books* 1 (1984)
35. E. Glaeser et al., Growth in cities. Journal of Political Economy **100**, 1126–1152 (1992). https://doi.org/10.1086/261856
36. R. Martin, P. Sunley, On the notion of regional economic resilience: conceptualization and explanation. Journal of economic geography **15**(1), 1–42 (2015). https://doi.org/10.1093/jeg/lbu015
37. R.L. Moomaw, Productivity and city size: a critique of the evidence. The Quarterly Journal of Economics **96**(4), 675–688 (1981). https://doi.org/10.2307/1880747
38. T. Tabuchi, Urban agglomeration, capital augmenting technology, and labor market equilibrium. Journal of Urban Economics **20**(2), 211–228 (1986). https://doi.org/10.1016/0094-1190(86)90008-2
39. S.S. Rosenthal, W.C. Strange, Evidence on the nature and sources of agglomeration economies, in *Handbook of regional and urban economics*. Vol. 4. Elsevier, 2004, pp. 2119–2171. https://doi.org/10.1016/S1574-0080(04)80006-3
40. E.L. Glaeser, M.G. Resseger, The complementarity between cities and skills. Journal of Regional Science **50**(1), 221–244 (2010). https://doi.org/10.1111/j.1467-9787.2009.00635.x
41. J. Lobo et al., Urban scaling and the production function for cities, in *PLoS One* 8.3 (2013), e58407. https://doi.org/10.1371/journal.pone.0058407
42. W.C. Wheaton, M.J. Lewis, Urban wages and labor market agglomeration. Journal of Urban Economics **51**(3), 542–562 (2002). https://doi.org/10.1006/juec.2001.2257
43. D. Pumain et al., An evolutionary theory for interpreting urban scaling laws, in *Cybergeo: European Journal of Geography;* (2006). https://doi.org/10.4000/cybergeo.2519
44. M. Batty, *The new science of cities*. Cambridge, London: MIT press, 2013. ISBN: 978-0-262-01952-1
45. F.L. Ribeiro, J. Meirelles, Meta-analysis of the categories of urban variables related to scaling, in *in preparation* (2024)
46. J. Meirelles et al., Evolution of urban scaling: Evidence from Brazil, in *PLOS ONE* 13.10 (Oct. 2018). Ed. by Celine Rozenblat, e0204574. https://doi.org/10.1371/journal.pone.0204574
47. M. Schläpfer et al., The scaling of human interactions with city size, in Journal of the Royal Society. Interface / the Royal Society **11**(98), 20130789 (2014). https://doi.org/10.1098/rsif.2013.0789
48. R. Florida, Cities and the creative class. City & community **2**(1), 3–19 (2003). https://doi.org/10.1111/1540-6040.00034
49. P. Gordon, S. Ikeda, Does density matter?" In: *Handbook of creative cities*. Edward Elgar Publishing, 2011
50. A.J. Stier, M.G. Berman, L.M.A. Bettencourt, COVID-19 attack rate increases with city size, in *arXiv preprint* (2020). https://doi.org/10.48550/arXiv.2003.10376. eprint: `2003.10376`
51. E. Strano, V. Sood, Rich and poor cities in Europe. An urban scaling approach to mapping the European economic transition, in PLoS ONE **11**(8), e0159465 (2016). https://doi.org/10.1371/journal.pone.0159465
52. H.V. Ribeiro, D. Rybski, J.P. Kropp, Effects of changing population or density on urban carbon dioxide emissions, in Nat. Commun. **10**, 3204 (2019). https://doi.org/10.1038/s41467-019-11184-y

53. J.R. Burger et al., Global city densities: Re-examining urban scaling theory, in *Front. in Conserv. Sci.* 3 (2022), p. 879934. https://doi.org/10.3389/fcosc.2022.879934
54. F.L. Ribeiro et al., On the relation between Transversal and Longitudinal Scaling in Cities, in *PLoS ONE* (2020), pp. 1–20. https://doi.org/10.1371/journal.pone.0233003
55. G. Xu et al., Scaling laws in intra-urban systems and over time at the district level in Shanghai. China, in Physica A: Statistical Mechanics and its Applications **560**, 125162 (2020). https://doi.org/10.1016/j.physa.2020.125162
56. C. Alexander, *Notes on the Synthesis of Form*. Vol. 5. Harvard University Press, 1964
57. V. Moraes Netto, *The social fabric of cities*. Vol. 250. (Routledge New York, 2017)
58. C. Cottineau et al., Diverse cities or the systematic paradox of Urban Scaling Laws. Computers, Environment and Urban Systems **63**, 80–94 (2017). https://doi.org/10.1016/j.compenvurbsys.2016.04.006
59. E. Arcaute et al., Constructing cities, deconstructing scaling laws, in *Journal of The Royal Society Interface* 12.102 (2015), p. 20140745. https://doi.org/10.1098/rsif.2014.0745
60. S.G. Ortman et al., The pre-history of urban scaling, in *PLoS One* 9.2 (2014), e87902. https://doi.org/10.1371/journal.pone.0087902
61. D. Rybski et al., Cities as nuclei of sustainability?" In: *Environ. Plan. B* 44.3 (2017), pp. 425–440. https://doi.org/10.1177/0265813516638340
62. N.Z. Muller, A. Jha, Does environmental policy affect scaling laws between population and pollution? Evidence from American metropolitan areas, in PloS one **12**(8), e0181407 (2017). https://doi.org/10.1371/journal.pone.0181407
63. R. Louf, M. Barthelemy, Modeling the polycentric transition of cities, in *Physical Review Letters* 111.19 (2013). https://doi.org/10.1103/PhysRevLett.111.198702
64. R. Louf, M. Barthelemy, How congestion shapes cities: from mobility patterns to scaling, in Sci. Rep. **4**(1), 5561 (2014). https://doi.org/10.1038/srep05561
65. C. Molinero, S. Thurner, How the geometry of cities determines urban scaling laws, in. J. R. Soc. Interface **18**(176), 20200705 (2021). https://doi.org/10.1098/rsif.2020.0705
66. G. Spadon et al., Reconstructing commuters network using machine learning and urban indicators. Scientific Reports **9**(1), 1–13 (2019). https://doi.org/10.1038/s41598-019-48295-x
67. L.G.A. Alves, D. Rybski, H.V. Ribeiro, Commuting network effect on urban wealth scaling, in *Scientific Reports* 11.1 (2021), p. 22918. https://doi.org/10.1038/s41598-021-02327-7
68. E.G. Altmann, Spatial interactions in urban scaling laws, in *PloS one* 15.12 (2020), e0243390. https://doi.org/10.1371/journal.pone.0243390
69. A. Gomez-Lievano, H. Jin Youn, L.M.A. Bettencourt, The statistics of urban scaling and their connection to Zipf's law, in *PLoS ONE* 7.7 (2012), e40393. https://doi.org/10.1371/journal.pone.0040393
70. R. Florida, *The Flight of the Creative Class: The New Global Competition for Talent*. Harper Business, 2007. ISBN: 978-0060756918
71. M. Keuschnigg, S. Mutgan, P. Hedström, Urban scaling and the regional divide, in *Science Advances* 5.1 (2019). https://doi.org/10.1126/sciadv.aav0042
72. C. Alexander, *A city is not a tree (1965)*. 50th Anniversary Edition. The Dalles, OR: Sustasis Press, 2015. ISBN: 9789463864817
73. B. Hillier, *Space is the Machine* (Cambridge University Press, Cambridge, 1996)
74. L. Martin, L. March, *Urban space and structures*. Vol. (282). (Cambridge University Press, Cambridge, 1972)
75. M. Batty, Urban Modelling, in (2009). URL: http://www.casa.ucl.ac.uk/rits/batty-urban-modelling-2009.pdf
76. D. Rybski, E. Arcaute, M. Batty, Urban scaling laws, Environment and Planning B: Urban Analytics and City. Science **46**(9), 1605–1610 (2019). https://doi.org/10.1177/2399808319886125
77. C. Cottineau et al., Paradoxical Interpretations of Urban Scaling Laws, in *arXiv preprint* arXiv:1507.07878 July (2015), pp. 1–21. arXiv: arXiv:1507.07878v1
78. V.M. Netto, E. Brigatti, C. Cacholas, From urban form to information: Cellular configurations in different spatial cultures. Environment and Planning B: Urban Analytics and City Science **50**(1), 146–161 (2023). https://doi.org/10.1177/23998083221107382

79. A. Coutrot et al., Entropy of city street networks linked to future spatial navigation ability. Nature **604**(7904), 104–110 (2022). https://doi.org/10.1038/s41586-022-04486-7
80. E. Arcaute, J.J. Ramasco, Recent advances in urban system science: Models and data, in *PLoS ONE* 17.8 August (2022), pp. 1–16. https://doi.org/10.1371/journal.pone.0272863
81. R. Louf, M. Barthelemy, Scaling: lost in the smog. Environment and Planning B: Planning and Design **41**(5), 767–769 (2014). https://doi.org/10.1068/b4105c
82. J.C. Leitão et al., Is this scaling nonlinear?" In: *Royal Society Open Science* 3 (2016). https://doi.org/10.1098/rsos.150649
83. F.L. Ribeiro, Física das Cidades, in *Revista de Morfologia Urbana* 8.1 (2020), e00159. https://doi.org/10.47235/rmu.v8i1.159
84. L. Bettencourt, G. West, A unified theory of urban living." In: *Nature* 467.7318 (Oct. 2010), pp. 912–3. https://doi.org/10.1038/467912a
85. J. Lobo et al., Urban science: Integrated theory from the first cities to sustainable metropolises, in *Report submitted to the NSF on the Present State and Future of Urban Science* (2020). https://doi.org/10.2139/ssrn.3526940
86. F.L. Ribeiro, J. Lobo, D. Rybski, Zipf's law and urban scaling: Hypotheses towards a Unified Urban Theory, in *arXiv preprint* (2021), pp. 1–3. arXiv:2107.11782. URL: http://arxiv.org/abs/2107.11782

Chapter 3
The Benefits and Costs of Agglomeration: Insights from Economics and Complexity

Andres Gomez-Lievano and Michail Fragkias

Abstract There are many benefits and costs that come from people and firms clustering together in space. Agglomeration economies, in particular, are the manifestation of centripetal forces that make larger cities disproportionately more wealthy than smaller cities, pulling together individuals and firms in close physical proximity. Measuring agglomeration economies, however, is not easy, and the identification of its causes is still debated. Such association of productivity with size can arise from interactions that are facilitated by cities ("positive externalities"), but also from more productive individuals moving in and sorting into large cities ("self-sorting"). Under certain circumstances, even pure randomness can generate increasing returns to scale. In this chapter, we discuss some of the empirical observations, models, measurement challenges, and open question associated with the phenomenon of agglomeration economies. Furthermore, we discuss the implications of urban complexity theory, and in particular urban scaling, for the literature in agglomeration economies.

3.1 Introduction

The study of *agglomeration economies* explores the economic benefits that arise when individuals live and work *in close physical proximity*. The topic originates in the observation of the existence of strong and persistent geographical concentrations in economic life: production and consumption activities in particular. The idea of agglomeration economies has been one of the most central topics studied by urban economists and economic geographers, if not *the* most studied one [1].

A. Gomez-Lievano (✉)
Analysis Group, Boston, MA, USA
e-mail: agomezlievano@asu.edu

Growth Lab, Harvard University, Cambridge, MA, USA

M. Fragkias
Department of Economics, Boise State University, Boise, ID, USA
e-mail: michailfragkias@boisestate.edu

D. Rybski (ed.), *Compendium of Urban Complexity*, Understanding Complex Systems,
https://doi.org/10.1007/978-3-031-82666-5_3

The reaping of agglomeration economies has historically been identified as a central issue in our quest for increasing economic productivity and eventual prosperity at the local, regional and global scale. But shocks in the way we live and work may have a significant effect in the trajectory of those economies from spatial concentration. In some cases, we can ask whether these economies will continue to benefit humanity in the long term. For example, the COVID-19 pandemic has shown that households and firms are able to renegotiating life/work arrangements, affecting the spatial patterns of living and working. Depending on the outcome of this process, and the economies present in complementary technologies for the exchange of ideas (e.g. remotely), we can expect important differentials in the level of agglomeration economies experienced and resulting economic prosperity.

There is a tendency to think that cities exist *because* of agglomeration economies. The reasoning goes that since cities are "the absence of space between people and companies" (as the urban economist Edward L. Glaeser said [2, p. 6]), if there are benefits to agglomerating, people will tend to seek each other's company, and cities will naturally form. According to this reasoning, cities are the physical manifestation of the gains from being close to one another. However, there is a certain circularity here, leading in turn to a misconception: that one concept (cities) cannot exist without the other (agglomeration economies). That both are synonyms. But they are not. So to understand what agglomeration economies are, it is useful to clarify this difference.

On the one hand, cities can exist because of reasons not associated with any economic benefits from living and working in high densities. For example, people can come together inside some walls for religious reasons, to protect themselves against enemies or nature, or to be next to some natural resource like a river, forest, mine, or port [3]. On the other, there are many disadvantages from living in cities, like traffic, pollution, high rents, crime, and disease. Thus, agglomeration economies are only an aspect of cities.

The benefits and costs from agglomerating act as two opposing forces. The former as centripetal forces that pull people (and companies) together, and the latter as centrifugal forces that push them away. In recent times, cities have become less dependent on being close to certain natural resources. At the same time, evidence tell us that the benefits from agglomeration are dominating its costs, and this has been reflected in the fast pace of urbanization and the continued growth of cities that we see today around the world. Hence, agglomeration can be considered one of the most important forces of our current world, and we better understand its effects (i.e. agglomeration economies).

In what follows, we address several dimensions of agglomeration economies and several open research questions in this area. Section 3.2 reviews the basic and advanced notions surrounding agglomeration economies–how they have been theoretically conceptualized, empirically measured, and mathematically modeled; the challenge of causal inference and differing experiences across parts of the world. Section 3.3 discusses ideas from economics and complexity science that provide a more nuanced understanding of agglomeration phenomena: the notion of spatial equilibrium from economics (in contrast to the idea of disequilibrium emerging complexity science); the microfoundations of agglomeration economies; artificial increasing returns to scale; and urban scaling theory. Section 3.4 explores the possibility of blending mainstream economic analysis of agglomeration with the complexity science perspective. Section 3.5 reviews agglomeration in the context of sociology, cultural and evolutionary anthropology, pointing to the possibility of a grand synthesis regarding agglomeration phenomena across fields. Finally, Sect. 3.6 concludes and discusses open questions that still remain.

3.2 The Empirics of Agglomeration Economies

3.2.1 *Conceptual and Technical Frameworks*

Agglomeration economies, broadly construed, are the benefits that emerge from the clustering of (economic) activity closely in space. In order to quantify and estimate them, we will use a more precise definition: *agglomeration economies* are the *increasing returns to scale* (IRS), as well as *external economies specific to a location*, related to economic productivity and well-being, occurring in specific locations, such as cities, as they get larger and/or denser. In the field of economic geography, increasing returns to scale play a significant role in the location choice of firms, leading to the formation of core-periphery patterns [4].

Agglomeration in economics has been a central theme of study in urban and regional economics and the new economic geography [5]. Specifically in economics, agglomeration economies are economies that are external to firms, but internal to a location [6]. Four types of agglomeration economies based on scale are conceptualized in this framework: (a) economies internal to the firm at a given location – that is economies from mass production, (b) economies external to a firm at a specific location, (c) localization economies (internal to an industry at that location) and (d) urbanization economies (external to both firm and industry at a particular location).

The "returns to scale" of a quantity F (e.g. the total well-being in a settlement) with respect to a variable x (e.g. the size of the settlement) is given by the change in the function when x is multiplied a constant $\lambda > 1$.[1] Increasing returns occur when $F(\lambda x) > \lambda F(x)$. Let $f(x) = F(x)/x$ be the *per capita* value. Then, it is easy to verify that IRS implies that $f(\lambda x) > f(x)$. The quantity f can be wages, patenting rates, or other measures of individual productivity in a given city, and the quantity x is the total employment, the working age population, the total population size or even can be the population density of the city. The discussion of increasing returns to scale leads into the topic of economies of scale when the payment of factors of production is introduced.

Economists perceive external economies of agglomeration as resulting from a shift in a production function of a firm due to the effects of a change in scale of

[1] Since the returns to scale concept in economics originates in production theory, we utilize this concept in line with a "city as a firm" analogy [5]. That is, in this exposition of the concept, we can think of a city as represented by a single firm.

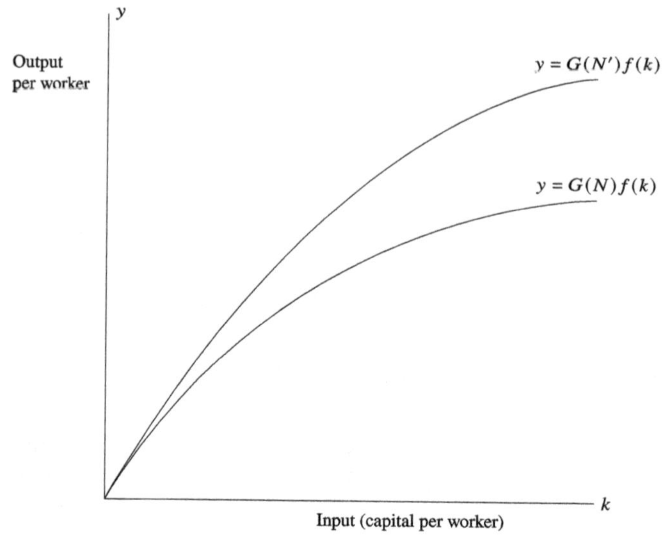

Fig. 3.1 External economies of agglomeration as experience by a firm through a shift in the production function due to an increase in the size of a city (N to N')

the area that a firm exists in. In the simplest form, this concept can assume $Y = G(N)F(K, L)$, where N is population size, K is capital and L is labor. Expressed in terms of output per worker–dividing by L, we get $y = G(N)f(k)$.[2] The function $G(N)$ is referred to as a 'Hicks-neutral' externality since we make the assumption that it does not affect the optimal choice of labor and capital in the production process.[3] Formally, if $dG/dN > 0$, we observe economies of agglomeration while if $dG/dN < 0$, diseconomies of agglomeration. Figure 3.1 shows economies of agglomeration by plotting a hypothetical production function of a firm existing in a larger agglomeration (N') vs a smaller agglomeration (N). Sveikauskas (1975) [7] was among the first to explore the empirical relationship between city size and labor productivity from this point of view. His paper concluded that average wages were disproportionately higher in large cities because Hicks-neutral productivity was substantially higher in large cities.

Before going into models of agglomeration economies, the statistical technicalities, or the underlying causes, let us look at some data so that we understand why there is so much interest from urban economists about this phenomenon. As an example, let us start with data from Metropolitan Statistical Areas (MSAs)[4] in the

[2] Here, we are assuming the function $F(K, L)$ to have *constant* returns to scale. That is, $F(\lambda K, \lambda L) = \lambda F(K, L)$ for any positive constant λ. In particular, if $\lambda = 1/L$, we get that $(1/L)F(K, L) = F(K/L, 1) \equiv f(k)$, a function of capital per worker $k = K/L$.

[3] A Hicks-neutral externality raises the marginal product of labor and capital proportionally at a specific capital to labor ratio.

[4] MSAs are the most common unit of analysis in urban studies, and the US methodology for how to define them has been adopted by many other countries. Their delineation is not based

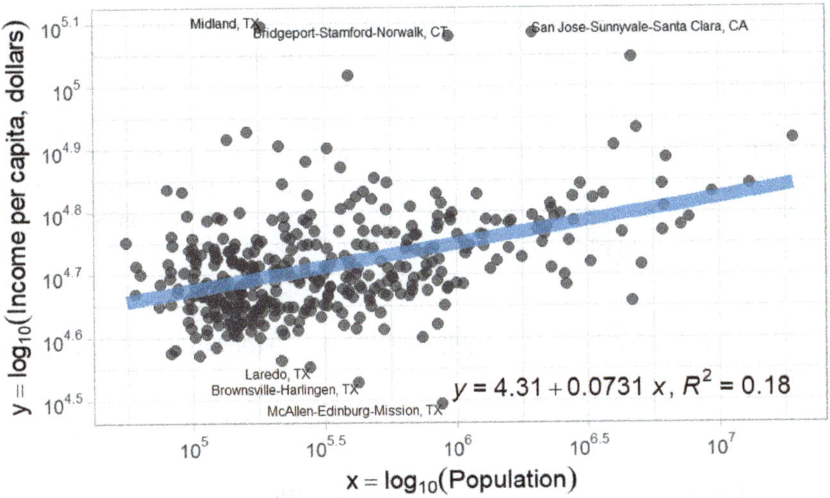

Fig. 3.2 Association between income per capita and population size across US MSAs in 2020. The name of the MSAs with the highest and lowest personal income are indicated in the figure. All dollar estimates are of current dollars (not adjusted for inflation). Source of data: Bureau of Economic Analysis (BEA) [8]

United States (US) for the year 2020 in Fig. 3.2. There are 384 dots in the plot, each dot representing an MSA, and shows how the logarithm of average personal income correlates with the logarithm of population size. The use of a log-log plot to reveal agglomeration economies is not accidental. This is because the "returns to scale" of a function f can be quantified by the ratio between the rate of increase of f (defined as $\frac{1}{f}\partial f/\partial x$) and the rate of increase in x (defined as $\frac{1}{x}\partial x/\partial x$). This is equivalent to comparing changes between logarithms, as $\partial \ln(f)/\partial \ln(x)$. As a consequence, a typical step into assessing the presence of agglomeration economies is plotting how $\ln(f)$ relates to $\ln(x)$.

As can be seen in Fig. 3.2, per capita income tends to be higher in larger cities. The coefficient of the regression is approximately 0.073. Coefficients between logarithmic variables denote "elasticities". An elasticity of 0.073 implies that a 1 % change in population size is approximately associated with a 0.073 % change in personal income.[5]

on administrative or political boundaries (like those of counties, municipalities, provinces, states, countries, etc.), but on commuting patterns. The idea behind using commuting patterns to define urban boundaries is that cities are the places where people *live and work*. MSAs account for approximately 86 % of the total US population.

[5] The actual formula is that an ϵ percent change in the independent variable (e.g. city size) leads to a $[(1+\epsilon/100)^{\beta}-1]\times 100$ percent change in the dependent variable (e.g. income). When ϵ is small, then $(1+\epsilon)^{\beta}-1\approx\beta\epsilon$, which means that a "$\epsilon$ percent change in the independent variable corresponds to a $\beta\epsilon$ percent change in the dependent variable", as was done above in the text. Sometimes, however, researchers interpret the elasticities with a 100 % change in the independent variable (e.g. doubling city size). In that case the "$\beta\epsilon$" formula will overestimate the effect on the

It is worth noting that the relationship in Fig. 3.2 between log-population and log-income seems to be linear, and that the 7.3 % elasticity seems to apply across all city sizes. Such linear relationships in the form of $\ln(y) = A + B\ln(x)$, where the coefficient B is a constant, describe a *power-law* relationship between quantities x and y, of the form $y = ax^B$ (where $a = e^A$).[6] At the level of cities, agglomeration economies are often (though not always) represented with these power-law relationships. Mathematically, we therefore talk of evidence of agglomeration economies at the aggregate level when

$$y = cN^{\delta}, \tag{3.1}$$

where c is a constant, N is city size, and, crucially, $\delta > 0$ for variables y that measure of economic well-being at a per-capita level.

3.2.2 Measurement Issues: Productivity and Scale

Going back to Fig. 3.2, the power-law relationship between city size and personal income suggests the presence of agglomeration economies in US MSAs. Personal income, however, includes earnings not only from job salaries, but also from interests, dividends, transfers, and other sources,[7] which do not necessarily reflect the benefits accrued specifically due to living at a given location. For example, an individual might live in San Francisco but their income may come from renting an apartment in New York. This income would still be counted in San Francisco, even though the economic activity generating such income is from New York. So how can we measure the "direct" benefits from living in a particular city? A better quantity would be a measure of *productivity*.

In economics, *productivity* is an abstract measure of the performance in a production process. It has to do with the efficiency by which an agent (person, firm, industry, city, region, or country) converts inputs into output. Quantifying productivity is challenging because one must account for increases in output that are not trivial. For instance, if doubling the inputs of a particular production process ends up doubling the output produced, then the productivity of that process has not changed. This suggests productivity should be quantified as a ratio of $\frac{\text{output}}{\text{inputs}}$. At times, $\frac{\text{sales}}{\text{worker}}$ act as a proxy of productivity. But it is not that simple. There are more subtle effects to think about. For example, rather than considering doubling the inputs, consider substituting some type of inputs with another type (e.g. substituting workers with machines). A substitution may or may not induce a change in productivity. Which

dependent variable. With an elasticity of 7.3 %, doubling the size of a city corresponds to income increasing by 5.2 %.

[6] If the relationship uses $\log_{10}$ instead of natural logarithms, then $a = 10^A$.

[7] See https://www.bea.gov/resources/learning-center/what-to-know-income-saving (last accessed: 15.05.2022).

substitutions would suggest an increase in productivity? Should productivity be measured as $\frac{\text{profits}}{\text{costs of inputs}}$?

Figure 3.2 also raises the question of whether total population size N is the right measure of agglomeration. Would it be better to use *population density* (that is, population per unit area)? Duranton and Puga (2020) [9] argue that neither is a great choice: "For instance, a highly concentrated but tiny cluster of economic activity is unlikely to generate strong agglomeration economies. On the other hand, workers located at the edge of large metropolitan areas are unlikely to benefit from their full scale in the job-matching process" (p. 7). They also suggest that theory is not going to resolve this issue. Currently, the main practice in economics is to utilize total population, but increasingly, with the availability of better data, local density is also utilized

Thus, the economics of density literature moves away from exploring agglomeration economies as the relationship that emerges between measures of productivity and city size, and focuses on the alternative approach exploring the relationship of productivity against *density*. Duranton and Puga (2020) show that urban density and population are highly correlated, assuming non-naive measures of urban density (e.g. that do not include extensive rural areas in the calculation). That is, proper density measures should reflect how density is experienced by the typical urban resident of the MSA.

The answers to the choice of measurement are not easy and change depending on the specific production process under consideration, and many other assumptions. We will not delve in such discussions here.[8] Instead, we will use *nominal wages*, as is customarily done in the literature, to reflect the productivity of a place. (We will review the conceptual arguments in Sect. 3.3.1). We will mainly adopt scale as a measure of spatial concentration. As we will show, these choices allow us to discuss the literature from urban economics and urban scaling uniformly (see, however, [11] and their use of density in the context of the urban scaling literature). Given these modeling choices, we can ask: are wages higher in larger cities?

3.2.3 Empirical Estimation

Let us define approximate productivity by taking a measure of total aggregate output generated by a single metropolitan area (i) during a given year (e.g. total annual wage bill), Y_i and dividing it by the number of workers in the city, N_i,

$$y_i = Y_i/N_i \,. \tag{3.2}$$

[8] There is a whole field for how to measure productivity, see e.g. [10]. Similarly, for measuring density.

The ratio in Eq. (3.2) can be interpreted as the average productivity of workers in the city.[9]

Given data for y_i and N_i across several cities within a same urban system (like a country), the relationship between productivity and scale can be assessed using ordinary least squares (OLS) regression analysis. From the economics perspective, the mathematical model specification is derived by a competitive firm-level profit maximization problem solution [12]. The statistical specification usually takes the form of a log-log model, given that researchers are exploring the relationship in terms of elasticities or estimating a power law (like in Fig. 3.2).

$$\ln(y_i) = \ln(c) + \delta \ln(N_i) + \epsilon_i, \tag{3.3}$$

where δ is the elasticity, and $\ln(c)$ is interpreted as the constant in standard regression. Some debate exists on whether ordinary least squares is the appropriate methodology in estimating power-laws, or whether maximum likelihood estimation or Bayesian econometrics should be utilized (see, e.g. [13–15]).

Alternatively, given the land area of cities (A_i), one can also estimate the same model but replace the size of the city with a population density measure, $\rho_i = \frac{N_i}{A_i}$.

$$\ln(y_i) = \ln(c) + \delta \ln(\rho_i) + \epsilon_i \, , \tag{3.4}$$

Typical results are shown in Fig. 3.3. Our results show a positive association between population size and well-being as measured by average wage. In this case, the estimated elasticity is 0.057, implying that doubling the size of a city in this cross-section is associated with an increase of 4.0 % in average wages. This elasticity is often referred to as the *urban wage premium*.

Similar results are found in the economics of density literature. Considering the agglomeration benefits when concentration is measured in terms of density, the literature provides similar results: Ahlfeldt and Pietrostefani [17], in a meta-analysis, measure the elasticity of productivity with respect to density at 0.04.

Estimates from simple bivariate regressions like Eqs. (3.3) and (3.4) have many limitations. They do not actually identify the *causal* effect of what being close to other people does to productivity and well-being. They are aggregate correlations and, as such, they do not rule out other alternative causes. For example, the same correlation could emerge if people who are more productive prefer to be close to each other, and as a consequence, tend to move and *self-sort* into (large) cities. As we mentioned in the introduction, another cause of both agglomeration (i.e. population growth) and higher productivity would be the presence of some natural advantages (potentially unobserved).

With the above in mind, next, we review the empirical challenges of accurately measuring agglomeration economies.

---

[9] In general, when you take any measure of total output (e.g. total number of patents, total number cars produced, total value of good exported) and you divide it by the labor used to produce it, you get a measure of *labor* productivity. As mentioned in the previous section, caution needs to be taken in the interpretation.

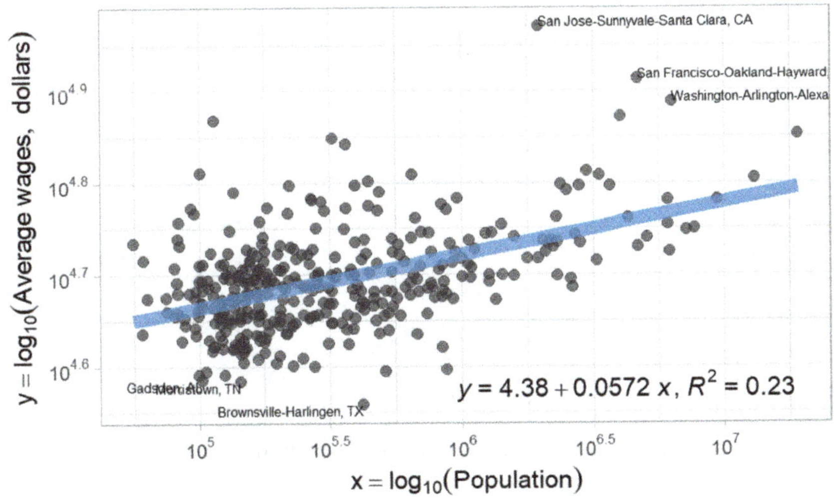

Fig. 3.3 Association between average wages and population size across US MSAs in 2020. The name of the MSAs with the highest and lowest average wages are indicated in the figure. Source of data: Occupational Employment and Wage Statistics (OEWS), Bureau of Labor Statistics (BLS) [16]

3.2.4 *Empirical Challenges: Causal Inference*

Quantifying agglomeration economies consists of being able to single out the *causal effect* that physical proximity has on productivity. We have discussed this, concretely, in the case of the effect city size has on wages. The challenges involved can be illustrated in the causal diagram depicted in Fig. 3.4. Each box in the figure represents a variable, and each arrow represents a causal effect. Thus, measuring agglomeration economies amounts to quantifying the strength of the green arrow.

The main challenge in measuring the green arrow is that other arrows exist that also induce correlations between city size and productivity. These issues are represented by the top and bottom boxes in Fig. 3.4. On the one hand, the top box and its yellow arrows represent variables that simultaneously cause cities to get larger *and* firms to have higher productivities. An example is a local natural advantage, such as a river or a sea port, which may attract firms to operate there (causing city size to increase), and may be an important source of resources for the city (causing rises in productivity). These variables are called "confounders". They are a problem if they are unobserved. In general, however, the problem of confounders can be minimized by including available variables for such local urban characteristics as "controls" in regressions [18]. For example, one could modify Eq. (3.3) to

$$\ln(y_i) = \ln(c) + \delta \ln(N_i) + \phi X_i + \epsilon_i \,,$$

where X_i can be variables that stand for local characteristics of the city.

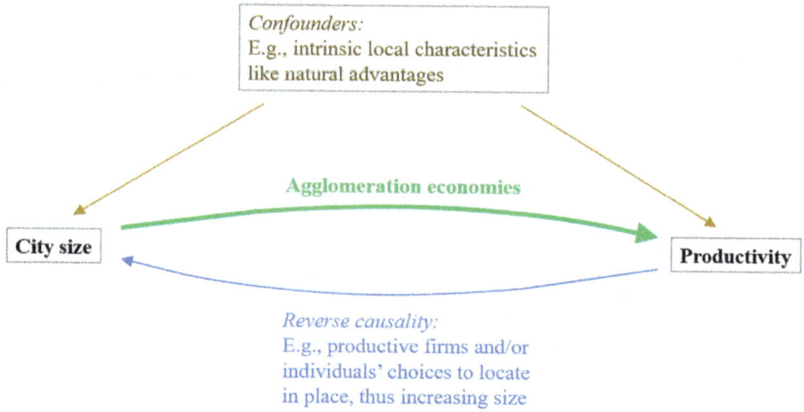

Fig. 3.4 Diagram of causes and effects that could explain correlations between city size and wages

The real challenge comes from the bottom part of the diagram. That is, the blue arrow in Fig. 3.4 in which the direction of causality runs in the opposite direction as the one associated with the concept of agglomeration economies (green arrow). The fact that the arrow points in the opposite duration is not meant to suggest that there is something that counteracts or reduces the effect of agglomeration economies. Instead, the arrows just represent which variable are affected by changes in other variables. In particular, the green and blue arrows illustrate the fact that a positive association between city size and productivity might have two possible explanations: either city size causes higher productivity or individual cities that are already more productive tend to systematically grow to large sizes. The bottom part of the diagram represents the issue of so-called "reverse causality", in which economic agents decide to locate themselves close to other agents that are inherently more productive, thus increasing the size of cities [19, 20]. As a consequence, choices will create a correlation between size and productivity. According to this direction of causality, the correlation will emerge not because larger sizes cause economic activities to be more productive, but the other way around: city size will be the *consequence* of the choices that productive individuals, firms and industries make. If high-productivity firms (or industries) tend to cluster in cities (increasing their size), the urban wage premium found in large cities could instead reflect firms' specific production processes rather than any benefits derived from geographical proximity.

The issue of reverse causality illustrated in Fig. 3.4 is sometimes also referred to as "self-sorting", and it turns out to be incredibly hard to adjust for in statistical analyses that seek to measure the causal effect of agglomeration economies. In other words, one step in measuring the presence and strength of the green arrow is to "shut down" (statistically speaking) the effect of the blue arrow. However, difficulties include the fact that the production function of firms is often hard to observe, as well as the fact that it is hard to identify the reasons why firms and people migrate from city to city.

A full review of the details involved in separating the causal effect of agglomeration economies from other causal sources like confounding and self-sorting can be found in Combes and Gobillon (2015) [21]. The literature in urban economics keeps investigating new methodologies to address these empirical challenges, and the sophistication of the methods keeps increasing steadily, but most of the solutions have converged on a specific strategy: using data that tracks single individuals over time. This strategy was first proposed by Glaeser and Maré in 2001 [19] (see also [22]).

With data at the level of individuals, the basic idea is to compare the productivity of individual workers before and after moving to a large city. This strategy needs several additional refinements to control for other simultaneous effects that can interfere with the causal effect of city size on productivity, but makes the analysis of agglomeration economies more transparent and interesting than when one is analyzing aggregate data at the level of cities. The difficulty becomes finding such data. Data at the level of individuals is either administrative data (like insurance, tax, or social security data, which are often private and hard to access) that tend to suffer from issues of statistical representation (because these are samples from selected groups of people that are not representative of the overall population), or census data, which are statistically representative, but offer only a cross-sectional view of the population and are not longitudinal.

3.2.5 Agglomeration Economies in Developed and Developing Countries

Bearing in mind the methodological caveats discussed before, we review here what researchers have found about the magnitude of agglomeration economies around the world. Overall, the consensus is that the typical magnitude of agglomeration economies is approximately 4 % (see [17, 23, 24]). There is, however, great variability around the world, in part due to the methodological choices that different studies have taken. Unfortunately, most of the research in urban economics has been disproportionately focused on the developed countries. This fact can in part be explained by the wider availability of metro-level data in OECD countries.

Hence, the value of 4 % is mostly representative of high-income countries. In the US, for example, accepted estimates range between 3.6 % and 4.3 %, while in France is around 3 %. The few studies that have been conducted in developing countries tend to find higher elasticities. Thus, researchers have estimated elasticities of 19 % in China, 17 % in Africa, 6 % – 16 % in Latin American countries (see [17, 23, 24] for recent reviews and meta-analyses of the literature). This is important, because it means that developing countries derive huge benefits from agglomeration. It urges decision makers to understand the underlying causes for the strong need in those countries for density, and the consequent pitfalls.

In summary, while researchers agree that empirical evidence supports the idea that agglomeration economies in dense urban areas boost productivity, there is a significant knowledge gap due to a lack of understanding of agglomeration economies in developing countries. Meta-analyses and literature reviews [9, 17, 23, 24] rely heavily on estimates from wealthy countries, and no systematic attempt has been made to quantify the disparities in estimations between advanced and poor nations. Filling this knowledge gap is critical since the majority of people on Earth is projected to live and work in developing-country cities, which are rising quickly in population.

3.3 Models and Explanations from Economics and Complexity

3.3.1 Economics: The Idea of Spatial Equilibrium

Spatial equilibrium is the organizing idea in urban economics for thinking about cities. Here, we will focus on the inter-urban implications of spatial equilibrium (i.e. what happens across cities) rather then the intra-urban implications (i.e. what happens within cities). The literature of the former was developed by Rosen (1979) and Roback (1982), which is why economists sometimes talk about the Rosen-Roback model of urban spatial equilibrium [1].

Imagine there exists a place where the amenities are so nice that everyone wants to live there (e.g. for many of us, that would be a city with warm weather throughout the year, beautiful sunsets, fertile soil, next to both beautiful mountains and a warm beach, where people can swim in the sea and have plenty of fish they can eat with little effort). Through the exchange of information about the available options for establishing a life, people would start moving to that place. From being close to each other, the right matches between people form to create happy marriages and successful enterprises. The place would start producing amazing innovations, many of which could start being exported to the rest of the world. Wealth would accumulate, the best firms and the brightest people would want to establish themselves there, wages would increase, and all this economic success would attract more people and firms to go there. It is not hard to imagine that there is at least one place in each country on Earth that is not too far from this description. Why are we not all living in that place? Why do we live in many cities as opposed to a single (gigantic) city? What are the underlying trade-offs? Something has to give. But what? In economics, the concept of a *spatial equilibrium* formalizes and captures the essential elements at play here.

Mainstream economics is centered around the idea of equilibrium–perhaps to a degree that actually hurts economic inquiry. In contrast, complexity scientists tend to think of cities as entities in disequilibrium [25]. However, thinking about systems in equilibrium does help with developing some useful insights, even amid the recognition that agents do not act as perfect rational optimizers (and are, more likely, 'boundedly rational' [26]). In urban economics, the main organizing idea is that individuals can move freely within a nation, akin to the 'no friction' assumption in physics, which is an idealization that simplifies reality. It is not true in reality, given that myriad factors tether individuals to their birth cities (e.g. family, private properties they need to take care of, health, or simple lack of resources to travel), but it is a good approximation supported by the fact that migration has been one of the universals of human nature [27]. Indeed, despite its costs, moving across space in search of better opportunities is one of the most effective ways to improve one's economic well-being [28].

Assuming that individuals are free to move in space (between cities) implies that people are ultimately *indifferent* between locations. This may seem counterintuitive, but such implication arises because we assume individuals have taken into account all benefits and costs, and have all made their final decision (i.e. if they moved and didn't like it, they are free to move again and find another location, until they are happy). Thus, the concept of a *spatial equilibrium*. Let us develop some additional intuitions from this key idea.

From the point of view of individuals, three things matter (again, as a simplification): wages, rent, and amenities. When spatial equilibrium applies, an individual cannot have a good wage, a low rent, and great amenities, all three simultaneously. For example, in the beautiful city we pictured above, wages and amenities were great. However, demand to live there would make housing very expensive. Moreover, the increasing density of people may in fact damage the amenities that were the original appeal of the place (e.g. due to environmental degradation, or visual, sound, and air pollution). In this simplified version of the world, things will reach an equilibrium. In this equilibrium, not all people will decide to move to the most beautiful city because some people will choose to live in places with fewer amenities if the cost of housing is cheaper, or if wages are higher.

The assumption of spatial equilibrium therefore implies that high wages in a place are, actually, an indication of poor amenities, high rent prices, or both. High wages and high rents do not equilibrate across cities (that is, wages and rents can continue being different between cities) because high wages in a given city are, in turn, an indication that the firms have decided to locate there for a reason. And the reason must be that, by locating themselves in that city, firms are being able to be more productive (and therefore more profitable).[10] So, at the end, higher wages are an indication that the area is benefiting from agglomeration.[11]

---

[10] Recall, in standard economic theory, assuming perfect competitive markets, the wage a firm pays to a worker is determined by the *additional* value he or she brings to the firm. In economic language, wage is the *marginal product of labor*.

[11] This is formalized in a theorem first stated by Starrett in 1978 [29], called the Spatial Impossibility Theorem, restated by Ottaviano and Thisse (see, e.g. [30]). The theorem implies that there are three sufficient conditions for economic activities to agglomerate: (i) heterogeneous space, (ii) positive externalities, or (iii) imperfect markets. The concept of agglomeration economies is particularly associated with (ii), but there are mechanisms in which (i) and (iii) also create agglomeration economies. We discuss the specific mechanisms in Sect. 3.3.2.

Through this line of reasoning we have arrived at an important insight: to reveal agglomeration economies, we should look at *nominal* wages and check how they correlate with the density or the size of a city. Looking at *real* wages, as opposed to nominal wages, would defeat the whole purpose. By assumption, we have said that wages are offset by the costs of living. Therefore, adjusting nominal wages to account for the costs of living would simply reveal that all real wages are equal across all locations, and we would destroy the information generated by agglomeration economies.

3.3.2 Economics: Causes of Agglomeration Economies–Sharing, Matching and Learning

Economists have been wondering for a long time about the mechanisms that would cause economic agents to benefit from crowding together in close physical proximity. The original classification for the causes of agglomeration economies is usually attributed to Alfred Marshall–a foundational figure in economics–first published in 1890 in his work *Principles of Economics*.[12] There, Marshall proposed that agglomeration was a consequence of reducing the transportation costs for goods, people, and ideas. According to Marshall, by locating near each other, firms will reduce the shipping costs of materials (reducing, for example, the interaction costs between buyers and suppliers), the costs associated with the labor market (e.g. the search and employment of appropriate workers), and finally, the costs of finding new ideas (e.g. firms will benefit from "knowledge spillovers").

Duranton and Puga (2004) [32] re-conceptualized this classification. According to them, once a firm in a particular industry is established in a city, it paves the way for other firms to get established there because the firm has assumed some initial fixed costs, has created a network of customers and suppliers, and has revealed important knowledge about how to operate successfully in the city. Thus, once the first firm arrives into a city, other firms will benefit, and this will show up as increases in productivity. In Duranton and Puga's classification, close physical proximity would entail benefits to the firms because of better

- *sharing* of local infrastructure, indivisible facilities, suppliers and customers, leading to savings on fixed costs, as well as firms sharing risk,
- *matching* between employers and employees, or sellers and buyers, and
- *learning* opportunities in terms of generation, spread and accumulation of knowledge.

Since Marshall's time, the causes for the gains from agglomeration have been thoroughly researched. However, the relative impact of each of these mechanisms has been hard to quantify. This is due, in part, to the so-called "Marshallian equivalence" [32], which states that all agglomeration mechanisms provide the same prediction:

[12] Citations sometimes refer the 1920's 8th edition of Marshall's book [31].

firms involved in similar activity will tend to cluster since they produce benefits for one another. The addition of gains due to the three causes makes determining which mechanism has the most weight very challenging. Identifying how each of these mechanisms impact agglomeration, however, is important not only from a policy perspective, but also to anticipate how technological advances and socioeconomic change will affect the growth of cities.

The current understanding is that these mechanisms affect different industries in distinct ways, and that these effects have been changing over time. Diodato et al. (2018) [33] carried out an extensive study of these observations. They show that value chain connections used to be dominant for most of the twentieth century for most industries. Currently, however, there has been a shift in the economy into services and skill-sharing has become the most important motivator driving firms in that sector, while value chain connections remains the main cause for co-location patterns in manufacturing.

Are Duranton and Puga's views the final word on the origins of agglomeration economies? There are reasons to suppose that city size has more to it than their three mechanisms. For example, there are additional urban phenomena that do not represent economic advantages but are still amplified in larger cities (like diseases and crime) in the same manner that economic gains are. Yet, the mechanisms of sharing, matching, and learning are unlikely to also explain those other non-economic phenomena. From a complexity point of view, however, a simple set of mechanisms should be able to explain all these phenomena, economic and non-economic. One element that could explain why larger cities potentiate all human activities is the fact that they are more diverse (Sect. 3.3.4 will discuss this idea).

3.3.3 Complex Systems: A Criticism of the Economics Approach–Artificial Increasing Returns to Scale

There is an empirical problem with measuring agglomeration economies that merits a section on its own here, separate from the empirical challenges that we have summarized so far. The issue we want to discuss here is one in which the statistical association that emerges between size and productivity does not arise from difficulties in disentangling economic causes and effects (i.e. cannot be represented in Fig. 3.4) or from one the three economic mechanisms explained in the previous section. It is an effect from over relying on the concept of "average" that one often does when analyzing complex systems.

The diagram in Fig. 3.4 and the three mechanisms for agglomeration economies (sharing, matching and learning) both make an assumption that, at first sight, would be unobjectionable. That in order to quantify the sources of correlations between city size and productivity using data, one can safely compute *averages*. For example, in Fig. 3.3 we looked at the *average* wage across cities. Or, when researchers take care of the issues of confounding and self-sorting, they aim to measure how size will influence the productivity of individuals *on average*. In all these efforts, researchers assume the Law of Large Numbers (LLN) applies: that taking the average over a finite sample of observations approximates the true underlying expected value. This assumption is fundamental to doing science in general. However, quantities in complex systems sometimes have the nasty property that their distributions either lack an expected mean, or their mean is very hard to estimate. Science in these cases must advance with caution, being mindful of this issue (see, e.g. [34–36]). How does this problem affect the analysis of cities?

In simple words, the challenge is that cities are fundamentally unequal, heterogeneous, places. Income, productivity, wages, and other economic variables, tend to be unevenly distributed across firms and individuals in cities. The distributions that describe these quantities are often power-laws, log-normals, or more generally, heavy-tailed distributions, which limit the validity of the LLN [37]. If the validity of the LLN is in question, one cannot take averages.

To take an average of a set of observed numbers, one needs to guarantee, first, that their distribution has a finite variance, and second, that the sample size is large enough. Andrés Gómez-Liévano et al. [38] calculated that if urban quantities are log-normally distributed (as wages in cities tend to be), then averages can be taken only if $\ln(N) \geq \sigma^2/2$, where N is the city size and σ^2 is the variance of log-wages. This relation means that the larger the variance of log-wages, the larger the size must be in order to guarantee that the sample average $\frac{1}{N}\sum_j y_j$ of wages across individuals $j = 1, \ldots, N$ makes sense. An interesting effect occurs to the sample average of wages when $\ln(N) < \sigma^2/2$, that is, when LLN does not apply: the sample mean (e.g. average wage) and the sample size N become positively correlated. If such situation exists in real data, higher average wages will be observed in larger cities, *even if the true underlying expected mean is constant across cities*. In other words, under certain situations in which the LLN does not apply, an artificial effect emerges that can be mistakenly interpreted as evidence for agglomeration economies. However, pure independent and identically distributed random variables can generate this effect, as illustrated in Fig. 3.5 (we encourage the reader to simulate this phenomenon and replicate this simple figure).

This effect stands as a cautionary tale about how we think about cities. Many mathematical models assume representative agents and apply averages to data to make statements about such agents. However, the heavy-tailed distributions in cities imply that such representative agents may not exist, and that computing empirical averages may be mistaken. Conceptually, our models should instead start from the assumption that heterogeneity is a fundamental characteristic of cities. Methodologically, this effect implies that our null hypothesis should be that observing a positive association between size and productivity ought to be expected even if there are no *real* causes correlating size and productivity.

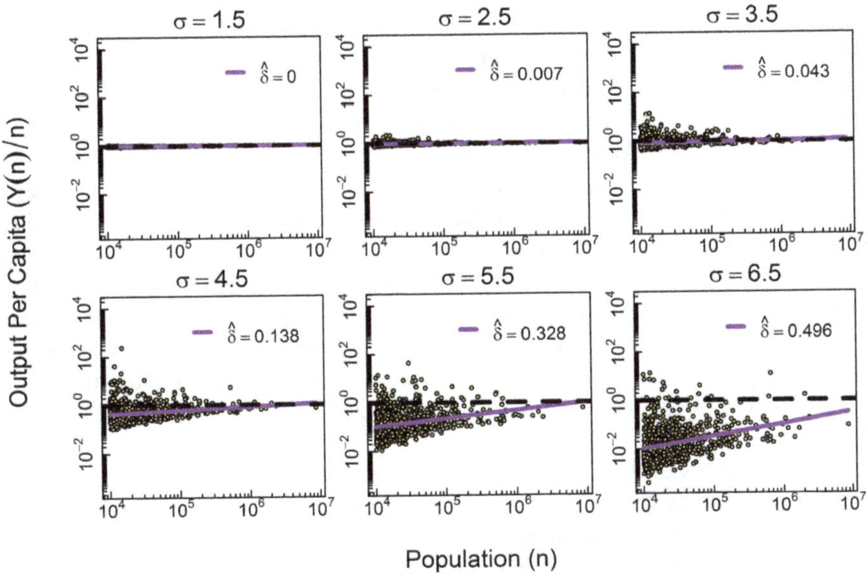

Fig. 3.5 Correlations between city size and measures of output per capita like wages emerge without agglomeration economies or self-sorting for pure statistical reasons due to wages being lognormally distributed. In these simulations, each dot represents a hypothetical city, and within each city individuals are assigned a wage sampled independent and identically from a lognormal distribution with same mean and standard deviation across all cities. Each panel plots the relationship that emerges for different standard deviations of log-wages σ. *Source* figure from [38], with permission

3.3.4 *Complex Systems: Urban Scaling Theory and Its Relationship to Agglomeration Economies*

As has been discussed elsewhere in this book, cities have been studied using the lens of Scaling Theory. The idea behind scaling is that associations of the form $Y = aX^b$ between quantities of interest and the scale of the system may indicate a property of the phenomenon of fundamental importance: scale-invariance. That is, the rule that explains the phenomenon acts the same across scales. The success of using scaling in physics and biology to derive theories for how systems work (see, e.g. [34, 39, 40]) encouraged the study of cities with the goal of finding a unified theory for how cities work [41].

Using scaling analysis, complex system scientists have observed that the relationship between many urban quantities Y and size N is characterized by a power-law function $(Y/Y_0) = (N/N_0)^{\beta}$, suggesting a universal mechanism at work (for an overview of the field see Chap. 2, as well as [42]). Indeed, urban metrics following this law span several different activities and have even been observed even in archaeological data (see, e.g. [43, 44]). Of particular interest for our current chapter are phenomena for which exponents β have been observed to be larger than 1. When expressing Y in per capita terms, $y = Y/N$, these *super-linear* power laws indicate that larger cities have higher y, because $y \propto N^{\delta}$, with $\delta = \beta - 1 > 0$.

What is the connection between the scaling relationships that have been identified by complexity scientists, and the relationship between productivity and population size that characterizes the phenomenon behind agglomeration economies, which is also described by a power-law (see Eq. (3.1)? The fact that there are other phenomena, different from income and wages, that occur with higher intensity in larger cities raises the question of whether the economic arguments described in Secs. 3.3.1 and 3.3.2 (ideas of spatial equilibrium, sharing, matching, or learning) are actually correct. Criminal activities, such as homicides, or the prevalence of diseases, such as cases of syphilis, also increase on a per capita basis with population size, following power-laws (see, e.g. [14, 45]). Would economic arguments make sense in order to explain why homicides, or cases of syphilis, scale with size? At a first look, it does not seem likely.

Some would argue that the theory on agglomeration economies emerging from economics should be abandoned given that a more general theory may exist. Surely, a theory that can explain a large number of urban properties has a significant advantage. Alternatively, however, future work could examine the possibility that specific economic theories, such as the ones in the space of agglomeration economies, are a special case of urban scaling theory.

The promise of a unified theory of cities derived from scaling laws has not been fulfilled yet. Nevertheless, a first parsimonious explanation for the super-linear scaling of economic variables from a complex systems point of view was proposed by Bettencourt (2013) and used the idea that larger populations are inherently more interconnected [46].

The argument is as follows. Assume that the output of a given individual in the city is proportional to the density of interactions that such person has with other people in the city, $y \propto \rho_{\text{social}}$. The density of interactions, in turn, is defined to be proportional to N/A_{social}, where population count is divided not by the surface area of the city, but by the "social area" in which interactions between people occur (e.g. the infrastructural network), A_{social}. Now, let people be evenly distributed throughout the city's surface are, A_{surface}, and let the average distance between individuals be d. Given this, the surface area is proportional to the population size multiplied by the area between them, $A_{\text{surface}} \propto Nd^2$. The social area where interactions occur represents only a small fraction of the city's surface area. The model assumes that the social area is actually defined by the *linear* space between people, which yields $A_{\text{surface}} \propto Nd$. Thus, if $\rho = N/A_{\text{surface}}$ is the population density, Bettencourt's assumption implies that the population density is proportional to the density of interactions squared, $\rho \propto \rho_{\text{social}}^2$. Going back to the output per capita, we get that

$$y \propto \rho_{\text{social}} \propto \rho^{1/2} . \tag{3.5}$$

According to Eq. (3.5), output per capita scales with the square root of city population density. In this model, this relationship holds regardless of the type of output, the physical makeup of the city, or the budget constraints of individuals. To derive a relation between y and N that does not depend on the area A_{surface}, however, one does need to make further assumptions. Thus, Bettencourt (2013) assumes that the minimum output per capita allowed in the city must be enough to pay for the cost of traversing the whole city. The minimum output is assumed to be proportional to the population density, $y_{\min} \propto \rho$ (not ρ_{social}), while the cost of traversing the city is assumed to be proportional to the linear extent of the city, $c \propto L \propto A_{\text{surface}}^{1/2}$. The condition $y_{\min} = c$ implies that $(N/A_{\text{surface}}) \propto A_{\text{surface}}^{1/2}$, such that one can write $A_{\text{surface}} \propto N^{2/3}$. Putting everything together yields

$$y \propto N^{1/6}\,. \tag{3.6}$$

In words, the model predicts an elasticity between productivity and size of $\delta = 1/6 \approx 16.7\,\%$, which is far from the estimates for developed countries we mentioned in Sect. 3.2.5, but not too far from the estimates for developing countries.

Note that Bettencourt's model is agnostic about how interactions are converted into output. The model does not indicate whether interactions represent instances of *sharing* resources, or *matches* between employers and employees (or suppliers and buyers), or cases of *learning* due to the interchange of information between individuals when they meet. In that sense, it is more general. Note, however, that the model has similar limitations in explaining phenomena like disease or homicides. It assumes social interactions are necessary and sufficient conditions to generate the different types of output (cases of crime and disease), and it either predicts a universal relationship with population density with an exponent of 0.5 that should hold across all activities (Eq. (3.5)), or a relationship with city size arising from economic assumptions that involve a 'minimum output per capita' equating a 'cost of transportation' (Eq. (3.6)).

Some improvements on Bettencourt's model have been proposed in which interactions are necessary but not sufficient (e.g. [47, 48]). These models also account for the cognitive constraints that limit the number of social links that humans can maintain [49–51], and consider individuals to be *combiners* of information, rather than conduits (for a review of the evidence in this area see [52]).

We highlight the model by Gomez-Lievano et al. (2017) [47], for example, in which people derive value not from the quantity of social links, but from the diversity that comes from them. In such model, individuals in larger cities are exposed to more diverse cultural factors, other people's know-how, and different pieces of information, which can be combined to generate value [53, 54]. In this model, therefore, larger sizes lead to more diversity, and more diversity leads to higher output.[13] Mathematically, a power-law scaling emerges as the combined outcome of exponentiating a logarithm, as $Y \propto e^{diversity} \propto e^{\beta \ln(N)+c} \propto N^{\beta}$. This model is well supported

[13] It is worthwhile pointing out that this finding aligns with the concept of urbanization economies, also known as the urbanization economies as opposed to the concept of localization economies. The former involves the spatial concentration of firms due to across-industry advantages (the richness of economic activity in cities). The latter involves the clustering of similar firms due to within-industry advantages.

by theory and data (e.g. see [55–58]). One of the predictions of this model is that socio-economic outputs that are more complex (i.e. those that require a higher diversity of inputs or ingredients to be produced) will display larger scaling exponents, will be less prevalent in general, and will be more variable from city to city. These predictions were recently verified by Balland et al. (2020) [59].

Fabiano and Diego [60] conducted a survey of the main mathematical models of urban scaling [60]. In general, urban scaling researchers tend to depict cities as more messy, more governed by stochasticity, intricate interactions, and serendipitous complementarities among heterogeneous actors, than mainstream urban economists, who like to emphasize cities' orderly structure, the economic forces of supply and demand, and the optimal decisions from rational agents. The actual world is likely to be a blend of these different perspectives, and a discussion between academics in urban economics and complex systems science should generate useful insights.

3.3.5 Complex Systems: Urban Scaling Theory Criticisms

Since the formulation of the urban scaling hypothesis [61–63], several researchers from within the complexity field have offered criticisms of the framework and the approach. The criticisms are mostly directed at the statistical methods (see, e.g. [13, 15, 64]), but some are aimed at the concept itself (e.g. [65–67]). Literature showing counterexamples or raising contradictions in statistical analyses can be found Chap. 2. We mostly focus below on conceptual issues related to urban scaling as they relate to agglomeration economies as discussed in previous sections.

Shalizi (2011) [13], for example, argued that the relationship between economic output and city size is not a power law. Shalizi suggested that other functions of population N can fit the relationship equally well. If this is true, then the relationship of urban characteristics and population is not scale invariant. Bettencourt et al. (2013) [68] respond to this criticism by stating that there is room for other explanations of urban scaling: "The fact that these other functions of N can fit urban data well (but not better) than power-laws in most observed cases [...] merely means that other models may be viable explanations of the scaling behavior of urban indicators, along with scale invariant functions". When statistics is not enough to distinguish among different functional forms, the authors argue one must rely on theory: "[B]ecause the scale-adjusted logarithmic function does not derive from theory to the best of our knowledge or fit the data manifestly better, we believe the hypothesis of urban scaling in terms of a scale invariant function, predicted by theory, remains at this point the better explanation" (p. 12).

Other criticisms, such as those of Leitão et al. (2016) [15] and Altmann (2020) [64], have indicated that a misspecification of the distribution of the dependent variable in urban scaling analysis, or the independence assumption between observations conventionally made in linear regressions, lead to wrong conclusions about the functional form relating outcomes Y with city sizes N. These are valid concerns, and researchers will need to rely on theory to decide which statistical methods and functions to use to fit the data, as Bettencourt et al. (2013) [68] argue.

To these methodological criticisms, we can add the empirical challenges we mentioned in Sect. 3.2.4, in which the statistical approaches used by urban scaling researchers to estimate the power-law exponents do not address the issues of confounders and reverse causality. Complexity scientists might argue that they are not concerned about reverse causality, since the question in urban scaling is not necessarily why size *causes* productivity, but why size and productivity maintain a very special relationship with one another across scales. The problem of confounders, however, still holds, since confounding adds spurious correlations between variables (see [45] for an example of urban scaling analysis that does take into account confounders).

But what if the very premise of urban scaling analysis, that aggregate urban quantities are a function of total size, is mistaken? Such conceptual questions started with Arcaute et al. (2015) [65], who noted that cities are not objects with clear physical boundaries, and therefore there is no definitive way of quantifying their size, or many other urban quantities. Thus, the premise of urban scaling may be misplaced. These authors showed how different ways of delineating the boundaries of cities return power-laws with different exponents, putting in doubt the very concept of agglomeration economies. Other studies have followed (e.g. [66, 67]) arguing that researchers should be more cautious about computing aggregate measures like total population size, or total output. Instead, they should include in their models the specific intra-city composition of economic and social activity and physical infrastructure in order to understand urban economic performance.

Note that the criticisms raised above have been addressed by mainstream urban economists by having a strict definition of cities as "labor markets" (i.e. places were people live and work), and by focusing their empirical work to studying the effects of agglomeration at the micro level of individual firms and workers. Given this focus, both their mathematical models and their statistical methods can be fairly detailed about mechanisms, causes and effects.

3.4 Complexity and Economics: Convergence for a Better Understanding of Cities?

Since at least the 1980s, several efforts of bringing a diverse set of economists, physicists and other scholars in fields related to complexity science have produced important research [69]. The field of complexity economics emerged from that work: assumptions of perfect rationality, well-defined problems and optimal behavior were relaxed. Heterogeneous agents, each with imperfect information and capable of adaptation to their environments were introduced. The approach brought about models of economies that are "organic, always creating itself, alive and full of messy vitality." [70]. Unfortunately, its impact in the mainstream of the economics profession is debatable.

The original challenge for the new field involved established research "cultures" and methodological or modeling choices or traditions. Mainstream economists have been modeling social phenomena utilizing representative agents (e.g. households, firms and developers) who solve optimization problems with the end outcome of the system arriving at an equilibrium. Complexity science-insprired economics models social phenomena utilizing heterogeneous agents, with localized knowledge, following rules and adapting to strategies of neighbors. While both approaches lead to emergent aggregate phenomena, the complexity economics approach is broader in scope and does not rely on limiting assumptions made in mainstream economics (representative agents, rationality, complete markets, etc). Below we summarize and contrasts the major features typically found in both mainstream urban economics and the complexity-oriented urban scaling theory approaches (Table 3.1).

Thisse (2014) [71] offers one of the few attempts made from mainstream economists, exploring the possibility of a synthesis between the field of urban economics and urban geography (which embodies several themes of complexity). Reviewing the work by Batty (2013) [72], he reaches a few conclusions that are revealing. Thisse focuses on the fundamental aspect of the methodological choices an economist makes and the associated belief system: "macroscopic phenomena can best be understood by examining how they arise from the motivated actions and interactions of individual agents in a world where various types of uncertainty prevail." This is not a point that a complexity scientist would contest. But he takes issue with an urban complexity view "formed by purposeless agents whose aggregate behavior is governed by stochastic laws." (Mainstream) economists model cities "as emerging from the decisions of many rational agents–firms, households, developers, and local governments–whose actions are often coordinated by markets." He correctly

Table 3.1 Summary of features of urban economics and urban scaling theory approaches. *Source* adapted from [70]

Feature	Urban economics	Urban scaling
Agents	Representative (but N different types)	Diverse
Behavior	Optimizing (depending on agent type–utility, profit, rents)	No goals other than interacting locally
Rationality	Perfect	Not defined
Interactions	Homogeneous (impersonal markets, no network formation)	Channeled by networks (social and physical) of differing strength and types
Interaction extent	Global	Localized
Theme	Optimal resource allocation and spatial equilibrium	Formation of structures, emergence of socioeconomic aggregates
Structural change	Equilibrium shifts	Endogenous re-organization

suggests that this difference in approaches is not "an insurmountable obstacle to interactions between urban geographers and urban economists".

While interactions can occur without significant effort, for unifying our understanding of the functioning of cities, we argue that what is needed is parallel shifts in urban modeler methodological positions. For example, on the one hand, economists need to rely less on representative agents and rational expectations and allow for agent heterogeneity. The strong emphasis on microfoundations leading to macro-outcomes has required unrealistic assumptions to the detriment of whole fields of study in economics [73]. Furthermore, network structure of interactions between people is important. Consumer and producers do not only interact in anonymous market through the price system, but develop relationships of differing strength in networks. Also, moving away from analytical models and adopting agent-based simulation modeling approaches in the context of economics could help bridge some of the gaps that exist. Some progress towards this goals already exists ([74]) but additional efforts will assist in the convergence of these knowledge domains.

On the other hand, urban scaling researchers (complexity scientists, physicists, statisticians, urban geographers, etc.) should attempt to further elaborate the existence of markets and distinct mechanisms of production and exchange in their models. The emergence of markets or other economic institutions from interactions within population (who adopt roles of producers and consumers) would be even more valuable. We are also faced with a paradox in urban scaling theory. In some ways, the theory appears to not allow for enough depth in idiosyncratic dimensions. Some people are more talented than others and could be attracted to bigger cities based on their individual characteristics. Some cities become diversified and some specialized in terms of their economic base. The migration of population from rural to urban areas is also not part of urban scaling theory. Business cycles–periods of economic booms and busts–are not explained. This richness of the urban experience is currently somewhat lost in a general theory of urbanization. Several attempts to close this gap are underway (e.g. [47, 48, 75, 76]).

Finally, it's worthwhile pointing out that differences between the two fields are also due to the way that research questions are asked. In urban scaling theory, the main question asked is whether the scaling of urban phenomena (an aggregate outcome) can be attributed to localized agent interactions (micro behavior). In economics, the questions are much more focused and directed: what type of economies do we observe? what are their causes within the operation of a market system? If scaling theory is indeed a 'General Theory of Urbanization', we should have observed similar scaling phenomena across distinct economic structures (for example, were systems of cities in the USSR experiencing similar scaling phenomena?) Some evidence exists on scaling of settlements in ancient societies – which were admittedly much simpler in terms of the functioning of markets, exchange and production. We argue that more research is needed when considering the potentially differential effects of modern economic structure.

3.5 Assessing Agglomeration Effects Beyond Economics and Complexity

We have summarized the topic of agglomeration economies from the economics and the complex systems perspectives. However, economists and complexity scientists are not the only ones who have studied the effects of agglomeration, and these other perspectives merit a brief mention.

We pointed out in Sect. 3.3.2 that the economist Alfred Marshall had speculated about the causes and the effects of physical proximity back in the 1890's. He wrote about this question after noticing that industries specialize geographically. As we explained, he contended that spatial proximity promotes the transmission of knowledge, reduces transport costs of inputs and outputs, and allows firms to benefit from a more efficient labor market. At around the same time, the French sociologist Émile Durkheim also remarked the effects of agglomeration from a sociological point of view. These and many other of Durkheim's ideas were published as part of his 1893 doctoral dissertation, *The Division of Labour in Society* [77]. Durkheim is seldom cited by economists and complexity scientists, but his influence might be as important, or more, than Marshall's. It is interesting to observe that the central issue of population scale and density have been addressed by foundational figures in both economics and sociology.

While Marshall emphasized the *immediate* effects of agglomeration on *firms*, Durkheim underlined the *collective* effects on *society*. In Durkheim's view, population density and growth were the crucial drivers of social and economic development. Durkheim's theory is interesting in that increasing the number of people in a location would increase the number of social interactions, but he did not interpret the latter to lead directly to prosperity. Instead, in Durkheim's account, the increase in the number of interactions would lead to more competition, division of labor, specialization, diversity, and ultimately, conflict. These, in turn, would force society to develop norms and behaviors to generate, sustain, and improve social integration. This vision was encapsulated in Durkheim's concept of "dynamic density".

Durkheim's ideas were further developed by the American sociologist Louis Wirth (see [78]), who wrote explicitly about cities, with clarity and insight. These ideas, however, were not formalized into mathematical models until the 1970s. Once sociologists started to connect models with data, a heated debate was sparked about the effects of urban agglomeration (see, e.g. [49, 79–84]). On the one hand, some researchers contended that Durkheim and Wirth's theories were unnecessarily complicated. The fact that social interactions scale as the square of population size, $I \propto N^2$, was enough to explain the positive relationships between socio-economic rates and the size of cities. Mayhew and Levinger (1976) [81], for example, used this argument to explain the higher rates of crime in large cities. On the other, such mathematical arguments were viewed as too simplistic, missing "substantive understanding" of how social systems work. It is instructive to read, for example, the criticism raised by the sociologist Claude S. Fischer (1977) [82]. For Fischer, in brief, the fact that *potential* interactions grow super-linearly with size does not imply that *actual* interactions do ("why do they [the proponents of the interactions-based theory] not predict increasing marriage or birth rates with city size?", p. 453). Fischer condemns such models, because they "assume that people in cities are like numbered marbles in a large urn, constantly being shaken up and randomly 'contacting' one another. [These models are] not even a heuristically useful approximation of the truth. Casual reading of urban ethnographies—even a casual look out the window—will show that urbanites' lives cannot be described this way. Most city dwellers lead sensible, circumscribed lives, rarely go downtown, hardly know areas of the city they neither live nor work in, and see (in any sociologically meaningful way) only a tiny fraction of the city's population."

Finally, a similar debate about the effects of population size on social systems has taken place, almost in parallel and with no connection to economics, complexity, or sociology, in the fields of cultural and evolutionary anthropology, where the issue is referred to as the "population size hypothesis" (see, e.g. [55, 56, 85–89]). In anthropology, one of the central concerns is to explain how cultures have evolved, allowing humans to occupy almost all ecological habitats on Earth (without subspeciation, which is the typical mechanism for all other organisms to adapt to new environments) [90]. Given the limits to what they can measure from archaeological and ethnographic data, anthropologists have studied the "cultural complexity" of societies (rather than productivity), defined by them as the number of cultural traits (literally, the count of distinct artefacts, social roles, etc).

When data started to accumulate enough to be amenable to statistical analysis, a clear relationship between population size and cultural complexity across several societies was found in 1960s [85].[14] That is, societies that had larger population sizes were observed to have more technologies, deeper divisions of labor, and more political, religious, and organizational units (for recent analysis of this association see [56, 89]). Since then, the question has been whether population size leads to higher rates of cultural innovation, vice versa, or whether there are other variables causing both in tandem (i.e. the same questions we ask in Fig. 3.4).

One of the current theories to explain why larger population sizes give rise to higher rates of cultural complexity is Joseph Henrich's "collective brain hypothesis" [54]. The argument is very different from those of economists, complexity scientists, or sociologists that we have reviewed so far in that Henrich's explanation is rooted in evolutionary theory (the field itself is called "cultural evolution"). The basis on which this theory is built are the recent observations that show that humans are super-imitators (for a review, see [93]). This unique cognitive feature, unmatched across the animal kingdom, allows innovations to spread and pass down across generations without deterioration [55]. As a result of this unique ability to mimic our close peers, cultures can accumulate an ever-expanding repertory of practices, conventions, tools, and know-how. The cultural brain hypothesis suggests that as culture

[14] The anthropologist Robert L. Carneiro was a strong supporter of adapting insights from the physical and natural sciences into anthropology. Incidentally, he introduced an idea from psychology called "scale analysis" [91], which should not be confused with "scaling analysis" as used in the urban scaling literature. His methodologies pre-dated the development of the field of economic complexity by almost 50 years (see, e.g. [92]).

evolves and expands, it exerts evolutionary forces that adjust sociality, imitation, and transmission variance to cope with the increased complexity of tools, practices, beliefs, and behaviors. In this view, the characteristics of individuals remain more or less constant across populations of different sizes, but not the "collective brain" of those societies. Instead, collective brains get larger as populations get larger. Hence, agglomeration leads to more innovation, higher productivity, and even higher IQ scores, but not because of smarter individual brains, but because of smarter collective brains.

How do these multi-disciplinary perspectives enrich our understanding of agglomeration economies in cities? Complexity scientists ought to be in the position to synthesize the key ideas and propose new ones. We end this chapter with some final conclusions and questions for further thought.

3.6 Conclusions and Open Questions

We argue that understanding cities is a very important priority for humanity, as we increasingly become an urban civilization. This understanding passes through the understanding of both benefits and costs of agglomerating.

Researchers exploring this fundamental aspect of the function of cities should be aware of the multiplicity of issues related to measurement of productivity and scale (population size vs. population density), as well as the definition of the boundaries of a city or a metropolitan area. Conducting sensitivity analysis across alternative measurement choices should be an aim of empirical research.

The economics of density, both in terms of the benefits through agglomeration economies, and the costs–congestion, crime, etc.-can be viewed both complimentary but also as a subset of the work occurring through complexity science. Agglomeration economies, from the economics perspective, is a phenomenon that relates to firms and the production of goods and services. The explanations offered by scaling theory could be viewed as a more abstract level of explanation of agglomeration economies; still, explanations offered by mainstream economists add a level of detail that is not addressed yet in the broader scaling context. Economic theory conceptualizes agglomeration economies as an economic production and cost specific concept while urban scaling theory discusses it as a broader concept, with application beyond the economics conceptualization. Unifying the approaches will, in the foreseeable future, be a challenge due to methodological differences between the fields–mainstream economists are averse to the idea of agents who are not driven by some degree of self interest and the lack of modeling of markets and exchange.

Perhaps what would be beneficial for a possible unification would be that the complexity approach adopts some micro-foundations that align with the economics approach. This could be amended by developing spatially- and network-explicit agent-based models of urban economies across scales- modeling consumption and production processes within and across cities as well as other dimensions of urban life. These models are expected to produce emergent power/scaling laws for various urban phenomena in systems of cities. In this way, explanations of the observed phenomena across the distinct modeling traditions could be more directly linked.

Urban scaling theory has done a substantial amount of work on the costs of scale or density–what Glaeser (2011) called the "demons of density". While some of the obvious challenges were identified early in the economics literature [94], comparatively, very little attention has been placed on the microfoundations of costs of agglomeration as compared to the benefits of agglomeration. Complementarities could be present in this space, and urban scaling theory could offer guidance. More attention should be placed in uniting social economics theories to those of urban economics, perhaps supplemented by social psychology theories on crowding.

It is worthwhile noting that as more detailed models of urban scaling come about, the connections between the complexity approach and other fields can begin to emerge. For example, modeling people as derive value not from the quantity of social links, but from the diversity that comes from them [47] aligns with the concept of urbanization economies, emerging in part from what is known as Jacobs knowledge spillovers (for a review of evidence for Jacobs externalities, see [95]). Still, urban scaling researchers need to address the concept of localization economies and identify whether scaling theory has a contribution to make in the discussion of localization vs. urbanization economies.

The topic of this chapter presents several opportunities for novel research and synthesis of the existing approaches. Thus, we end with some open questions and proposed topics for future research.

Questions for further discussion

- Are the causes of agglomeration economies as explained by mainstream economists a subset of a complexity science based urban scaling theory? For example, can we express sharing, matching and learning as production theory concepts that are special cases of broader concepts in scaling theory? Would those be: quantity of linkages, quality of linkages and adaptation? Or other features of scaling/complexity theory?
- Can urban scaling theory incorporate distinct types of agents interacting in networks in a dynamic setting, through evolving norms and institutions? E.g. firm formation, land development activity, evolving markets and economic institutions, etc.
- What are the possible pathways through which mainstream urban economists can integrate concepts from complexity science into their theory and models?
- How to enrich our understanding of the advantages of agglomeration as arising from evolutionary processes that involve history and path-dependence?

References

1. E.L. Glaeser, *Cities, Agglomeration, and Spatial Equilibrium*. (OUP Oxford, 2008)
2. E.L. Glaeser, *Triumph of the city: how our greatest invention makes us richer, smarter, greener, healthier, and happier*. (Penguin Press, New York, 2011). ISBN: 9781594202773
3. A. O'Sullivan, The first cities, in *A Companion to Urban Economics*. (John Wiley & Sons, Ltd, 2006). Chap. 3, pp. 40–54. ISBN: 9780470996225. https://doi.org/10.1002/9780470996225.ch3
4. P. Krugman, Increasing returns and economic geography, in *Journal of Political Economy* 99 (3 1991), pp. 483–499. https://doi.org/10.1086/261763
5. M. Fujita, P.R. Krugman, A. Venables, *The Spatial Economy: Cities, Regions, and International Trade* (MIT press, 1999)
6. G.S. Goldstein, T.J. Gronberg, Economies of scope and economies of agglomeration, in *Journal of Urban Economics* 16.1 (1984), pp. 91–104. https://doi.org/10.1016/0094-1190(84)90052-4
7. L. Sveikauskas, The productivity of cities, in *The Quarterly Journal of Economics* 89.3 (1975), pp. 393–413. URL: http://www.jstor.org/stable/1885259
8. Bureau of Economic Analysis (BEA). *To access the data, go to "PERSONAL INCOME AND EMPLOYMENT BY COUNTY AND METROPOLITAN AREA", click "Personal Income, Population, Per Capita Personal Income (CAINC1)", select "Metropolitan Statistical Area" and download the data for "All areas". Last accessed: May 15, 2022.* URL: https://apps.bea.gov/iTable/iTable.cfm?reqid=70%5C&step=1%5C&acrdn=6 (visited on 05/15/2022)
9. Gilles Duranton and Diego Puga. "The economics of urban density, in *Journal of economic perspectives* 34.3 (2020), pp. 3–26. https://doi.org/10.1257/jep.34.3.3
10. Jesus Felipe and Franklin M Fisher. "Aggregation in production functions: what applied economists should know, in *Metroeconomica* 54.2-3 (2003), pp. 208–262. https://doi.org/10.1111/1467-999X.00166
11. Haroldo V Ribeiro, Quentin S Hanley, and Dan Lewis. "Unveiling relationships between crime and property in England and Wales via density scale-adjusted metrics and network tools, in *PLoS One* 13.2 (2018), e0192931. https://doi.org/10.1371/journal.pone.0192931
12. Ronald L Moomaw. "Productivity and city size: a critique of the evidence, in *The Quarterly Journal of Economics* 96.4 (1981), pp. 675–688. https://doi.org/10.2307/1880747
13. Cosma Rohilla Shalizi. "Scaling and hierarchy in urban economies, in *arXiv preprint* arXiv:1102.4101 (2011). https://doi.org/10.48550/arXiv.1102.4101
14. Andres Gomez-Lievano, HyeJin Youn, and Luis M. A. Bettencourt. "The Statistics of Urban Scaling and Their Connection to Zipf's Law, in *PLoS ONE* 7.7 (July 2012), e40393. https://doi.org/10.1371/journal.pone.0040393
15. Jorge C Leitao et al. "Is this scaling nonlinear?" In: *Royal Society open science* 3.7 (2016), p. 150649. https://doi.org/10.1098/rsos.150649
16. Bureau of Labor Statistics (BLS). *Occupational Employment and Wage Statistics (OEWS), May 2020, 'Metropolitan and nonmetropolitan area'. Last accessed: May 15, 2022.* URL: https://www.bls.gov/oes/special-requests/oesm20ma.zip (visited on 05/15/2022)
17. Gabriel M. Ahlfeldt and Elisabetta Pietrostefani. "The economic effects of density: A synthesis, in *Journal of Urban Economics* 111 (2019), pp. 93–107. ISSN: 0094-1190. https://doi.org/10.1016/j.jue.2019.04.006
18. Pierre-Philippe Combes et al. "Estimating agglomeration economies with history, geology, and worker effects, in *Agglomeration economics*. University of Chicago Press, 2010, pp. 15–66. https://doi.org/10.7208/chicago/9780226297927.003.0002
19. Edward L Glaeser and David C Mare. "Cities and skills, in *Journal of Labor Economics* 19.2 (2001), pp. 316–342. https://doi.org/10.1086/319563
20. Pierre-Philippe Combes, Gilles Duranton, and Laurent Gobillon. "The identification of agglomeration economies, in *Journal of Economic Geography* 11.2 (2011), pp. 253–266. https://doi.org/10.1093/jeg/lbq038

21. Pierre-Philippe Combes and Laurent Gobillon. "Chapter 5–The Empirics of Agglomeration Economies, in *Handbook of Regional and Urban Economics*. Ed. by J. Vernon Henderson Gilles Duranton and William C. Strange. Vol. 5. Handbook of Regional and Urban Economics. Elsevier, 2015, pp. 247–348. https://doi.org/10.1016/B978-0-444-59517-1.00005-2
22. Pierre-Philippe Combes, Gilles Duranton, and Laurent Gobillon. "Spatial wage disparities: Sorting matters!" In: *Journal of Urban Economics* 63.2 (2008), pp. 723–742. https://doi.org/10.1016/j.jue.2007.04.004
23. Patricia C. Melo, Daniel J. Graham, and Robert B. Noland. "A meta-analysis of estimates of urban agglomeration economies, in *Regional Science and Urban Economics* 39.3 (2009), pp. 332–342. https://doi.org/10.1016/j.regsciurbeco.2008.12.002
24. Arti Grover, Somik V. Lall, and Jonathan Timmis. *Agglomeration Economies in Developing Countries*. Policy Research Working Papers. World Bank, Washington, DC, 2021. URL: http://hdl.handle.net/10986/36003
25. Michael Batty. "Cities in Disequilibrium, in *Non-Equilibrium Social Science and Policy*. Ed. by Jeffrey Johnson et al. Series Title: Understanding Complex Systems. Springer International Publishing, 2017, pp. 81–96. ISBN: 978-3-319-42422-4 978-3-319-42424-8. https://doi.org/10.1007/978-3-319-42424-8_6. (Visited on 11/01/2023)
26. Herbert A Simon. "A behavioral model of rational choice, in *The Quarterly Journal of Economics* (1955), pp. 99–118. https://doi.org/10.2307/1884852
27. David Reich. *Who we are and how we got here : ancient DNA and the new science of the human past*. eng. First edition. New York: Pantheon Books, 2018. ISBN: 1-101-87033-8
28. Peter J Richerson and Robert Boyd. "Migration: An engine for social change, in *Nature* 456.7224 (2008), pp. 877–877. https://doi.org/10.1038/456877a
29. David Starrett. "Market allocations of location choice in a model with free mobility, in *Journal of economic theory* 17.1 (1978), pp. 21–37. https://doi.org/10.1016/0022-0531(78)90120-5
30. Gianmarco Ottaviano and Jacques-François Thisse. "Chapter 58–Agglomeration and Economic Geography, in *Cities and Geography*. Ed. by J. Vernon Henderson and Jacques-François Thisse. Vol. 4. Handbook of Regional and Urban Economics. Elsevier, 2004, pp. 2563–2608. https://doi.org/10.1016/S1574-0080(04)80015-4
31. Alfred Marshall. *Principles of Economics*. Macmillan and Company, 1890
32. Gilles Duranton and Diego Puga. "Micro-foundations of urban agglomeration economies, in *Handbook of regional and urban economics* 4 (2004), pp. 2063–2117. https://doi.org/10.1016/S1574-0080(04)80005-1
33. Dario Diodato, Frank Neffke, and Neave O'Clery. "Why do industries coagglomerate? How Marshallian externalities differ by industry and have evolved over time, in *Journal of Urban Economics* 106 (2018), pp. 1–26. https://doi.org/10.1016/j.jue.2018.05.002
34. Manfred Schroeder. *Fractals, chaos, power laws: Minutes from an infinite paradise*. Dover, 1991
35. Didier Sornette. *Critical Phenomena in Natural Sciences: Chaos, Fractals, Selforganization and Disorder: Concepts and Tools*. 2nd. Springer Series in Synergetics. Heidelberg: Springer, 2006
36. M. E. J. Newman. "Power laws, Pareto distributions and Zipf's law, in *Cont. Phys.* 46.5 (Sept. 2005), pp. 323–351. https://doi.org/10.1080/00107510500052444
37. Paul Embrechts, Claudia Klüppelberg, and Thomas Mikosch. *Modelling extremal events: for insurance and finance*. Vol. 33. Springer, 1997
38. Andrés Gómez-Liévano, Vladislav Vysotsky, and Jose Lobo. "Artificial increasing returns to scale and the problem of sampling from lognormals, in *Environment and Planning B: Urban Analytics and City Science* 48.6 (2021), pp. 1574–1590. https://doi.org/10.1177/2399808320942366
39. Geoffrey B West, James H Brown, and Brian J Enquist. "A general model for the origin of allometric scaling laws in biology, in *Science* 276.5309 (1997), pp. 122–126
40. Grigory Isaakovich Barenblatt. *Scaling*. Vol. 34. Cambridge University Press, 2003
41. Luís M. A. Bettencourt et al. "Urban Scaling and Its Deviations: Revealing the Structure of Wealth, Innovation and Crime across Cities, in *PLoS ONE* 5.11 (Nov. 2010), e13541. https://doi.org/10.1371/journal.pone.0013541

42. Luís MA Bettencourt. *Introduction to urban science: evidence and theory of cities as complex systems*. MIT Press, 2021
43. Scott G Ortman et al. "The pre-history of urban scaling, in *PloS one* 9.2 (2014), e87902. https://doi.org/10.1371/journal.pone.0087902
44. Jose Lobo et al. "Settlement scaling theory: Bridging the study of ancient and contemporary urban systems, in *Urban Studies* 57.4 (2020), pp. 731–747. https://doi.org/10.1177/0042098019873796
45. Oscar Patterson-Lomba et al. "Per capita incidence of sexually transmitted infections increases systematically with urban population size: a cross-sectional study, in *Sexually Transmitted Infections* (2015), pp. 1–5. https://doi.org/10.1136/sextrans-2014-051932
46. Luís M. A. Bettencourt. "The Origins of Scaling in Cities, in *Science* 340.6139 (June 2013), pp. 1438–1441. https://doi.org/10.1126/science.1235823
47. Andres Gomez-Lievano, Oscar Patterson-Lomba, and Ricardo Hausmann. "Explaining the prevalence, scaling and variance of urban phenomena, in *Nature Human Behaviour* 1 (2017), p. 0012. https://doi.org/10.1038/s41562-016-0012
48. Andres Gomez-Lievano and Oscar Patterson-Lomba. "Estimating the drivers of urban economic complexity and their connection to economic performance, in *Royal Society open science* 8.9 (2021), p. 210670. https://doi.org/10.1098/rsos.210670
49. Stanley Milgram. "The Experience of Living in Cities, in *Science* 167.3924 (1970), pp. 1461–1468. ISSN: 00368075, 10959203. URL: http://www.jstor.org/stable/1728966 (visited on 10/16/2022)
50. Robin IM Dunbar. "The social brain hypothesis, in *Evolutionary Anthropology: Issues, News, and Reviews: Issues, News, and Reviews* 6.5 (1998), pp. 178–190. https://doi.org/10.1002/(SICI)1520-6505(1998)6:5%3C178::AID-EVAN5%3E3.0.CO;2-8
51. Bruno Gonçalves, Nicola Perra, and Alessandro Vespignani. "Modeling users' activity on twitter networks: Validation of dunbar's number, in *PloS one* 6.8 (2011), e22656. https://doi.org/10.1371/journal.pone.0022656
52. Laurel Fogarty, Nicole Creanza, and Marcus W Feldman. "Cultural evolutionary perspectives on creativity and human innovation, in *Trends in ecology & evolution* 30.12 (2015), pp. 736–754. https://doi.org/10.1016/j.tree.2015.10.004
53. Martin L Weitzman. "Recombinant growth, in *The Quarterly Journal of Economics* 113.2 (1998), pp. 331–360. https://doi.org/10.1162/003355398555595
54. Michael Muthukrishna and Joseph Henrich. "Innovation in the collective brain, in *Philosophical Transactions of the Royal Society B: Biological Sciences* 371.1690 (2016), p. 20150192. https://doi.org/10.1098/rstb.2015.0192
55. Joseph Henrich. "Demography and cultural evolution: how adaptive cultural processes can produce maladaptive losses: the Tasmanian case, in *American Antiquity* (2004), pp. 197–214. https://doi.org/10.2307/4128416
56. Michelle A Kline and Robert Boyd. "Population size predicts technological complexity in Oceania, in *Proceedings of the Royal Society B: Biological Sciences* 277.1693 (2010), pp. 2559–2564. https://doi.org/10.1098/rspb.2010.0452
57. Hyejin Youn et al. "Scaling and universality in urban economic diversification, in *Journal of The Royal Society Interface* 13.114 (2016). ISSN: 1742-5689. https://doi.org/10.1098/rsif.2015.0937
58. James M Zumel Dumlao et al. *Innovation Dynamics of Cultural Production: Evidence in Rap Lyrics*. Tech. rep. San Francisco, CA: University of San Francisco, 2020. URL: https://repository.usfca.edu/thes/1315
59. Pierre-Alexandre Balland et al. "Complex economic activities concentrate in large cities, in *Nature human behaviour* 4.3 (2020), pp. 248–254. https://doi.org/10.1038/s41562-019-0803-3
60. Fabiano L Ribeiro and Diego Rybski. "Mathematical models to explain the origin of urban scaling laws: a synthetic review, in *arXiv preprint* arXiv:2111.08365 (2021). https://doi.org/10.48550/arXiv.2111.08365

61. Deborah Strumsky, José Lobo, and Lee Fleming. *Metropolitan Patenting, Inventor Agglomeration and Social Networks: A Tale of Two Effects*. SFI Working Paper 2005-02-004. Santa Fe, NM: Santa Fe Institute, 2005
62. Fabien Paulus et al. "An evolutionary theory for interpreting urban scaling laws, in *Cybergeo: Revue europ é enne de g é ographie/European journal of geography*.343 (2006), p. 20. https://doi.org/10.4000/cybergeo.2519
63. Luis M. A. Bettencourt, José Lobo, and Deborah Strumsky. "Invention in the city: Increasing returns to patenting as a scaling function of metropolitan size, in *Research Policy* 36.1 (Feb. 2007), pp. 107–120. ISSN: 0048-7333. https://doi.org/10.1016/j.respol.2006.09.026
64. Eduardo G Altmann. "Spatial interactions in urban scaling laws, in *Plos one* 15.12 (2020), e0243390. https://doi.org/10.1371/journal.pone.0243390
65. Elsa Arcaute et al. "Constructing cities, deconstructing scaling laws, in *Journal of the royal society interface* 12.102 (2015), p. 20140745. https://doi.org/10.1098/rsif.2014.0745
66. Clémentine Cottineau et al. "Diverse cities or the systematic paradox of urban scaling laws, in *Computers, environment and urban systems* 63 (2017), pp. 80–94. https://doi.org/10.1016/j.compenvurbsys.2016.04.006
67. Somwrita Sarkar et al. "Evidence for localization and urbanization economies in urban scaling, in *Royal Society Open Science* 7.3 (2020), p. 191638. https://doi.org/10.1098/rsos.191638
68. Luís M. A. Bettencourt, José Lobo, and Hyejin Youn. "The hypothesis of urban scaling: formalization, implications and challenges". arXiv:1301.5919v1 [physics.soc-ph] 24 Jan 2013. 2013. arXiv:1301.5919
69. Philip W. Anderson, Kenneth J. Arrow, and D. Pines. *The economy as an evolving complex system*. Redwood City, CA: Addison-Wesley, 1988
70. W.B. Arthur. "Foundations of complexity economics, in *Nature Reviews: Physics* 3 (2021), pp. 136–145. https://doi.org/10.1038/s42254-020-00273-3
71. Jacques-François Thisse. "The new science of cities by Michael Batty: The opinion of an economist, in *Journal of Economic Literature* 52.3 (2014), pp. 805–19. https://doi.org/10.1257/jel.52.3.805
72. Michael Batty. *The new science of cities*. MIT press, 2013
73. Joseph E Stiglitz. "Where modern macroeconomics went wrong, in *Oxford Review of Economic Policy* 34.1-2 (2018), pp. 70–106. https://doi.org/10.1093/oxrep/grx057
74. Kristian Behrens and Frédéric Robert-Nicoud. "Agglomeration theory with heterogeneous agents, in *Handbook of regional and urban economics* 5 (2015), pp. 171–245. https://doi.org/10.1016/B978-0-444-59517-1.00004-0
75. Denise Pumain. "An evolutionary theory of urban systems, in *Rozenblat C., Pumain D., Velasquez E. International and Transnational Perspectives on Urban Systems*. Springer, 2018, pp. 3–18
76. Luís MA Bettencourt. "Towards a statistical mechanics of cities, in *Comptes Rendus Physique* 20.4 (2019), pp. 308–318. https://doi.org/10.1016/j.crhy.2019.05.007
77. Émile. Durkheim, *De la division du travail social* (Félix Alcan, Paris, 1893)
78. Louis Wirth, "Urbanism as a Way of Life. American journal of sociology **44**(1), 1–24 (1938)
79. John D Kasarda. "The structural implications of social system size: A three-level analysis, in *American Sociological Review* (1974), pp. 19–28. https://doi.org/10.2307/2094273
80. John D Kasarda and Patrick D Nolan. "Ratio measurement and theoretical inference in social research, in *Social Forces* 58.1 (1979), pp. 212–227. https://doi.org/10.1093/sf/58.1.212
81. Bruce H. Mayhew and Roger L. Levinger. "Size and the Density of Interaction in Human Aggregates, in *American Journal of Sociology* (1976), pp. 86–110. https://doi.org/10.1086/226271
82. Claude S. Fischer. *Comment on Mayhew and Levinger's "Size and the Density of Interaction in Human Aggregates"*. 1977
83. Kazuo Seiyama. "A review of mathematical models of formal organizations–why and how they failed –, in *Journal of Mathematical Sociology* 12.1 (1986), pp. 71–96. https://doi.org/10.1080/0022250X.1986.9990005

84. Thomas Rotolo and Charles R Tittle. "Population size, change, and crime in US cities, in *Journal of Quantitative Criminology* 22.4 (2006), pp. 341–367. https://doi.org/10.1007/s10940-006-9015-x
85. Robert L. Carneiro. "On the relationship between size of population and complexity of social organization, in *Journal of Anthropological Research* 42.3 (1967), pp. 355–364. https://doi.org/10.1086/jar.42.3.3630039
86. Ester Boserup, *Population and technological change: A study of long-term trends* (University of Chicago Press, Chicago Ill, United States, 1981)
87. Ronald Demos Lee. "Induced population growth and induced technological progress: Their interaction in the accelerating stage, in *Mathematical Population Studies* 1.3 (1988), pp. 265–288. https://doi.org/10.1080/08898488809525278
88. Stephen Shennan. "Demography and cultural innovation: a model and its implications for the emergence of modern human culture, in *Cambridge archaeological journal* 11.1 (2001), p. 5. https://doi.org/10.1017/S0959774301000014. (Visited on 05/28/2017)
89. Mark Collard, Briggs Buchanan, and Michael J O'Brien. "Population size as an explanation for patterns in the Paleolithic archaeological record: more caution is needed, in *Current Anthropology* 54.S8 (2013), S388–S396. https://doi.org/10.1086/673881
90. Joseph Henrich. *The secret of our success: how culture is driving human evolution, domesticating our species, and making us smarter*. Princeton University Press, 2015
91. Robert L Carneiro. "Scale analysis as an instrument for the study of cultural evolution, in *Southwestern Journal of Anthropology* 18.2 (1962), pp. 149–169. https://doi.org/10.1086/soutjanth.18.2.3629014
92. César A. Hidalgo and Ricardo Hausmann. "The building blocks of economic complexity, in *PNAS* 106.25 (June 2009), pp. 10570–10575. https://doi.org/10.1007/s10887-011-9071-4
93. Kevin N. Laland. *Darwin's unfinished symphony: how culture made the human mind*. eng. Princeton: Princeton University Press, 2017. ISBN: 9780691151182
94. Israel Pressman and Arthur Carol. "Crime as a diseconomy of scale, in *Review of Social Economy* 29.2 (1971), pp. 227–236. https://doi.org/10.1080/00346767100000035
95. Catherine Beaudry and Andrea Schiffauerova. "Who's right, Marshall or Jacobs? The localization versus urbanization debate, in *Research policy* 38.2 (2009), pp. 318–337. https://doi.org/10.1016/j.respol.2008.11.010

Chapter 4
Urban Mobility

Laura Alessandretti and Michael Szell

Abstract In this chapter, we discuss urban mobility from a complexity science perspective. First, we give an overview of the datasets that enable this approach, such as mobile phone records, location-based social network traces, or GPS trajectories from sensors installed on vehicles. We then review the empirical and theoretical understanding of the properties of human movements, including the distribution of travel distances and times, the entropy of trajectories, and the interplay between exploration and exploitation of locations. Next, we explain generative and predictive models of individual mobility, and their limitations due to intrinsic limits of predictability. Finally, we discuss urban transport from a systemic perspective, including system-wide challenges like ridesharing, multimodality, and sustainable transport.

4.1 Introduction and History of Quantitative Mobility Studies

From commuting to work, running errands, to going on a leisure trip, mobility is an integral part of our daily lives. Humans allocate a significant amount of time, money and energy to travel, with the average US household spending on mobility a non-negligibly high fraction of around 16% of its budget [1], and with commuting accounting on average for around 56 minutes per day [2]. Mobility underlies some of the most critical challenges of our times, including the design of sustainable urban transportation and the containment of epidemics. For example, a badly designed transport system can drive critical problems such as congestion, road deaths, and pollution. Transportation is estimated to account for around 27% of greenhouse gas emissions in the US, out of which around 57% comes from light-duty on-road

---

L. Alessandretti
DTU Compute, Technical University of Denmark, Lyngby, Denmark
e-mail: laualc@dtu.dk

M. Szell (✉)
NEtwoRks, Data, and Society (NERDS), ITU Copenhagen, Copenhagen, Denmark
e-mail: misz@itu.dk

D. Rybski (ed.), *Compendium of Urban Complexity*, Understanding Complex Systems,
https://doi.org/10.1007/978-3-031-82666-5_4

vehicles (cars, motorbikes, vans) [3]. People on the move drive social mixing which in turn can facilitate the spread of infectious diseases [4]. Important societal issues are deeply intertwined with mobility patterns, and it has become critical to quantify and understand human movements. Understanding human mobility is important not only from a theoretical standpoint but also crucial to inform policy makers on how to improve urban transport and infrastructure, with the ultimate goal to invest into policies that improve the movement of people or that reduce the need for movement in the first place.

The study of human movements has a long history. The first rigorous approaches to quantify human movement patterns can be traced back to the 19th century, for example to Eduard Lill [5], statistician and chief inspector of the imperial-royal Austrian north-west railway company, who noticed that the number of people traveling from a place decreases as a hyperbola with distance, eventually giving rise to the gravity laws of travel discussed in Chap. 6. From this statistical perspective, the aggregate of many people can be described via simple mathematical laws, despite each individual having their individual, unique reasons for movement. This perspective is taken to the extreme in statistical physics, where large numbers of particles are easily described via their emergent properties, for example the temperature and pressure of a gas. This reductionist Ansatz, and the corresponding tools developed in statistical physics, are powerful ways to understand large numbers of interacting or moving entities *statistically*, whether they are particles or humans [6].

From these early studies on human mobility, we have come a long way. As an empirical science, the availability of high-resolution data has been crucial to drive evidence-based understanding. Such data have started to become available at the start of the 21st century, with first the diffusion of mobile phones, smartcards, and GPS positioning systems installed on vehicles, then smartphones and wearable devices. Concurrently with the development of increasingly powerful computational tools, the data revolution has enabled a richer and richer understanding of human mobility patterns. First, scientists could shift focus from the collective to the individual level, by studying single trajectories that have increasingly high temporal and spatial resolution, to the point that we can now trace actual route choices. Second, we are achieving a better and better understanding of the mechanisms underlying mobility behavior, through enriching mobility data with additional information, such as the features of the visited locations, individuals' social interactions, environmental and weather data, and features of the built environment [7].

In this chapter, we give an overview on how the field of complex systems, shaped considerably by statistical physicists, makes use of and adapts their tools in the context of urban mobility. Our chapter does not claim completeness but is an introductory overview of the fundamental concepts with hand-picked highlights; for a comprehensive review see, e.g., Barbosa et al. [8]. We outline the main aspects of the field by following largely the historical co-evolution of data and research. In Sect. 4.2 we discuss sources of data and issues of data quality. In Sect. 4.3 we introduce metrics and descriptive results about the statistical properties of individual empirical trajectories. In Sect. 4.4 we summarize the main statistical and mechanistic models about urban mobility patterns. Although we explicitly do not cover other approaches to urban mobility, for example from transport engineering, we follow up with an outlook in Sect. 4.5 on sustainable transport and urban livability due to the urgency of this topic in the context of the climate crisis, asking to which extent some of the long-range or high-speed movements in cities of today are actually necessary and how they could be avoided. Finally, we provide a list of important tools and further materials in Sect. 4.6 and end with conclusions in Sect. 4.7.

4.2 Data Sources

Since the early 2000s, the development and widespread adoption of technologies that record the positions of individuals over time has driven a rapid growth of the field of human mobility, through enabling access to data [7]. Here, we describe some of the most widely used data sources.

4.2.1 Data from Mobile Network Operators

Mobile network operators gather location data for billing and operational purposes, through monitoring the cell towers that user devices connect to. These data have been widely used for human mobility research. Data collected by network operators has inherently limited spatial resolution, because the coverage of cell towers is quite wide, ranging from some tens of meters in urban areas up to tens of kilometers in rural areas. In terms of temporal resolution, there has been a rapid evolution over the last 20 years. Before the diffusion of smartphones, research was largely based on Call Detail Records (CDR), which captures the closest cell tower to individuals as they issue or receive calls or SMS. More recently, researchers have gained access to higher-frequency data. eXtended Detail Records (XDR) are collected when users explicitly request an http address or when the phone downloads content from the Internet (e.g., emails, messages, app updates). Control Plane Records (CPRs) are network-triggered (e.g., assigning a new antenna, connecting new devices) and are used to monitor the cellphone network status. CDR, XDR, and CPR vary significantly in their time granularity and data sparsity [9].

4.2.2 Data from Smartphone Applications

Other useful data sources are represented by services that collect Global Positioning System (GPS) positions through smartphone applications or services. In the early years of human mobility research, these data would mostly be collected by Location Based Social Networks (LSBN) services, such as Foursquare or Twitter (now "X"), which would gather location data only when individuals used the service.

Nowadays, other applications, including navigation apps or apps that perform targeted advertising, collect GPS positions from user devices at higher frequency. In turn, these applications are often equipped with software development kits (SDK) that can send the GPS data from the user's device to the company that produced the SDK. Through using several smartphone applications, SDK companies can thus collect comprehensive geolocation data. Especially since the COVID-19 pandemic, SDK data has become widely used for mobility research [10, 11].

4.2.3 Transport and Mobile Sensor Data

The digitization of private and public transport services now allows tracking of citizens in the public transportation system, such as through public transport travel card, and analyzing/visualizing entire taxi systems and transportation fleets. Further, detailed records are being generated by novel mobility sharing systems, from car and bicycle sharing to e-scooter and ride sharing. Custom sensors, installed on vehicles, can provide the potential to sense ecological urban variables and the sentiments of city dwellers in unprecedented detail.

Most data sources come with a wide range of biases and measurement problems, such as missing (sub)trajectories, zigzagging, or GPS jitter. Mobility data usually originate from applications that were not developed for research purposes, leading to typically low data quality. Any serious movement data analysis therefore has to account for adequate data preprocessing, including assessment and mitigation of data quality issues, which is often an extensive task [12]. Further, stop and home detection are not trivial [13, 14]. For example, several factors are crucial in the correct estimation of home locations, such as spatial aggregation or localization errors.

4.3 Statistical Properties of Individual Trajectories

Characterizing the statistical properties of individual trajectories is necessary to understand the underlying dynamics of human mobility and to design reliable predictive models. Generally, one can think of the movements of an individual as a trajectory consisting of *displacements* between locations and *pauses* at locations where the individual stops and spends time. Depending on the sampling frequency, the availability of spatio-temporal details, and the goal or scope of the analysis, trajectories can be studied either as sequences of locations, as sequences of locations embedded in space and time, or as high-frequency trajectories. In this section, we review some of the fundamental metrics that have been used in the literature to characterize individual trajectories and the resulting empirical findings. We will introduce richer and richer representations of individual trajectories. The entropy measures that we introduce here swiftly can be well motivated from statistical mechanics, and can be extended or applied to many additional questions in urban science, see Chap. 12.

4.3.1 *Trajectories as Sequences of Locations*

In the simplest form, a trajectory can be represented as a sequence of locations, discarding aspects related to the position of locations in physical space. A large stream of literature has focused on the characteristics of these sequences of places.

One first key question is: How predictable are these sequences of locations? Or, in other words, to what extent is human mobility repetitive? This question is important for modeling purposes, to understand how simple we can make our models, and for privacy considerations. A key metric to address these questions is entropy, which can be thought of as a measure of uncertainty.

The most naive way of assessing predictability of an individual i is by examining their visited number of unique locations N_i. In the trivial case $N_i = 1$ there is no uncertainty about the individual's location and it is fully predictable. However, the larger N_i gets, the more uncertain the location and the less predictable the individual becomes. A measure for the uncertainty of an individual's location which considers only N_i was introduced by Song et al. [15] as the *random entropy*

$$S_i^{\text{rand}} = \log_2 N_i$$

of an individual i's trajectory, given their visited number of unique locations N_i. For example, a person 1 who visited 2 locations has $S_1^{\text{rand}} = \log_2 N_1 = \log_2 2 = 1$, a person 2 who has visited 4 locations has $S_2^{\text{rand}} = \log_2 N_2 = \log_2 4 = 2$.

Coming from thermodynamics and statistical physics, entropy more formally measures the degree to which the probability of a system is spread out over different possible microstates. In information theory it is called *Shannon entropy*

$$H(X) = -\sum p(x) \log_2 p(x)$$

of a discrete random variable X, where $\sum$ denotes the sum over all possible values x of X. In mobility research, the Shannon entropy of an individual i's trajectory has been called *temporal-uncorrelated entropy* [15]

$$S_i^{\text{unc}} = -\sum_{j=1}^{N_i} p_i(j) \log_2 p_i(j) ,$$

where $p_i(j)$ denotes the fraction of individual i's visits to location j.

As an example, consider a person 1 who visited a location A 270 times and a location B 30 times. Then $p_1(A) = 0.9$, $p_1(B) = 0.1$, and

$$\begin{aligned} S_1^{\text{unc}} &= -\sum_{j=1}^{N_1} p_1(j)\log_2 p_1(j) \\ &= -\big(p_1(A)\log_2 p_1(A) + p_1(B)\log_2 p_1(B)\big) \\ &= -\big(0.9\log_2 0.9 + 0.1\log_2 0.1\big) \\ &\approx 0.47\,. \end{aligned}$$

In another example, imagine person 2 visited location A 200 times, location B 300 times, location C 300 times, location D 200 times, meaning $p_2(A) = 0.2$, $p_2(B) = 0.3$, $p_2(C) = 0.2$, and $p_2(D) = 0.2$. Then $S_2^{\text{unc}} \approx 1.97$.

The random and temporal-uncorrelated entropies of an individual are identical, $S^{\text{rand}} = S^{\text{unc}}$, when all locations are visited with equal probability. However, S^{unc} becomes smaller the more unequal, or skewed, the distribution probabilities of different locations are. From the examples above, $S_2^{\text{rand}} = 2$ which is almost the same as $S_2^{\text{unc}} = 1.97$ because all four locations were visited with similar frequency. However, for individual 1, $S_1^{\text{rand}} = 1$ is much larger than $S_1^{\text{unc}} = 0.47$ because in this case the visitation frequencies were much more skewed, 270 to 30. These two concepts of entropy, random and temporal-uncorrelated, are thus useful to assess how much the skew in visiting probabilities affects the uncertainty of locations.

Extending the concept of entropy one step further adds the aspect of ordering, asking: How heterogeneous are visitations not only across locations but also in their time-ordering? The *real entropy* S_i^{real} of an individual i's trajectory is given by

$$S_i^{\text{real}} = -\sum_{T_i' \subset T_i} P(T_i')\log_2 P(T_i')\,,$$

where $P(T_i')$ is the probability of finding a particular time-ordered subsequence T_i' in the trajectory T_i [15]. Thus, an individual with a trajectory of repeating or regular patterns, such as ABACDCABACDCABACDC will have a much lower real entropy than an individual who has an identical number and visitation distribution of locations but a less regular order of visitations, for example ABCDACABCDCABACACD. Similarly as S^{rand} was an upper limit for S^{unc}, here S^{unc} is an upper limit for S^{real}. The identity $S^{\text{unc}} = S^{\text{real}}$ is given when the order of locations in the trajectory is randomized, thereby destroying all time-correlated information.

A measure associated to the entropy S is the predictability Π, derived from information theory [15], which is the probability that an optimal predictive algorithm can predict correctly the individual's future whereabouts. Every type of entropy has a respective predictability measure, for which the inequality relation between entropies $S^{\text{rand}} \geq S^{\text{unc}} \geq S^{\text{real}}$ is inverted to $\Pi^{\text{rand}} \leq \Pi^{\text{unc}} \leq \Pi^{\text{real}}$. The more uncertainty, the less predictability. As Song et al. [15] have shown, a significant share of predictability is encoded in the time ordering which make the trips of most individuals highly predictable.

This high predictability of human trajectories has strong privacy implications. For example, it was shown that for the trajectories of 1.5 million individuals, with hourly sampling and spatial aggregation to 6,500 cell phone towers, only 4 random spatio-temporal points are needed to identify 95% of individuals [16]. Such insights have profound legal and ethical consequences on the sharing and tracking of micro-mobility data, and have spurred research on privacy-enhancement techniques such as cloaking, suppression, aggregation, swapping, or differential privacy [17].

While the statistical characterization of trajectories, either individually or in aggregate, is crucial for fundamental predictability and privacy considerations, two important practical questions remain: 1) What are those regular sequences of locations, and 2) What is the relation with other people's movements in the city? The first question can be tackled via motif analysis, where the most common trips of individuals are interpreted as a network of trips between locations [18]. Motifs in complex networks are generally small subgraphs that appear with significantly higher than expected probability [19]. Analyzing both surveys and mobile phone data in different cities, Schneider et al. [18] found that half of the daily mobility networks are just described by two trivial motifs, consisting of one node (no movement) or of two nodes (back and forth movement, typically home-work). The vast majority of all other daily movements are described by remaining 15 motifs. The second question–the relation with other people's movements–has been investigated for example in the context of bus transit patterns of 5 million individuals in Singapore [20]. Due to the striking regularity of daily commutes, the emergence of "familiar strangers" in a tight-knit contact network was observed, with crucial consequences on the impact of human behavior on diffusion/spreading processes. More recent studies have compared explicitly social relations with mobility behavior [21, 22]. For example, the Copenhagen network study [21] has found striking differences in geospatial versus social entropy in a population of 1,000 university students, uncovering "party nights" that are characterized by geospatial exploration but conservative social behavior.

The reason why sequences of human locations are highly predictable is rooted in the high level of regularity of day-to-day routines, with mobility highly dominated by home-work commuting. Moreover, the way humans visit places and allocate time among them is characterized by universal properties, as shown by several empirical studies. Some notable properties include:

Heterogeneous distribution of visits. Humans allocate their time in a heterogeneous way across places, implying they spend most of their time within a small set of favorite locations. Ranking locations from the most visited to the least visited, the number of visits for a location with rank L follows a power-law $P \sim L^{-\alpha}$ with exponent $\alpha \sim 1.2$ [23, 24].

Sublinear exploration. The number of distinct locations S visited by an individual grows over time (t) as $S(t) \sim t^{\beta}$, with $\beta \sim 0.6$, implying that there is a decreasing tendency to explore new locations [23].

Periodicity. The probability to return to a location visited n hours before is characterized by peaks at 24 h, 48 h and 72 h, due to the recurrence and temporal periodicity inherent to human mobility, which is driven by circadian and weekly patterns. [23]

Conservation of the number of visited places. It was shown that, despite the fact that our routines change and evolve over long time-scales, some properties of the set of visited locations, including its size, remain approximately conserved at the level of the single individual [24].

4.3.2 Trajectories in Space and Time

If the data are available, realism can be increased by considering trajectories as sequences of locations embedded in space and time. Two key properties that characterize these sequences are the waiting time and the displacement distributions, see Fig. 4.1.

The *waiting time distribution* $P(\Delta t)$ describes probabilistically the time Δt that an individual pauses after a move. The *displacement distribution* $P(\Delta r)$ describes probabilistically the distance Δr that an individual moves after a pause. The distribution of waiting times (or pause durations), Δt, between movements and the distribution of distances, Δr, travelled between pauses are useful to quantitatively assess the dynamics of human mobility. For example, specific probability distributions of distances and waiting times characterise different types of diffusion processes.

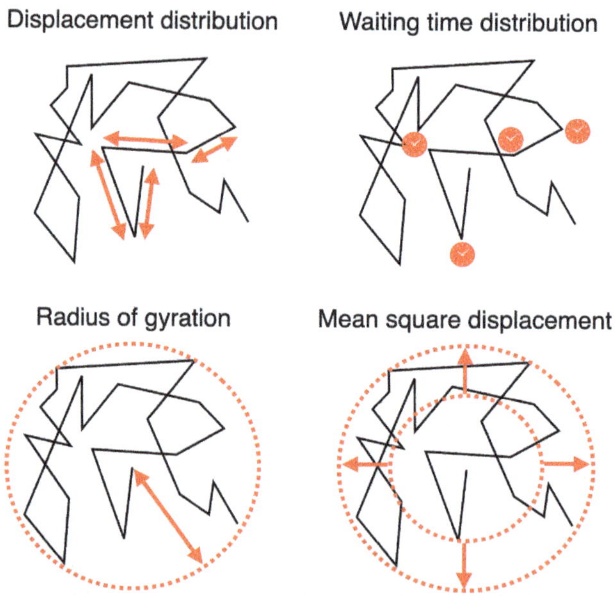

Fig. 4.1 To quantitatively study the movements of humans, the field of complex systems adapts the random walk approach from statistical physics originally developed for describing particle movements, with metrics such as displacement distribution, waiting time distribution, radius of gyration, mean square displacement. The black connected line segments denote a trajectory starting in the center, the red markers illustrate the different metrics

Thanks to the recent availability of data used as proxy for human trajectories, the characteristic distributions of distances and waiting times between consecutive locations have been widely investigated. There is no agreement, however, on which distribution best describes these empirical datasets. Pioneer studies, based on CDR [23, 25] and banknote records [26], found that the distribution of displacement Δr is well approximated by a power-law, $P(\Delta r) \sim (\Delta r)^{-\beta}$, (or "Lévy distribution" [27], typically with $1 < \beta < 3$), and that an exponential cut-off in the distribution may control boundary effects [25]. These findings were confirmed by studies based on GPS trajectories of individuals [28–30] and vehicles [31, 32], as well as online social networks data [33–35]. It has been noted, however, that power-law behaviour may fail to describe intra-urban displacements [36]. Other analyses, based on online social network data [37–39] and GPS trajectories [40–43] showed that the distribution of displacements is well fitted by an exponential curve, $P(\Delta r) \sim e^{-\lambda \Delta r}$, in particular at short distances. Finally, analyses based on Taxi GPS [44, 45] suggested that displacements may also obey log-normal distributions, $P(\Delta r) \sim (1/\Delta r) \cdot e^{-(\log \Delta r - \mu)^2/2\sigma^2}$. Another study found that this is the case also for single-transportation trips [29].

Fewer studies have explored the distribution of waiting times between displacements, Δt, as trajectory sampling is often uneven; e.g., in CDR data location is recorded only when the phone user makes a call or sends an SMS, and LBSN data include the positions of individuals who actively "check-in" at specific places. Analyses based on evenly sampled trajectories from mobile phone call records [18, 23], and individuals GPS trajectories [28, 30] found that the distribution of waiting times can also be approximated by a power-law. A recent study based on GPS trajectories of vehicles, however, suggests that for waiting times larger than 4 hours, this distribution is best approximated by a log-normal function [46]. Several studies have highlighted the presence of natural temporal scales in individual routines. Distributions of waiting times display peaks that correspond to the typical times spent home on a typical day (around 14 hours) and at work (3-4 hours for a part-time job and 8-9 hours for a full-time job) [18, 47, 48].

The datasets considered have different *spatial resolution and coverage*, and few studies have so far considered the whole range of displacements occurring between 10^1 m and 10^7 m (10,000 km). Another issue for comparability concerns the *temporal sampling* in the datasets analysed so far. Uneven sampling typical of CDR and LBSN data (i) does not allow to distinguish phases of *displacement* and *pause*, since individuals could be active also while transiting between locations, and (ii) may fail to capture patterns other than regular ones [49, 50], because individuals' voice-call/SMS/data activity may be higher in certain preferred locations. Finally, studies focusing on displacements effectuated using one or several *specific transportation modalities* (private car [46, 51], taxi [43], public transportation [52], or walking [30]) capture only a specific aspect of human mobility behaviour.

Nevertheless, an aggregate measure of diffusion over many individuals, which describes how fast an area is explored, is given by the *mean square displacement* MSD(t), see Fig. 4.1. It measures the deviation of the position of N individuals from a reference position over a timespan t as an (ensemble) average over all the individuals j,

$$\mathrm{MSD}(t) = \langle \mathbf{r}_j(t) - \mathbf{r}_j(0) \rangle = \frac{1}{N} \sum_{j=1}^{N} \left(\mathbf{r}_j(t) - \mathbf{r}_j(0) \right)^2 ,$$

where $\mathbf{r}_j(t)$ is the position of individual j at time t and $\mathbf{r}_j(0)$ is their initial reference position. The mean squared displacement is a measure introduced in statistical mechanics to study diffusion processes, and can be thought of as measuring the portion of the system "explored" by an ensemble of individuals. Empirical research has shown that human mobility patterns follows a slow diffusion process, in which the MSD grows logarithmically [23].

To distinguish whether individuals all sample from the same displacement or waiting time distributions, or whether individuals have different movement behaviors, it is useful to define the *radius of gyration*, see Fig. 4.1. It is the typical distance travelled by an individual over a fixed timespan t,

$$r_g = \sqrt{\frac{1}{n} \sum_{i=1}^{n} |\mathbf{r}_i - \mathbf{r}_{\mathrm{cm}}|^2} ,$$

where n is the number of jumps within timespan t, $\mathbf{r}_i$ represents the position at step i, and $\mathbf{r}_{\mathrm{cm}} = \frac{1}{n} \sum_{i=1}^{n} \mathbf{r}_i$ is the center of mass of the trajectory. Gonzalez et al. [25] have introduced this metric to human mobility studies, showing that individuals have different typical travel distances. In particular, it has been shown that the distribution of the radius of gyration across individuals is broad and can be well described by a truncated power-law [25].

4.4 Modeling Urban Mobility Patterns

To study the movements of humans, the field of complex systems adapts models from statistical physics, which studies the movements of particles. Central to this approach is the concept of *random walk*, see Fig. 4.1. A random walk describes the random, Brownian motion of particles suspended in a liquid or gas, and can be taken as the simplest model for a moving individual. Generally defined, a random walk is a random process describing a path of random steps on a mathematical space. For example, a random walk on the integer number line $\mathbb{Z}$ could start at 0 and at each step move -1 or $+1$ with equal probability, decided by a coin flip. Thus, after two coin flips, the following outcomes are possible:

- -2, after $0 - 1 - 1$
- 0, after $0 - 1 + 1$ or $0 + 1 - 1$
- 2, after $0 + 1 + 1$

In the above example, the random walk describes a trajectory on $\mathbb{Z}$ with a fixed jump distance of 1 and an arbitrary waiting time–for repeated coin flipping this could be for example 3 seconds. Further, this random walk is memoryless, as the coin flip is a random process independent of previous outcomes. There are several ways to extend the behavior of random walkers, adding complexity to more realistically model human mobility. For example, in continuous-time random walk (CTRW) models, jump lengths and waiting times can follow arbitrary distributions. This model was suggested as one of the first models able to capture the probability distributions of displacements and waiting times, where the step-lengths and waiting times were calibrated from heavy-tailed probability distributions observed in human mobility [26].

However, human mobility is not completely random, and researchers quickly realized that more realistic mechanisms were needed to capture the properties of human trajectories. An ingredient to make trajectories more realistic is memory, i.e. individuals can remember and be influenced by visits to previous places. In 2010, Song et al. [23] proposed an agent-based model with memory that could capture many of the observed properties of mobility.

The *Exploration and Preferential Return* model (EPR) proposes that whenever individuals visit a location, they either (i) explore with probability $P_{\text{new}} = S^{-\gamma}$, where S is the number of distinct locations previously visited, or (ii) return to a previously visited location with complimentary probability $P_{\text{ret}} = 1 - P_{\text{new}}$ [23]. When returning to a previously visited location, the probability p_i to visit location i is proportional to the number of visits the individual previously had to that location, an effect known as preferential return. Several variations of the EPR model have been proposed to account for other aspects of human mobility including more memory and recency effects [24, 53, 54], social interactions [55], and heterogeneities across individuals [56].

An important development of the EPR model is the *Preferential Exploration and Preferential Return* (PEPR) model, developed by Schläpfer et al. [55]. This model couples the movements of agents, so that, when exploring new locations, they are preferentially attracted towards highly frequented areas, which can be seen as a manifestation of Central Place Theory. Importantly, this model is able not only to reproduce the individual mobility, but also collective patterns of mobility, thus linking the literature streams on individual and collective mobility (collective mobility is covered in Chap. 5). In particular, the PEPR model accounts for an important empirical discovery made by Schläpfer et al. [55], related to the frequency at which individuals visit different locations. By studying mobility traces extracted from phones the authors had found that the number of users who visit a location at distance r, exactly n times in a period of length T, decreases as $N \sim r^{-2} f^{-2}$, where $f = n/T$.

While capturing many properties of human movements, preferential return models are still unable to explain the power-law distribution of displacements in mobility data. The observation that mobility patterns are captured by power-law distributions seemed to show that human mobility is scale free in a truly fundamental sense. But concurrently, it was known that the natural and built environment is rich in spatial scales, from neighbourhoods, to cities, regions, countries and continents [57].

The notion of scale is fundamental within the fields of geography and spatial cognition, because we think of space in a hierarchical fashion characterized by typical scales [58]. In this sense, there appeared to be an important schism between traditional geography and the data-driven work on human mobility. This divide got explicitly addressed in 2020 by Alessandretti et al. [59] working with a very large dataset of GPS traces. This study showed that typical scales are indeed embedded within the same type of scale-free trajectories analyzed by previous researchers. The explanation is that scales manifest as containers of mobility behavior. Neighborhoods have typical sizes, but the distance between the neighborhoods a certain individual visits is unrelated to the size of neighborhoods (it is rather set by the city scale). Similarly with cities which also tend to have a typical geographical size, but the cities a certain individual tends to visit may be close–or located in opposite ends of their home country. A similar logic is true for scales across neighborhoods, cities, regions, countries, continents, and the entire world. The authors use a model-based approach to infer the typical sizes of these containers across millions of individuals.

Based on these ideas, the *container model* models physical space as a hierarchy of L levels, ordered from the smallest to largest (e.g. individual locations to countries). At any level l, space is partitioned into compact containers with a characteristic size. Each geographical location k can be identified as a sequence of nested containers, $k = (k_1, ..., k_l, ..., k_L)$ that contains it. The *level-distance* $d(j, k)$ between locations j and k is defined as the highest index at which the two sequences of containers describing j and k differ. For an agent located in j, the probability of moving to k is the product of two factors $P(j \rightarrow k) = p_{d(j,k),d(j,h)} \prod_{l \leq d(j,k)} a(k_l)$. The first factor, $p_{d(j,k)}$, is the probability of traveling at level-distance $d(j, k)$. The second factor $\prod_{l \leq d(j,k)} a(k_l)$ is the probability of choosing a specific location k at that level-distance, where $a(k_l)$ is the attractiveness of a container at level l including location k.

4.5 Improving Urban Transport Systems

The research described so far focuses mostly on understanding where and when people move, statistically describing individual trajectories or emerging aggregate patterns. The question of how people get to their destinations–which transport mode they use, which route they take and why, and the role of the underlying transport system–has remained largely uncharted from a quantitative perspective, also due to the limited availability of data with high enough resolution. For example, how individuals choose routes through balancing the interplay between public and private forms of transportation such as walking, driving and cycling remains poorly characterized. The existing empirical research using a *systemic* and *long-term* view on urban transport systems, including shared travel, multimodal travel, or system dynamics like induced demand is thus quite limited [60]. In this section we review first on the one hand some microscopic and mode-specific wayfinding mechanisms, on the other hand more systems-level approaches of how to think about the dynamics of urban transport systems with the aim to improve them. We discuss the importance of different urban transport modes, to help designing more efficient and sustainable urban transportation.

4.5.1 Wayfinding

Apart from a statistical, quantitative description of human mobility, it is important to understand the mechanisms of how humans are navigating in cities to find their ways. While pathfinding on a graph (such as a street network) is a solved problem in computer science via Dijkstra's shortest path algorithm and its modern replacements [61], the implementation of routing on multimodal networks can be challenging [62, 63], and an array of cognitive peculiarities (see also Chapter 1) have to be accounted for such as upbringing [64], vector-based navigation [65], information overload [66], or mode-specific travel behavior [67]. Recent findings have established that pedestrians, cyclists, and motorists do not tend to follow optimal routes, and are instead influenced by urban features and goal distance and direction [65, 68–70].

4.5.2 Shareability Networks

Existing urban transport systems are highly inefficient, such as the individualized transport mode of taxis. In New York City, for example, a large fleet of around 13,500 taxicabs serves individualized mobility needs, but with a high fraction of idle or low occupation runs. To quantify how much such a localized fleet could be improved, Santi et al. [71] introduced the concept of shareability network in 2014 using methods from network science.

The idea of shareability networks is to view a vehicle fleet from a global perspective and to optimize for the whole system with global knowledge instead of locally. For example, instead of having two taxis deliver two individuals in parallel from the same starting point to the same end point, the routes of the two passengers could have been bundled with just one taxi with small inconvenience to both. Taking this idea to the extreme, towards k-sharing with $k > 2$, would lead to a system of dynamically routed "taxi buses". As Santi et al. [71] showed, in New York City indeed a large fraction of trips could be shared with relatively low discomfort, with cumulative trip length cut by 40% or more.

Trip-sharing systems were introduced later by UBER (called UBERPool), and by similar ride hailing companies, leading to a claimed short-term reduction of driven kilometers [72]. However, in total, companies like UBER have later shown to have generated more traffic and congestion due to rebound effects and their competition with public transport [73–75]. This example shows how long-term systems thinking is necessary to understand the full impact and possible unintended consequences of short-term optimization, see Sect. 4.5.3.

The research on shareability networks was later expanded with a focus on fleet size and vehicle-sharing networks [76], finding a potential to reduce the New York City taxi fleet size by 30%. Such a potential reduction follows directly from a reorganization of taxi dispatching that could be implemented with a mobile phone app and does not assume ride sharing, nor requires changes to regulations, business models, or human attitudes towards mobility to become effective. Finally, further research uncovered scaling effects in ride-sharing [77, 78].

4.5.3 Sustainable, Multimodal Transport and Systems Thinking

Understanding urban mobility is impossible without understanding its history and the complex socio-technical system it is part of. This history shows a transition towards car-centricity in the 20th century for most cities, entailing critical consequences on the whole system. For example, over 1.3 Mio. people die on the road every year [79], vehicular pollution causes millions of more yearly deaths [80], and vehicular traffic noise has shown to cause dementia-related diseases on a large scale [81]. Further, cars come with massive inefficiencies due to their skewed space requirements and usage patterns [82]. Apart from empirical evidence that this is unsustainable [83], also mathematical models from complexity science show that cities as (car-centric) transport monocultures are not sustainable [84, 85], and that a mere replacement of fossil fuel cars with electric vehicles is not an adequate solution [86–89], especially when much more efficient and economic solutions such as mass transit or bicycles are available.

To fix such monocultures, the biggest question in urban transport is how to reverse car dependency [90], i.e. how to best replace unsustainable modes of transport and to prioritize more sustainable ones. In particular, active travel such as walking and cycling has the highest societal benefits [91]. Cost-benefit analysis which accounts for the environment and public health reveals that each kilometer walked or cycled in the European Union provides €0.37 or €0.18 to society, respectively, while each kilometer driven by car incurs a cost of €0.11 [92]. Overcoming transportation monocultures is also possible with a focus on more multimodal transport, i.e. the combination of multiple transport modes promising all of their benefits while avoiding their weaknesses. For an overview of state of the art complex systems based approaches to multimodal mobility see Alessandretti et al. [60].

Systems thinking is the appropriate method of understanding how (un)sustainable urban mobility emerges over time [93]. There is a complex co-evolution between urban transport infrastructure, land use, and socio-political/cultural factors with long-term feedback loops. Complex systems dynamics are crucially dependent on the *interactions* between its parts and not just on the parts themselves. For example, increased traffic volume creates pressure to invest into road infrastructure, which drives more traffic volume. This is the well-known phenomenon of *induced demand* [93]. Due to increased roadway capacities, catchment areas increase, leading to urban sprawl, which in turn increases traffic volume and erodes active modes of transport due to increased distances and lack of adequate transport infrastructure. In particular cultures such as in the US, this phenomenon has sometimes lead to extreme sprawl, which makes it all the harder to reverse car-dependence. This situation is also an example for path dependency and lock-in effects common in complex systems, where the more the system evolves towards a certain path, the harder it is to reverse.

The general phenomenon of car-dependence can be thought of as a dynamic reinforced through different socio-technical systems such as the automotive industry, land use, or car culture [94]. Therefore, long-term urban planning cannot detach transport planning from the planning of land-use, housing, parking [95], etc., but the whole system must be understood as being shaped by strong, non-linear, long-term interactions. Although the opposite of systems thinking, namely short-term engineering thinking, has its use cases, it is inadequate to apply to most problems in complex systems, because the optimization or control of sub-systems can lead to unintended consequences (such as induced demand or urban sprawl) [93]. Using short-term traffic engineering thinking has lead to the outdated "predict and provide" logic, where e.g. traffic volumes are predicted from past flows and the role of policy makers is to react by providing a short-term solution. On the contrary, following systems thinking, this logic is replaced with the more adequate "decide and provide" principle, which aims to pro-actively shape the future [96]. It acknowledges that e.g. the choice to drive a car is not solely the result of people's individual preferences, i.e. exogenous to the system, but determined largely by transport and urban systems organised around car driving, which can be changed with appropriate policies.

Such policies can leverage long-term systems dynamics, allowing to re-shape transport systems for the better [93]. For example, the reversal of induced demand, namely disappearing traffic, happens when public space is reallocated from private cars to space-efficient modes of transport. As research shows [90, 97], significant reductions in overall traffic levels can occur in the right conditions, because mode choice is more elastic than commonly believed: Many individuals are willing to give up traveling by car if there are appropriate alternatives. To succeed, it is also important to consider the right communication strategies and the strong psychological aspect of car attachment [98]. Apart from policies to develop multimodal networks and street redesign, a large array of further policies can address a variety of system dynamics, from market-based instruments such as carbon-prices, road pricing, to abolishing minimum parking requirements, or financial support to increase the attractiveness of micromobility [93]. For all these policies, it is important to prioritize correctly, for example following an "avoid-shift-improve" approach [99, 100]. First, focus on avoiding the need for mobility, for example through planning for proximity such as the 15-minute city [101]. Second, aim for a shift from unsustainable to more sustainable modes of transport. Only as a last third step aim to improve existing components of the system such as making vehicle technology more efficient.

4.6 Tools

Over the years, computational tools have become more and more important for studying urban systems and human mobility. For an overview of the available tools in geographic analysis and transport planning see Lovelace [102]. Here we focus on open-source tools as they allow fully reproducible research or analysis. There is a plethora of tools specialized on various purposes, including network analysis (single-layer or multilayer), routing and access (unimodal or multimodal), or mobility analysis. We classify these computational tools into three groups.

4.6.1 *Transport and Mobility Tools*

We first describe the most general tools for mobility and transport analysis.

To work with public transportation data, the Python library *transitfeed* [103] is suited to parse, validate and build General Transit Feed Specification (GTFS) files, which is a data specification that allows public transit agencies to publish their transit data in a usable form. This tool is particularly useful to those interested in the manipulation of the raw data. However, to convert the data into a network, some additional steps are needed, such as using *Peartree* (see below).

Movingpandas [104] is a Python package that provides trajectory data structures and functions for the analysis and visualisation of mobility data. In a similar sense, and also developed in Python, *scikit-mobility* [105] is a library that implements a framework for analyzing statistical patterns and modeling mobility, including functions for estimating movement between zones using spatial interaction models, and tools to asses privacy risks related to the analysis of mobility datasets.

4.6.2 *Transport Network Tools*

In this section we cover tools for studying transportation networks.

Multiple tools were developed to obtain data on transportation and multimodal infrastructures. One of the best known is *OSMnx* [106], a Python package that downloads street networks from OpenStreetMap into Python objects. *OSMnx* can further be used to download other transportation networks, and to build its multimodal transport networks.

Another reliable Python library to read data from OpenStreetMap and extract transportation networks is *Pyrosm* [107]. Differently from *OSMnx*, *Pyrosm* reads the data directly from OpenStreetMap's Protocol Buffer Format files (*.osm.pbf), while *OSMnx* downloads the data from the Overpass API [108]. For this reason *Pyrosm* is a particularly good alternative when working with large urban areas, states, and even countries, while *OSMnx* typically offers a more precise way to collect data from specific points in a city.

Finally, an alternative to *transitfeed*'s functionality of reading GTFS feeds is *Peartree* [109], a Python library allowing to convert GTFS feed schedules into the corresponding directed network graph.

4.6.3 Tools for Routing or Access on Transport Networks

In this section we cover tools that can be used for routing and navigation.

On one hand there are established general, high-performance tools for unimodal routing such as *graphhopper* [110] and *OSRM* [111]. These tools were developed for routing on one network type but could in principle be extended to multimodal routing. Explicit multimodal routing is provided by *OpenTripPlanner (OTP)* [112] and *R5* [113]. Both tools exist as fast R implementations, *r5r* [114] and *OTP for R* [115]; the Python implementation of *R5* is *R5py* [116]. The R package *dodgr* [117], thanks to an efficient algorithm for computing distances between pairs of nodes in a street network, further enables the aggregation of flows on network links from a set of origin and destination points, a matrix of pairwise flow densities, and the underlying street network.

Finally, several open source packages focus on computing accessibility metrics, e.g. the ease by which people can reach points of interests, such as those offering employment, shopping, medical care or recreation [118]. *Pandana* [119, 120] is a Python library that enables to compute the accessibility of places by retrieving points of interest and street network data from OpenStreetMap [121] and by efficiently computing shortest paths along the street network. *Access* [122] is a Python library that computes a wide range of spatial accessibility metrics from a set of origins and destinations, and travel times or distances between them. Built on top of *Pandana*, *UrbanAccess* [123] integrates the creation of multi-modal transport networks (transit and pedestrian) using GTFS data and the computation of accessibility metrics. Similar functionalities are offered in R and Python by *r5r* [114] and *R5py* [116], respectively.

4.7 Conclusions

Modelling urban mobility patterns is critical to address a wide range of societal challenges, from managing public health, to planning urban infrastructure and transportation. By gaining insights into how people move, we can make informed decisions towards developing transportation systems that are efficient and inclusive, reducing carbon emissions, and enhancing public safety.

The scientific study of human mobility dates back to the late 19th century and has been extensively studied in Transportation Science and Human Geography throughout the 20th century. Over the past few decades, the field has made significant advances, facilitated by the availability of large datasets and the development of Complexity Science, which has brought together methods for understanding interconnected, non-linear systems.

In this chapter, we provided a brief, introductory overview of recent developments, including hand-picked highlights. We started with a short history of the field, focusing on the complexity science perspective. We then described the wide range of data sources used in mobility studies and the empirical findings these data have enabled. We also presented a variety of modelling approaches developed to capture different aspects of individual movements and discuss how these models can be used to study urban transportation networks. Finally, we presented a set of computational open-source tools that can be used for urban mobility analysis and research.

Notwithstanding the important discoveries made thus far, many challenges remain. For example, issues surround the understanding of causal mechanisms underlying human spatial behavior in urban contexts. While research using passively collected data has suggested some potential mechanisms, integrating experimental or quasi-experimental approaches may be necessary to shed light on how the cognitive and behavioral mechanisms identified in laboratory settings drive real-world behavior and validate hypotheses brought forward by the field of Human Mobility.

Other challenges include the availability and quality of data, including mis-representation, spatial and temporal resolution differences, and data skews. These problems can affect insights derived from the data, and the literature is still sparse on how to address them, partly due to the lack of ground-truth data. Additionally, mobility datasets are sensitive because they contain detailed information regarding people's whereabouts and behavioral patterns, making it critical that this data is handled in safe and ethical ways. The difficulties of obtaining informed consent for data use and enabling result reproducibility do not have clear solutions yet.

Finally, from an applications perspective, new challenges surround the understanding of emerging technologies for urban mobility, from micro-mobility, to autonomous vehicles, route recommendation engines, vehicle-to-vehicle communication, and vehicle-to-infrastructure communication. Efforts are needed to understand the impact of these developing technologies on travel behaviour and to comprehensively integrate widely diverse travel modes into a unified framework.

All in all, our chapter could serve as a basis for teaching the topic to an advanced audience (master level or higher); much of the discussed material is covered in open teaching resources available at https://github.com/mszell/geospatialdatascience.

References

1. *Transportation Economic Trends: Transportation spending–average household*. https://data.bts.gov/stories/s/Transportation-Economic-Trends-Transportation-Spen/ida7-k95k/
2. C. Burd, M. Burrows, B. McKenzie, Travel time to work in the united states: 2019, in *American Community Survey Reports, United States Census Bureau* 2 (2021), p. 2021. http://large.stanford.edu/courses/2022/ph240/schutt2/docs/acs-47.pdf
3. https://www.epa.gov/ghgemissions/sources-greenhouse-gas-emissions

4. L. Alessandretti, What human mobility data tell us about COVID-19 spread. Nature Reviews Physics **4**(1), 12–13 (2022). https://doi.org/10.1038/s42254-021-00407-1
5. E. Lill, *Die Grundgesetze des Personenverkehrs: Eine Studie* (Hartleben's Verlag, Vienna, 1889)
6. P. Ball, The physical modelling of human social systems. Complexus **1**(4), 190–206 (2003). https://doi.org/10.1159/000082449
7. B. Resch, M. Szell, Human-Centric Data Science for Urban Studies, in *ISPRS International Journal of Geo-Information* 8.584 (2019). https://doi.org/10.3390/ijgi8120584
8. H. Barbosa et al., Human mobility: Models and applications. Physics Reports **734**, 1–74 (2018). https://doi.org/10.1016/j.physrep.2018.01.001
9. L. Pappalardo et al., Evaluation of home detection algorithms on mobile phone data using individual-level ground truth, in *EPJ data science* 10.1 (2021), p. 29. https://doi.org/10.1140/epjds/s13688-021-00284-9
10. A. Aleta et al., Modelling the impact of testing, contact tracing and household quarantine on second waves of COVID-19. Nature Human Behaviour **4**(9), 964–971 (2020). https://doi.org/10.1016/j.tra.2010.08.003
11. S. Chang et al., Mobility network models of COVID-19 explain inequities and inform reopening. Nature **589**(7840), 82–87 (2021). https://doi.org/10.1038/s41586-020-2923-3
12. G. Andrienko, N. Andrienko, G. Fuchs, Understanding movement data quality. Journal of location Based services **10**(1), 31–46 (2016). https://doi.org/10.1080/17489725.2016.1169322
13. F. Calabrese et al., Understanding individual mobility patterns from urban sensing data: A mobile phone trace example. Transportation research part C: emerging technologies **26**, 301–313 (2013). https://doi.org/10.1016/j.trc.2012.09.009
14. L. Pappalardo et al., scikit-mobility: A Python Library for the Analysis, Generation, and Risk Assessment of Mobility Data, in *Journal of Statistical Software* 103.4 (2022), pp. 1–38. https://doi.org/10.18637/jss.v103.i04. https://www.jstatsoft.org/index.php/jss/article/view/v103i04
15. C. Song et al., Limits of predictability in human mobility. Science **327**(5968), 1018–1021 (2010). https://doi.org/10.1126/science.1177170
16. Y.-A. De Montjoye et al., Unique in the crowd: The privacy bounds of human mobility. Scientific reports **3**(1), 1–5 (2013). https://doi.org/10.1038/srep01376
17. M. Fiore et al., Privacy in trajectory micro-data publishing: a survey, in *Transactions on Data Privacy* 13 (2020), pp. 91–149. https://orbit.dtu.dk/en/publications/privacy-in-trajectory-micro-data-publishing-a-survey
18. C.M. Schneider et al., Unravelling daily human mobility motifs, in *Journal of The Royal Society Interface* 10.84 (2013), p. 20130246. https://doi.org/10.1098/rsif.2013.0246
19. R. Milo et al., Network motifs: simple building blocks of complex networks. Science **298**(5594), 824–827 (2002). https://doi.org/10.1126/science.298.5594.824
20. L. Sun et al., Understanding metropolitan patterns of daily encounters. Proceedings of the National Academy of Sciences **110**(34), 13774–13779 (2013). https://doi.org/10.1073/pnas.1306440110
21. V. Sekara, A. Stopczynski, S. Lehmann, Fundamental structures of dynamic social networks. Proceedings of the national academy of sciences **113**(36), 9977–9982 (2016). https://doi.org/10.1073/pnas.1602803113
22. Z. Chen et al., Contrasting social and non-social sources of predictability in human mobility. Nature communications **13**(1), 1–9 (2022). https://doi.org/10.1038/s41467-022-29592-y
23. C. Song et al., Modelling the scaling properties of human mobility. Nature Physics **6**(10), 818–823 (2010). https://doi.org/10.1038/nphys1760
24. L. Alessandretti et al., Evidence for a conserved quantity in human mobility. Nature human behaviour **2**(7), 485–491 (2018). https://doi.org/10.1038/s41562-018-0364-x
25. M.C. Gonzalez, C.A. Hidalgo, A.-L. Barabasi, Understanding individual human mobility patterns. Nature **453**(7196), 779–782 (2008). https://doi.org/10.1038/nature07850
26. D. Brockmann, L. Hufnagel, T. Geisel, The scaling laws of human travel. Nature **439**(7075), 462–465 (2006). https://doi.org/10.1038/nature04292

27. A. Baronchelli, F. Radicchi, Lévy flights in human behavior and cognition. Chaos, Solitons & Fractals **56**, 101–105 (2013). https://doi.org/10.1016/j.chaos.2013.07.013
28. X.-W. Wang, X.-P. Han, B.-H. Wang, Correlations and scaling laws in human mobility, in *PloS one* 9.1 (2014), e84954. https://doi.org/10.1371/journal.pone.0084954
29. K. Zhao et al., Explaining the power-law distribution of human mobility through transportation modality decomposition, in *Scientific reports* 5 (2015). https://doi.org/10.1038/srep09136
30. I. Rhee et al., On the levy-walk nature of human mobility. IEEE/ACM transactions on networking (TON) **19**(3), 630–643 (2011). https://doi.org/10.1109/tnet.2011.2120618
31. B. Jiang, J. Yin, S. Zhao, Characterizing the human mobility pattern in a large street network, in *Physical Review E* 80.2 (2009), p. 021136. https://doi.org/10.1103/physreve.80.021136
32. Y. Liu et al., Understanding intra-urban trip patterns from taxi trajectory data. Journal of geographical systems **14**(4), 463–483 (2012). https://doi.org/10.1007/s10109-012-0166-z
33. M.G. Beiró et al., Predicting human mobility through the assimilation of social media traces into mobility models, in *arXiv preprint* arXiv:1601.04560 (2016). https://doi.org/10.1140/epjds/s13688-016-0092-2
34. Z. Cheng et al., Exploring Millions of Footprints in Location Sharing Services". ICWSM **2011**, 81–88 (2011). https://doi.org/10.1609/icwsm.v5i1.14109
35. B. Hawelka et al., Geo-located Twitter as proxy for global mobility patterns. Cartography and Geographic Information Science **41**(3), 260–271 (2014). https://doi.org/10.1080/15230406.2014.890072
36. A. Noulas et al., A tale of many cities: universal patterns in human urban mobility, in *PloS one* 7.5 (2012), e37027. https://doi.org/10.1371/journal.pone.0037027
37. L. Wu et al., Intra-urban human mobility and activity transition: evidence from social media check-in data, in *PloS one* 9.5 (2014), e97010. https://doi.org/10.1371/journal.pone.0097010
38. Y. Liu et al., Uncovering patterns of inter-urban trip and spatial interaction from social media check-in data, in *PloS one* 9.1 (2014), e86026. https://doi.org/10.1371/journal.pone.0086026
39. R. Jurdak et al., Understanding human mobility from Twitter, in *PloS one* 10.7 (2015), e0131469. https://doi.org/10.1371/journal.pone.0131469
40. H. Liu, Y.-H. Chen, J.-S. Lih, Crossover from exponential to power-law scaling for human mobility pattern in urban, suburban and rural areas. The European Physical Journal B **88**(5), 1–7 (2015). https://doi.org/10.1140/epjb/e2015-60232-1
41. X. Liang et al., The scaling of human mobility by taxis is exponential. Physica A: Statistical Mechanics and its Applications **391**(5), 2135–2144 (2012). https://doi.org/10.1016/j.physa.2011.11.035
42. L. Gong et al., Inferring trip purposes and uncovering travel patterns from taxi trajectory data. Cartography and Geographic Information Science **43**(2), 103–114 (2016). https://doi.org/10.1080/15230406.2015.1014424
43. K. Zhao, M.P. Chinnasamy, S. Tarkoma, Automatic City Region Analysis for Urban Routing, in *2015 IEEE International Conference on Data Mining Workshop (ICDMW)*. IEEE. 2015, pp. 1136–1142. https://doi.org/10.1109/icdmw.2015.176
44. W. Wang et al., A comparative analysis of intra-city human mobility by taxi. Physica A: Statistical Mechanics and its Applications **420**, 134–147 (2015). https://doi.org/10.1016/j.physa.2014.10.085
45. J. Tang et al., Uncovering urban human mobility from large scale taxi GPS data. Physica A: Statistical Mechanics and its Applications **438**, 140–153 (2015). https://doi.org/10.1016/j.physa.2015.06.032
46. R. Gallotti et al., A stochastic model of randomly accelerated walkers for human mobility, in *Nature Communications* 7 (2016), p. 12600. https://doi.org/10.1038/ncomms12600
47. S. Hasan et al., Spatiotemporal patterns of urban human mobility. Journal of Statistical Physics **151**(1–2), 304–318 (2013). https://doi.org/10.1007/s10955-012-0645-0
48. A. Bazzani et al., Statistical laws in urban mobility from microscopic GPS data in the area of Florence, in *Journal of Statistical Mechanics: Theory and Experiment* 2010.05 (2010), P05001. https://doi.org/10.1088/1742-5468/2010/05/p05001

49. S. Çolak et al., Analyzing cell phone location data for urban travel: current methods, limitations, and opportunities. Transportation Research Record **2526**(1), 126–135 (2015). https://doi.org/10.3141/2526-14
50. G. Ranjan et al., Are call detail records biased for sampling human mobility?" In: *ACM SIGMOBILE Mobile Computing and Communications Review* 16.3 (2012), pp. 33–44. https://doi.org/10.1145/2412096.2412101
51. R. Gallotti, A. Bazzani, S. Rambaldi, Understanding the variability of daily travel-time expenditures using GPS trajectory data, in *EPJ Data Science* 4.1 (2015), p. 1. https://doi.org/10.1140/epjds/s13688-015-0055-z
52. C. Roth et al., Structure of urban movements: polycentric activity and entangled hierarchical flows, in *PloS one* 6.1 (2011), e15923. https://doi.org/10.1371/journal.pone.0015923
53. M. Szell et al., Understanding mobility in a social petri dish. Scientific reports **2**(1), 1–6 (2012). https://doi.org/10.1038/srep00457
54. H. Barbosa et al., The effect of recency to human mobility. EPJ Data Science **4**(1), 1–14 (2015). https://doi.org/10.1140/epjds/s13688-015-0059-8
55. M. Schläpfer et al., The universal visitation law of human mobility. Nature **593**(7860), 522–527 (2021). https://doi.org/10.1038/s41586-021-03480-9
56. L. Pappalardo et al., Returners and explorers dichotomy in human mobility. Nature communications **6**(1), 1–8 (2015). https://doi.org/10.1038/ncomms9166
57. D. Pumain, Hierarchy in natural and social sciences. Springer (2006). https://doi.org/10.1007/1-4020-4127-6
58. S. Grauwin et al., Identifying and modeling the structural discontinuities of human interactions, in *Scientific Reports* 7.46677 (2017). https://doi.org/10.1038/srep46677
59. L. Alessandretti, U. Aslak, S. Lehmann, The scales of human mobility. Nature **587**(7834), 402–407 (2020). https://doi.org/10.1038/s41586-020-2909-1
60. L. Alessandretti et al., Multimodal urban mobility and multilayer transport networks. Environment and Planning B: Urban Analytics and City Science , 1–33 (2022). https://doi.org/10.1177/23998083221108190
61. D. Delling et al., Engineering route planning algorithms, in *Algorithmics of large and complex networks*. Springer, 2009, pp. 117–139. https://doi.org/10.1007/978-3-642-02094-0_7
62. F. Poletti et al., Public transit route mapping for large-scale multimodal networks, in *ISPRS International Journal of Geo-Information* 6.9 (2017), p. 268. https://doi.org/10.3390/ijgi6090268
63. J.-Q. Li et al., A multimodal trip planning system with real-time traffic and transit information. Journal of Intelligent Transportation Systems **16**(2), 60–69 (2012). https://doi.org/10.1080/15472450.2012.671708
64. A. Coutrot et al., Entropy of city street networks linked to future spatial navigation ability. Nature **604**(7904), 104–110 (2022). https://doi.org/10.1101/2020.01.23.917211
65. C. Bongiorno et al., Vector-based pedestrian navigation in cities. Nature Computational Science **1**(10), 678–685 (2021). https://doi.org/10.1038/s43588-021-00130-y
66. R. Gallotti, M.A. Porter, M. Barthelemy, Lost in transportation: Information measures and cognitive limits in multilayer navigation, in *Science advances* 2.2 (2016), e1500445. https://doi.org/10.1126/sciadv.1500445
67. J. Broach, J. Dill, J. Gliebe, Where do cyclists ride? A route choice model developed with revealed preference GPS data, in Transportation Research Part A: Policy and Practice **46**(10), 1730–1740 (2012). https://doi.org/10.1016/j.tra.2012.07.005
68. E.J. Manley, J.D. Addison, T. Cheng, Shortest path or anchor-based route choice: a large-scale empirical analysis of minicab routing in London". en. In: *Journal of Transport Geography* 43 (Feb. 2015), pp. 123–139. ISSN: 0966-6923. https://doi.org/10.1016/j.jtrangeo.2015.01.006
69. N. Malleson et al., The characteristics of asymmetric pedestrian behavior: A preliminary study using passive smartphone location data". en. In: *Transactions in GIS* 22.2 (2018), pp. 616–634. ISSN: 1467-9671. https://doi.org/10.1111/tgis.12336
70. A. Lima et al., Understanding individual routing behaviour, in *Journal of The Royal Society Interface* 13.116 (2016), p. 20160021. https://doi.org/10.1098/rsif.2016.0021

71. P. Santi et al., Quantifying the benefits of vehicle pooling with shareability networks. Proceedings of the National Academy of Sciences **111**(37), 13290–13294 (2014). https://doi.org/10.1073/pnas.1403657111
72. U. Editor, *Earth Day is Sunday. And every day. — uber.com*. https://www.uber.com/blog/earth-day-2018/. [Accessed 06-Oct-2022]. 2018
73. A. Henao, W.E. Marshall, The impact of ride-hailing on vehicle miles traveled. Transportation **46**(6), 2173–2194 (2019). https://doi.org/10.1007/s11116-018-9923-2
74. A. Tirachini, Ride-hailing, travel behaviour and sustainable mobility: an international review. Transportation **47**(4), 2011–2047 (2020). https://doi.org/10.1007/s11116-019-10070-2
75. M. Diao, H. Kong, J. Zhao, Impacts of transportation network companies on urban mobility. Nature Sustainability **4**(6), 494–500 (2021). https://doi.org/10.1038/s41893-020-00678-z
76. M.M. Vazifeh et al., Addressing the minimum fleet problem in on-demand urban mobility. Nature **557**(7706), 534–538 (2018). https://doi.org/10.1038/s41586-018-0095-1
77. R. Tachet et al., Scaling law of urban ride sharing, in *Scientific Reports* 7.42868 (2017). https://doi.org/10.1038/srep42868
78. N. Molkenthin, M. Schröder, M. Timme, Scaling laws of collective ride-sharing dynamics, in *Physical Review Letters* 125.24 (2020), p. 248302. https://doi.org/10.1103/physrevlett.125.248302
79. World Health Organization, Global status report on road safety, in (2018)
80. F. Caiazzo et al., Air pollution and early deaths in the United States. Part I: Quantifying the impact of major sectors in 2005, in *Atmospheric Environment* 79 (2013), pp. 198–208. https://doi.org/10.1016/j.atmosenv.2013.05.081.
81. M. Lech Cantuaria et al., Residential exposure to transportation noise in Denmark and incidence of dementia: national cohort study, in *bmj* 374 (2021). https://doi.org/10.1136/bmj.n1954
82. M. Szell, Crowdsourced quantification and visualization of urban mobility space inequality, in *Urban Planning* 3 (2018), pp. 1–20. https://doi.org/10.17645/up.v3i1.1209
83. D. Banister, Unsustainable transport: city transport in the new century. Routledge (2005). https://doi.org/10.4324/9780203003886
84. R. Louf, M. Barthelemy, Modeling the polycentric transition of cities, in *Physical review letters* 111.19 (2013), p. 198702. https://doi.org/10.1103/physrevlett.111.198702
85. R. Prieto Curiel et al., A paradox of traffic and extra cars in a city as a collective behaviour, in *Royal Society Open Science* 8 (201808 2021). https://doi.org/10.1098/rsos.201808
86. F. Creutzig et al., Transport: A roadblock to climate change mitigation?" In: *Science* 350.6263 (2015), pp. 911–912. https://doi.org/10.1126/science.aac8033
87. A. Milovanoff, I.D. Posen, H.L. MacLean, Electrification of light-duty vehicle fleet alone will not meet mitigation targets. Nature Climate Change **10**(12), 1102–1107 (2020). https://doi.org/10.1038/s41558-020-00921-7
88. C. Brand et al., The climate change mitigation effects of daily active travel in cities, in *Transportation Research Part D: Transport and Environment* 93 (2021), p. 102764. https://doi.org/10.21203/rs.3.rs-39219/v2
89. J. Henderson, EVs are not the answer: a mobility justice critique of electric vehicle transitions. Annals of the American Association of Geographers **110**(6), 1993–2010 (2020). https://doi.org/10.1080/24694452.2020.1744422
90. International Transport Forum. *Reversing Car Dependency*. 2021, p. 41. https://doi.org/10.1787/bebe3b6e-en
91. Walking and Cycling, *latest evidence to support policy-making and practice* (WHO, Tech. rep, 2022)
92. S. Gössling et al., The social cost of automobility, cycling and walking in the European Union. Ecological Economics **158**, 65–74 (2019). https://doi.org/10.1016/j.ecolecon.2018.12.016
93. *Transport Strategies for Net-Zero Systems by Design*. Tech. rep. (OECD Publishing, 2021)
94. G. Mattioli et al., The political economy of car dependence: A systems of provision approach, in *Energy Research & Social Science* 66 (2020), p. 101486. https://doi.org/10.1016/j.erss.2020.101486

95. D.C. Shoup, *The high cost of free parking*. (Routledge, 2021)
96. G. Lyons, C. Davidson, Guidance for transport planning and policymaking in the face of an uncertain future. Transportation Research Part A: Policy and Practice **88**, 104–116 (2016). https://doi.org/10.1016/j.tra.2016.03.012
97. S. Cairns, S. Atkins, P. Goodwin, Disappearing traffic? The story so far, in *Proceedings of the Institution of Civil Engineers-Municipal Engineer*. Vol. 151. 1. Thomas Telford Ltd. 2002, pp. 13–22. https://doi.org/10.1680/muen.151.1.13.38856
98. S. Gössling, Why cities need to take road space from cars-and how this could be done. Journal of Urban Design **25**(4), 443–448 (2020). https://doi.org/10.1080/13574809.2020.1727318
99. D. Bongardt et al., Sustainable Urban Transport: Avoid-Shift-Improve (ASI), in *Transformative Urban Mobility Initiative* (2019)
100. E. Holden et al., Grand Narratives for sustainable mobility: A conceptual review, in *Energy Research & Social Science* 65 (2020), p. 101454. https://doi.org/10.1016/j.erss.2020.101454
101. C. Moreno et al., Introducing the "15-Minute City": Sustainability, resilience and place identity in future post-pandemic cities. Smart Cities **4**(1), 93–111 (2021). https://doi.org/10.3390/smartcities4010006
102. R. Lovelace, Open Source Tools for Geographic Analysis in Transport Planning, in *Journal of Geographical Systems* (Jan. 2021). ISSN: 1435-5949. https://doi.org/10.1007/s10109-020-00342-2
103. Google, *transitfeed*. V.1.2.16 https://github.com/google/transitfeed. 2018. https://github.com/google/transitfeed
104. A. Graser, MovingPandas: efficient structures for movement data in Python. GIForum **1**, 54–68 (2019). https://doi.org/10.1553/giscience2019_01_s54
105. L. Pappalardo et al., *scikit-mobility: a Python library for the analysis, generation and risk assessment of mobility data*. 2021. arXiv: 1907.07062 [physics.soc-ph]
106. G. Boeing, OSMnx: New methods for acquiring, constructing, analyzing, and visualizing complex street networks, in *Computers, Environment and Urban Systems* 65 (Sept. 2017), pp. 126–139. ISSN: 01989715. https://doi.org/10.1016/j.compenvurbsys.2017.05.004
107. H. Tenkanen, *pyrosm*. V.0.5.0 https://github.com/HTenkanen/pyrosm. May 2020. https://doi.org/10.5281/ZENODO.3818244. https://github.com/HTenkanen/pyrosm
108. OpenStreetMap contributors, *Overpass API*. https://overpass-api.de/. (2017)
109. K. Butts, *Peartree*. Peartree, V0.6.4, https://github.com/kuanb/peartree/releases/tag/0.6.4 (2021)
110. GraphHopper, *GraphHopper*. https://www.graphhopper.com/ (2022)
111. OSRM, *Open source routing machine*. http://project-osrm.org/ (2021)
112. OTP, *OpenTripPlanner*. http://www.opentripplanner.org/ (2022)
113. M. Wigginton Conway, A. Byrd, M. van der Linden, Evidence-based transit and land use sketch planning using interactive accessibility methods on combined schedule and headway-based networks, in *Transportation Research Record* 2653.1 (2017), pp. 45–53. https://doi.org/10.3141/2653-06
114. R.H.M. Pereira et al., r5r: Rapid realistic routing on multimodal transport networks with R5 in R, in *Findings* (2021), p. 21262. https://doi.org/10.32866/001c.21262
115. M. Morgan et al., OpenTripPlanner for R, in *Journal of Open Source Software* 4.44 (2019), p. 1926. https://doi.org/10.21105/joss.01926
116. C. Fink et al., *r5py: Rapid Realistic Routing with R5 in Python*. 2022. https://doi.org/10.5281/zenodo.7060437
117. M. Padgham, Dodgr: An r package for network flow aggregation, in *Findings* (2019), p. 6945. https://doi.org/10.32866/6945
118. P. Rietveld, The accessibility of railway stations: the role of the bicycle in The Netherlands. Transportation Research Part D: Transport and Environment **5**(1), 71–75 (2000). https://doi.org/10.1016/s1361-9209(99)00019-x
119. F. Foti, P. Waddell, D. Luxen, A generalized computational framework for accessibility: from the pedestrian to the metropolitan scale, in *Proceedings of the 4th TRB Conference on Innovations in Travel Modeling. Transportation Research Board* (2012)

120. Pandana, *Pandana*. V.0.6.1 2021. https://github.com/UDST/pandana
121. OpenStreetMap contributors, *Planet dump retrieved from* https://planet.osm.org. https://www.openstreetmap.org (2017)
122. J. Saxon et al., An open software environment to make spatial access metrics more accessible, in *Journal of Computational Social Science* (2021), pp. 1–20. https://doi.org/10.1007/s42001-021-00126-8
123. S.D. Blanchard, P. Waddell, Urbanaccess: generalized methodology for measuring regional accessibility with an integrated pedestrian and transit network. Transportation research record **2653**(1), 35–44 (2017). https://doi.org/10.3141/2653-05

Chapter 5
Modeling Urban Population Dynamics and City-to-City Migration

Rafael Prieto-Curiel and Carmen Cabrera

Abstract Migration plays a crucial role in urban growth. Over time, individuals opting to relocate led to vast metropolises like London and Paris during the Industrial Revolution, Shanghai and Karachi during the last decades and thousands of smaller settlements. Here, we analyze the impact that migration has on population redistribution. We use a model of city-to-city migration as a process that occurs within a network, where the nodes represent cities, and the edges correspond to the flux of individuals. We analyze metrics characterizing the urban distribution and show how a slight preference for some destinations might result in the observed distribution of the population.

5.1 Introduction

Human mobility takes place at different spatiotemporal scales. Day-to-day mobility, commuting to work or school, leisure, or shopping, takes place at the smallest scale. This type of mobility is characterized by high-frequency and short-distance movements [1, 2]. As the frequency lowers and travel distances increase, we may consider visits and tourism. If the time spent at the destination is even more prolonged, the movement may then be a change of place of residence. Migration is observed when the time spent at the destination is prolonged or permanent. Typically, for a movement to be considered migration, a change of residence must occur for at least six months to a year [1, 2]. In addition, the new home must be at a long enough distance from the previous one—for example, changing address to a different neighbourhood within the same city is often not regarded as a migratory movement, but changing to a different city is (Fig. 5.1). Migration, whether international or internal (within the same country), has always been a central feature of humanity, from the first

R. Prieto-Curiel (✉)
Complexity Science Hub, Vienna, Austria
e-mail: prieto-curiel@csh.ac.at

C. Cabrera
Department of Geography and Planning, University of Liverpool, Liverpool, UK
e-mail: c.cabrera@liverpool.ac.uk

D. Rybski (ed.), *Compendium of Urban Complexity*, Understanding Complex Systems,
https://doi.org/10.1007/978-3-031-82666-5_5

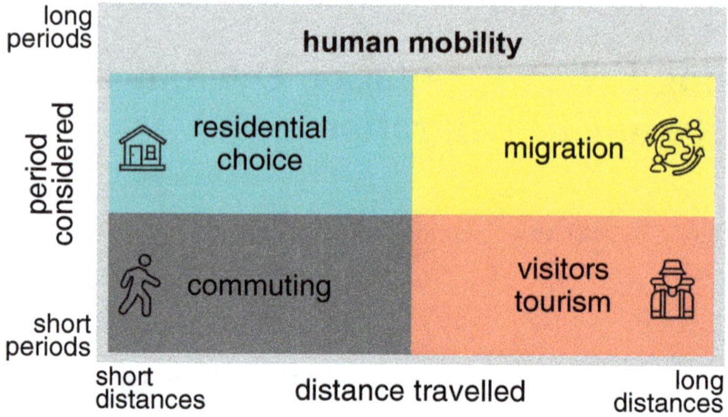

Fig. 5.1 Human mobility is a broad concept that includes aspects such as daily commuting, tourism, and migration. Movements that cover short distances and take place for periods shorter than a day are generally associated with daily commutes. Moving home happens over a long period or even permanently and may occur over a wide range of distances. Over short distances, moving home may be a change of residential choice. Over longer distances and long periods, we observe migration

crossing of the Bering Strait to the colonization of America or the rural flight that many countries experienced or are still experiencing during periods of the Industrial Revolution.

Detecting migration patterns formed by people moving between regions within or across countries has been an important research topic for decades. For example, it has been found that the physical distance between an origin and a destination negatively affects the probability of moving, meaning that migrants are more likely to move to destinations relatively closer to their origin [3, 4]. This pattern was already observed more than a century ago by Ernst Georg Ravenstein in the 1880s, described as one of the "Laws of migration" [5]. More generally, it has been found that the number of visitors to a given location scales with the distance travelled and the frequency of visitation [6]. As with many other study objects, the statistical significance of observed patterns increases with larger sample sizes [7]. Traditionally, data on human mobility was only available through administrative data sources such as censuses or surveys. Currently, we also have a range of novel location-tracking data sources that allow us to gather large amounts of information on human mobility and migration at unprecedented spatial and temporal resolution. Examples of data sources that have been used to analyze human mobility and migration include data from social media platforms as well as location tracker devices in mobile phones [8–13] or smart travel card data [14–17]. While these new data sources are very valuable, they also come with challenges, frequently arising from the fact that the data set design was originally not intended for researching human mobility patterns, see Chap. 4 and [18].

Migration can be a highly selective process. For example, economic migrants often migrate for better economic opportunities, such as higher salaries, improved job prospects, or access to resources. Those with the necessary skills, education, and financial means are more likely to migrate for economic reasons. Therefore, economic migrants may have specific characteristics or motivations that set them apart from those remaining in their home location. A similar effect happens with students who move abroad or skilled people moving to a specific industry (for example, Hollywood). As a result, migration shapes the demographics and composition of both the place of origin and the destination, impacting social, economic, and cultural dynamics in both regions. Furthermore, the choice of destination is dependent on the origin of the migrants, whereby international migrants are more likely to move to larger cities. In contrast, rural populations are more likely to move to smaller cities [3, 19].

Population change in a country or a region is driven by migration and natural change, the latter being the difference between births and deaths. While positive natural change is the primary mechanism driving population growth in the least developed countries, population change in the most developed countries is either a decline caused by negative natural change or a small growth dominated by incoming migrants. For example, in Djibouti, the total population change in 2021 was 1.37 %, and the population change due to natural change was 1.31 %. Therefore, natural change accounts for more than 95 % of the total growth. In contrast, in Austria, the total population change in 2021 was 0.17 %, but the population change due to natural change was -0.04 %, so the population would have declined if it was not for the incoming migrants [20]. While population change is only dominated by migration in some countries, migration always plays the main role in the redistribution of population at the national scale [21]. Internal and international migration can alter the population size of cities and, consequently, a country's urban structure [22, 23]. Similarly, in urban regions within a country, there is a tendency for population change to be dominated by incoming migrants, whereas natural change tends to be the major driver for population growth in rural regions [3].

Some places "push" parts of their population due to unfavourable conditions, whilst other locations attract (or "pull") people [24]. Then, if this process is sustained consistently over time, the origin and the destination will grow at different speeds. Cities have historically been attractive destinations for migrants, either coming from more rural environments in their relative proximity or from other countries. It is natural that, through the cumulative effect of migration, population growth has historically been faster in cities, leading to more economic activity (Chap. 3), infrastructure development, and, overall, increased city attractiveness, resulting in even more migration. This positive feedback cycle can continue for decades or centuries, shaping some cities into massive metropolises with a vast population [25].

For example, in 1970, the metropolitan area of Delhi had half the population of Kolkata. During the 1960s and 1970s, however, Kolkata was undergoing a time of political instability and financial stagnancy after the partitions of Bengal in 1905 and 1947, worsened by the rise of communal riots and the influx of East Bengali refugees during the Indo Pakistan war. Combined with the long-term effect of the change of capital city from Kolkata to Delhi in 1911, Kolkata lost its appeal, resulting in high inter-state migration from Eastern and North-Eastern India towards more prosperous regions [26]. Today, due to migration, Delhi has twice the population of Kolkata. Similarly, Mexico City has the lowest birth rate in the country, but its population represents a stable 17 % of the country's population for more than 50 years, primarily due to migration to the city. In Tanzania, Dar es Salaam has grown fast to 4.5 million people, and currently, it has one-third of the urban population of that country, but it also had one-third of Tanzania's urban population 50 years ago. Dar es Salaam has attracted migration, mainly from the rural parts of Tanzania and other small cities [27]. According to some predictions, Dhaka will have 80 million inhabitants by 2100, and most of this growth will be due to people moving into the city from other regions of Bangladesh [28]. If Nigeria's population continues to grow and people move to cities at the same rate as now, Lagos could become the world's largest metropolis, home to 85 or 100 million people [29].

Cities such as Paris, Mexico City, or Madrid are large metropolises in France, Mexico, and Spain, respectively, and cities such as Toulouse, Morelia, or Burgos are cities in the same countries of roughly the same age but with much smaller populations. This variation in city sizes is primarily related to migration. Cities may also shrink if they no longer offer good prospects for those moving there. For example, Detroit in the US went from a population of 1.9 million in the 1950s to 0.7 million in 2017. It is mainly due to migration that cities grow to become metropolises (or shrink).

Here, we will provide a general introduction to modelling the evolution of urban population size. Then, we will present a city-to-city migration model and its implications regarding city size.

5.2 A General Equation for the Evolution of City Population Size

Generally speaking, population change refers to the change in the number of people in a specified region during a specific period. Population change is usually attributed to two types of change: natural change and migration. Natural change is due to the difference between births and deaths, so if more people are born than dying, we observe positive natural change. Population change due to migration occurs due to the difference between the number of people moving into a location and the number of people leaving that location. So, if more people arrive than leave, there is a positive change due to migration.

When analyzing population change in the context of urban populations, a set of cities is typically considered as the regions of interest, and time is explored in yearly increments. Any locations outside of the group of cities are deemed to be outside of the system.

Formally speaking, population change is described by the so-called demographic equation. If we consider an urban system formed by N_c cities, where $N_i(t)$ is the population of city i at time t, then:

$$N_i(t + \Delta t) = N_i(t) + \Delta t \left[\upsilon_i N_i(t) + \sum_{j=1, j \neq i}^{N_c} (J_{ji}(t) - J_{ij}(t)) \right], \tag{5.1}$$

where υ_i is an effective random growth rate for city i, accounting for natural change, i.e. population change due to the difference between births and deaths, and for "out-of-system" growth, i.e. population change due to migratory movements between city i and other places outside the set of N_c cities in the urban system, such as rural regions or cities outside of the region of interest. The terms involving $J_{ij}(t)$ and $J_{ji}(t)$ account for the migratory flows between city i and the rest of the cities in the system. In particular, $J_{ij}(t)$ is the number of people moving from city i to city j between time t and time $t + \Delta t$, and vice-versa for J_{ji}. Like in [22], we will work in units where $\Delta t = 1$ for simplicity of notation.

The model proposed in Eq. (5.1) is very general and allows including new, emerging cities over time and excluding others if their size vanishes. In the continuous time limit, Eq. (5.1) is equivalent to the diffusion equation with noise, a well-known model with implications far beyond urban systems [23, 30] in phenomena such as finance, statistical mechanics, and the Kardar-Parisi-Zhang (KPZ) equation. However, as is often the case when using very general equations, it is challenging to use Eq. (5.1) for making predictions [23], as there are no known solutions for a general form of J_{ij} [30]. Various assumptions have been proposed to simplify the equation, leading to other well-known models for urban growth. For example, when we assume that there is an exact balance of migration flows, i.e. $J_{ij} = J_{ji}$, Eq. (5.1) becomes equivalent to Gibrat's model [31], which predicts a log-normal distribution of city population sizes. Similarly, when the mean-field approximation is considered, i.e. $J_{ij} = J/N_c$ for all $i, j = 1, \ldots, N_c$ with $J = \sum_{i,j=1}^{N_c} J_{ij}$, and we assume that J is small enough, then the solution of Eq. (5.1) leads to a power law distribution in equilibrium, which is consistent with the empirical Zipf's law . This suggests that the origin of Zipf's law lies in the interplay between internal random growth and exchanges between different cities. However, according to the mean-field approximation, as interurban mobility increases, the population sizes of all the cities in the system converge to the same value. Therefore, from a theoretical perspective, it is essential to consider not just the mean-field analysis but also other non-constant coupling J_{ij} alternatives [30] that are capable of accounting for random fluctuations in the migratory flows migrations.

In the remainder of this chapter, we discuss several models of city-to-city migration that go beyond the mean-field approximation. For simplicity, we will set $\nu_i = 0$ and focus only on urban population growth purely due to intraurban migration flows, so instead of working with Eq. (5.1), we will consider the following equation

$$N_i(t + 1) = N_i(t) + \sum_{j=1, j \neq i}^{N_c} (J_{ji}(t) - J_{ij}(t)) . \tag{5.2}$$

5.3 Modeling City-to-City Migration

Network analysis offers a simple way to conceptualize migratory movements between cities. Here, we represent the distinct origin and destination locations as network nodes, in our case, cities. The weights of the edges represent the probability that a person moves between the two nodes. This type of network has been used to model migration between California and other states in the US or people moving from and to Germany [32–35].

Using the same notation introduced in Sect. 5.2, we consider a region with N_c cities. Usually, to keep the notation simple, we assign the same index $i = 1, 2, \ldots, N_c$ to each node and to the city it represents so that node 1 corresponds to "city 1" and node 4 to "city 4". We add weighted edges between the pairs of nodes. The weight of an edge between nodes i and j corresponds to the probability $m_{ij} \geq 0$ that a person moves between cities i and j. Migration is a process that takes a long time, so in general, the movements between cities are considered over annual intervals as opposed to daily or weekly intervals. If $m_{ij} = 0$, we consider that the edge between i and j does not exist. Let m_{ii} be the probability that the person remains in the city i from one year to the next. This way, the sum of the probabilities satisfies $\sum_{j \in n} m_{ij} = 1$, for $i = 1, 2, \ldots N_c$. Here, we assume that the probability of moving is independent of past migrations, that it is the same for all individuals, and that it only depends on the origin and destination cities of a possible migratory movement.

Then, the number of people that move from i to j in a year, which we denote by $J_{ij}(t)$, can be regarded as a time-dependent random variable with an associated probability mass function. Even though the variables corresponding to the cities' populations and the intraurban population flows are time-dependent, we will not make this explicit for ease of notation unless it is necessary for clarity.

Since J_{ij} represents a count, a natural choice of distribution to model this quantity is a Binomial distribution. A Binomial distribution is often used to model the probability for a certain number of successes, given a number of random trials and the probability for success in each individual trial. We can think of the number of people in i, N_i, as the number of trials, and the probability of an individual migrating from i to j, m_{ij}, as the probability for success in each trial. Then, the number of people migrating from i to j, J_{ij}, can be regarded as the number of successes, distributed according to a Binomial distribution of the form

$$\begin{aligned}\Pr(J_{ij}; N_i, m_{ij}) &= \text{Bin}(J_{ij}, N_i, m_{ij}) \\ &= \binom{N_i}{J_{ij}} m_{ij}^{J_{ij}} (1 - m_{ij})^{N_i - J_{ij}} .\end{aligned}$$

The expected number of people moving from i to j is $E[J_{ij}] = m_{ij} N_i$, corresponding to the expected value of the Binomial distribution, and the variance is $Var[J_{ij}] = m_{ij}(1 - m_{ij})N_i$. Note that the expected value of J_{ij} depends on the population size of the origin node. Thus, considering the interaction of i only with node $j \neq i$, the expected population of city i at time $t+1$ is $N_i(t+1) = N_i(t) + m_{ji}N_j(t) - m_{ij}N_i(t)$.

If, instead, we consider the interaction of i with all the other nodes in the network (i.e. the flows between city i and all the other cities in the urban system), the model can be generalized from a Binomial to a Multinomial distribution. The mathematical properties are pretty similar [36]. The joint probability of the flows of people moving from node i to all other nodes in the network, given a population N_i at the origin and a set of probabilities $\mathbf{m}_i = (m_{i1}, ..., m_{iN_c})$ for moving from i to each j in the network, can be modelled as follows

$$\begin{aligned} Pr(\mathbf{J}_i; N_i, \mathbf{m}_i) &= \\ &= \text{Multinomial}(\mathbf{J}_i, N_i, \mathbf{m}_i) \\ &= \frac{N_i!}{J_{i1}! \dots J_{iN_c}!} m_{i1}^{J_{i1}} \times \dots \times m_{iN_c}^{J_{iN_c}}, \end{aligned} \tag{5.3}$$

where $\mathbf{J}_i$ represents the set of flows from i to every other node including itself, $(J_{i1}, \dots, J_{iN_c})$.

Underlying Eq. (5.2), there is a discrete-time Markov chain that models the states of a random variable over time. An important feature of Markov chains is that the current state of the variable depends only on the state at the previous time step [37]. In our case, the random variable is categorical and indicates the city where an individual is located. The possible states of this variable are the cities in the network. Hence, the Markov chain describes the sequence of cities in the network that the individual would visit at each time step. The probability of visiting city i at time t only depends on the city where they were located at the previous time step. Generally, a Markov chain can be characterized by a stochastic matrix, which contains the probabilities that the random variable transitions from one state to another in each time step [37]. Here, this matrix M is given by the probabilities to migrate between from city i to any other city j

$$M = \begin{pmatrix} m_{11} & m_{12} & \dots & m_{1N_c} \\ m_{21} & m_{22} & \dots & m_{2N_c} \\ \vdots & \vdots & \ddots & \vdots \\ m_{N_c1} & m_{N_c2} & \dots & m_{N_cN_c} \end{pmatrix}.$$

Note that, in general, these transition probabilities are not necessarily independent of time. If they are, we say that the Markov chain is time-homogeneous. The total population of the system, i.e. the sum of the population of all the N_c cities, is, on average, conserved over time. This is ensured by the fact that the elements in each row of M add up to 1. We can prove this by showing that, at any time t, the equality $E[\sum_i N_i(t+1)] = \sum_i N_i(t)$ is satisfied

$$E[\sum_{i=1}^{N_c} N_i(t+1)] = \sum_{i=1}^{N_c} E[N_i(t+1)] =$$

$$= \sum_{i=1}^{N_c} N_i(t) + \sum_{i=1}^{N_c} \sum_{\substack{j=1 \\ j \neq i}}^{N_c} m_{ij} N_j(t) - \sum_{i=1}^{N_c} \sum_{\substack{j=1 \\ j \neq i}}^{N_c} m_{ij} N_i(t) = \sum_{i=1}^{N_c} N_i(t) . \tag{5.4}$$

where the two last terms in the second last step add up to 0, as it can be derived from the fact that $\sum_i \sum_{j, j \neq i} m_{ij} N_j(t) = \sum_i \sum_j m_{ij} N_j(t) - \sum_i m_{ii} N_i(t) = \sum_i \sum_j m_{ij} N_j(t) = \sum_j N_j(t) [\sum_i m_{ji}] = \sum_j N_j(t)$ and equivalently for the subtracting term, $\sum_i \sum_{j, j \neq i} m_{ij} N_i(t)$.

It is possible to run simulations using an iterative algorithm, where each iteration represents a time step of the stochastic process. We set the initial time of this process to $t = 0$. At $t = 0$, the population size of all cities is known, and it is given by $\mathbf{N}(0) = (N_1(0), \ldots, N_{N_c}(0))$, and the matrix of transition probabilities M is also known. Let's assume the Markov process is time-homogeneous, so the matrix M remains constant over time. The algorithm involves the following steps, starting from $t = 0$:

1. For each node i, generate the number of people moving to every other node in the network, J_{ij}. This will be a set of random numbers drawn from a Multinomial distribution specified in Eq. (5.3), parametrised by N_i and m_i.
2. For each node i, update the corresponding city population to generate $N_i(t + 1)$. This is done by subtracting from $N_i(t)$ the total number of people that move to other nodes, i.e. $\sum_{j=1, j \neq i}^{N_c} J_{ij}$ and by adding to the resulting quantity the total number of people that move from other nodes, i.e. $\sum_{j=1, j \neq i}^{N_c} J_{ji}$.
3. Update t to $t + 1$ and repeat from step 1.

To obtain confidence intervals for the city population sizes at time t, we can run the simulations a sufficiently large number of times up to time t. Although there is no specific requirement for the number of iterations, at least 100 simulations are frequently used.

5.4 Characterizing the Disparity in the Distribution of City Population Sizes at Time t

The size distribution of many events in nature is characterized by a few huge observations and many smaller ones. For example, thousands of low-magnitude earthquakes have been recorded over a century, with very few earthquakes of magnitude eight or above. Similarly, there are only one or two massive cities in a country or region, whereas there are many smaller ones [38, 39], see also Chap. 1. Heavy-tailed probability distributions of event sizes, such as a power law or a lognormal distribution, are often used to account for this pattern. Zipf's law [40] is an empirical law that states that city population size follows a power law distribution with exponent -1, and it has repeatedly been reported to hold in the case of the distribution of urban population sizes (e.g. [41]). In the context of the distribution of wealth, Pareto's distribution is frequently used to describe that a small number of individuals concentrate a large proportion of the total wealth.

The Gini index is one of the most widely used measures to quantify the disparity in the distribution of wealth. It is possible to establish explicit functional relations between the parameters that characterize the probability distribution of the variable of interest and the Gini index [42]. However, this index may also be used to quantify the disparity in the sizes of other variables, such as city population size. Like the Gini index, other metrics summarize the disparity of a distribution. To be able to compare distributions across countries or regions and over time, disparity metrics need to satisfy the following properties [43]:

- Anonymity principle: permutations of "personal" labels within a given distribution should not affect the overall level of disparity, e.g. if the population of London and Manchester are swapped, the level of disparity for city population size in England should remain the same.
- Transfer principle (or Pigou-Dalton principle): disparity is reduced when a fixed amount is transferred from a larger to a smaller event, and the recipient is still smaller than the donor. For example, if people from the capital migrate to a smaller town, the level of disparity would be decreased.
- Population principle: the level of disparity should be independent of the size of the data set (e.g. number of cities in the sample). This ensures that it is possible to compare small and large regions or countries regardless of their total number of cities.
- Relative size principle: only relative values should matter. If the values for city population size increase by a constant proportion for all the cities in the dataset, then the level of disparity should remain the same. For example, if all cities in a country grew in size by the same proportion due to an overall increase in fertility rates, the level of disparity should remain the same.

Here, we focus on four metrics of disparity in city population size. These are (i) the Gini index, (ii) the metropolization index, (iii) the 90-10 inter-decile ratio, and (iv) the coefficient of variation. While the Gini index and the coefficient of variation use the whole distribution of city population sizes, the metropolization index and the 90-10 inter-decile ratio only look at specific distribution points.

Let us start by defining the Gini index. Let N_i be the population size of city i, with $i = 1, \ldots, N_c$, then the Gini index G is given by

$$G = \frac{\sum_{i=1}^{N_c} \sum_{j=1}^{N_c} |N_i - N_j|}{2 \sum_{i=1}^{N_c} \sum_{j=1}^{N_c} N_j}.$$

The Gini index's range of values spans from 0 to 1, with these extremes corresponding, respectively, to a situation where all the cities have the same population size and to a situation of maximal disparity in population sizes.

The metropolization index MI is defined here as the proportion of the population living in the largest city, i.e.

$$MI = \frac{max(\{N_i\}_{i=1}^{N_c})}{\sum_{i=1}^{N_c} N_i},$$

with values of MI varying between $1/N_c$ if all cities have the same population and a value close to 1 if one city has most of the region's population.

The 90-10 inter-decile ratio IR measures how many times larger the population size at the 90th percentile, denoted by $N_{\%90}$, is compared with the population size at the 10th percentile, denoted by $N_{\%10}$. Hence,

$$IR = \frac{N_{\%90}}{N_{\%10}}.$$

This measure takes a minimum value of 1, corresponding to the scenario where all the cities have the same population. It tends to infinity as the disparity in the distribution of city sizes increases.

Finally, the coefficient of variation CV equals the ratio of the standard deviation to the average of the city population size. If only a sample of empirical data is available, the CV can be estimated using the ratio of the sample standard deviation to the sample mean, i.e.,

$$CV = \frac{\sqrt{\frac{1}{N_c} \sum_{i=1}^{N_c} (N_i - \bar{N})}}{\bar{N}},$$

where $\bar{N} = \frac{1}{N_c} \sum_{i=1}^{N_c} N_i$ is the sample mean. The CV measures variability concerning the mean, so it is independent of the values of city population size and, therefore, satisfies the "population principle". While the Gini index is more sensitive to changes in the middle values of the distribution of sizes, the CV places more weight on the top end of the distribution. Therefore, if one of the largest cities in the region under study gains population faster than the others, the change in value in CV will be relatively greater than the Gini index.

5.5 Example 1: Migration Between Five Cities

Consider a network with five nodes corresponding to a system of five cities (Fig. 5.2). The matrix M in Table 5.1 gives the probability of moving between cities.

Migration from city A to city B differs from migration from B to A. This is reflected by the matrix M is not symmetric, meaning that m_{ij} is not always the same as m_{ji}. Some cities could attract more people than others. Note, however, that the total number of people in our system of cities is, on average, constant over time since we are not considering births, deaths or out-of-system population growth. This is ensured by the fact that the values in each row of matrix M add up to one, as demonstrated in Eq. (5.4). For example, in Table 5.1, the first row gives the

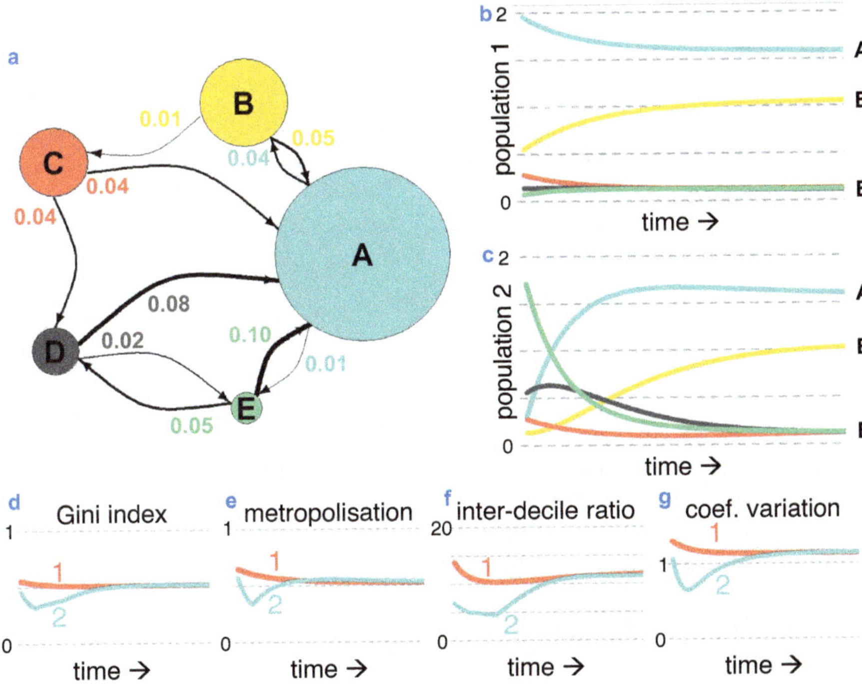

Fig. 5.2 **a** Five nodes represent five cities, labelled A, B, C, D and E. The edges have been given probabilities according to Table 5.1. The nodes' size on the left diagram corresponds to the long-run distribution of the population. **b,c** Evolution of city size, computed according to Eq. (5.1) for different initial distributions of population sizes $P(0)$. After a few time steps, both distributions **b** and **c** are roughly the same. Evolution of the Gini index **d**, the metropolisation index **e**, the 90–10 inter-decile ratio **f**, and the coefficient of variation **g** for the two initial distributions of populations of sizes

probabilities of moving from city A to each other city, and the sum is 1. The first column shows the probabilities of moving to city A from any other city, and these values are higher than for moving from A to any other city. That is, $m_{iA} > m_{Ai}$, with $i = B, C, D, E$, which means that city A pulls more people from cities B, C, D and E than it pushes to those cities.

Taking some initial population $P(0)$, we can analyze the evolution of the size of all cities (Fig. 5.2, right). With different initial populations (top and bottom of Fig. 5.2), the curves representing the evolution of city population sizes have different shapes. However, after enough time steps, the distribution of sizes becomes nearly the same. As time passes, the population flows between cities become more dominant than the initial populations in determining the long-term distribution of population sizes. In both cases depicted in Fig. 5.2, city A becomes the most populous, followed by city B. We can attribute this to the fact that, according to the transition probabilities in matrix M, people are more attracted to city A and are less likely to move outside that city. Moreover, the values in the diagonal of M are relatively high, corresponding to

Table 5.1 Probability of moving between two cities. The origin corresponds to the rows of the table, and the destination corresponds to the columns. Thus, the probability of moving between A and E, for example, is $m_{AE} = 0.01$

	A	B	C	D	E
A	0.95	0.04	0	0	0.01
B	0.05	0.94	0.01	0	0
C	0.04	0	0.92	0.04	0
D	0.08	0	0	0.90	0.02
E	0.10	0	0	0.05	0.85

the probability of a person staying in the same city from one year to the next. This will be the case in general as most people will remain in the same city from one year to the next, and in relative terms, almost no one will move. Many people will never move from their hometown, and those who do will likely remain in their new location for years, or even decades, before moving again [34].

The impact of the initial size of the cities becomes less relevant as time passes. Even though there are differences in the distribution of the populations in the first stages, the long-run distribution of the population is the same for both sets of initial city population sizes (Fig. 5.2). Therefore, the Gini index, the metropolization index, the inter-decile ratio, and the coefficient of variation also converge to similar values. In the example, for the initial city population sizes plotted on top (population 1), the Gini index is relatively high, meaning that a few cities have most of the country's population; the metropolization index is above 0.5, meaning that the largest city has more than half of the population; the inter-decile ratio is above 10, and the coefficient of variation is above 1. The initial city population sizes plotted on the bottom (population 2) are more mixed, so there are some short-term fluctuations in the four metrics under consideration. However, in the long run, the values converge to those obtained with the first set of initial city population sizes.

5.6 Example 2: Modeling Internal Migration in Madagascar

Here, we construct an example in Madagascar, a country occupying an island in Southeast Africa with roughly the same surface as France. Madagascar has approximately 30 million inhabitants, with its capital being Antananarivo, with 2.5 million people. The Africapolis dataset gives an estimate of the population of urban agglomerations in the continent, and it will be used to measure population and distance between cities [44]. Madagascar has 68 urban agglomerations, but here we focus on the top 25 most populous cities (Fig. 5.3).

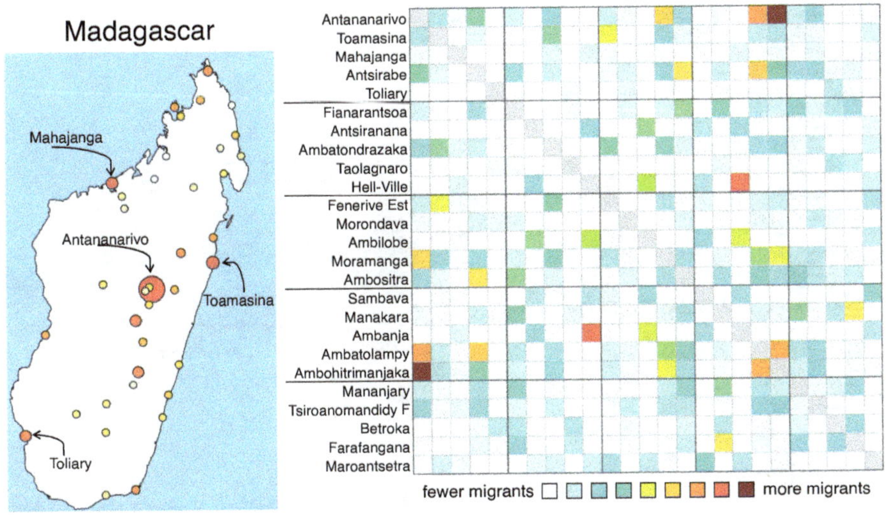

Fig. 5.3 Madagascar example. The map of Madagascar and its 25 most populous cities is shown on the left. The disk's size corresponds to each city's population in 2015. On the right, the probability of moving between two cities m_{ij} considering only the distance between city i and j, with $\alpha = 0.04$ and $\beta = 1/100$, is illustrated. Columns and rows are sorted according to the population of cities

First, let us consider what happens if people move to cities depending only on the distance between their city and the possible destinations. The physical distance between city i and j reduces the probability of moving. One way to capture this process is by considering the distance between the two cities, D_{ij}, and an exponential decay probability. That is, we can capture the effect of distance by expressing the probability of moving by

$$m_{ij} = \alpha \exp(-\beta D_{ij}),$$

where $\alpha > 0$ and $\beta > 0$ are model parameters. Notice that due to the negative sign inside the exponential, the expression gives a smaller probability of moving for longer distances. For example, with values of $\alpha = 0.04$ and $\beta = 1/100$, we obtain the matrix M in Fig. 5.3. For a given i, the distance from i to itself is $D_{ii} = 0$, which means that $\alpha = m_{ii}$. Furthermore, according to a Multinomial distribution, the values m_{ij} for $j = 1, \ldots, N_c$, must add up to 1. Therefore, the value of β must be such that for any i, $\sum_{j=1}^{N_c} m_{ij} = 1$.

The case where the probability of moving between cities is modelled using only distance has some interesting properties. First, the matrix M is symmetric (since distances are symmetric). Hence, if city i has more population than j, i.e. $N_i(t) > N_j(t)$, then the expected number of people moving between cities i and j, $E[J_{ij}]$ exceeds the expected number of people moving between city j and i, $E[J_{ji}]$ since according to the Multinomial distribution $E[J_{ij}] = m_{ij} N_i$, therefore:

$$E[J_{ij}] = m_{ij} N_i(t) > m_{ji} N_j(t) = E[J_{ji}].$$

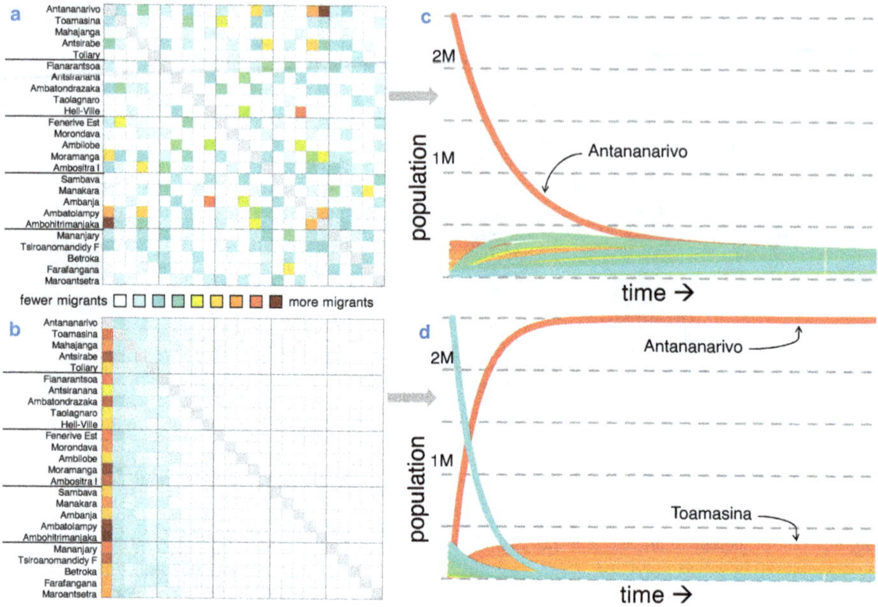

Fig. 5.4 Modeled probability of moving between cities (left) and distribution of city size as time evolves (right). The top panel **a** shows the scenario where the distance gives the probability of moving between cities, i.e. $m_{ij} = \alpha \exp(-\beta D_{ij})$. The bottom panel **b** shows the scenario where the probability is dependent on destination size too, i.e. $m_{ij} = \gamma N_j \exp(-\delta D_{ij})$. In both cases, the country's total population remains constant over time, so $\sum_{i=1}^{N_c} N_i(t) = \sum_{i=1}^{N_c} N_i(t+1)$ for any $t \geq 0$

The expected population flow between cities i and j would be the same as the flow between cities j and i only if both cities have the same population. Otherwise, more people will move from large to small cities. In the long run, an equilibrium will be reached when all cities have the same population. This model is illustrated in the top panel of Fig. 5.4).

While this model works mathematically, it leads to an equilibrium distribution that fails to capture the disparity in city population sizes in the long run. A more realistic approach (as we will see below) is to model the transition probabilities m_{ij} as a function of the distance between cities and the population of N_j rather than N_i, hence placing more emphasis on the characteristics of the possible destination locations that an individual might decide to migrate to.

Considering $m_{ij} = \gamma N_j \exp(-\delta D_{ij})$, with $\gamma > 0$ and $\delta > 0$, the long-run distribution of the population is similar to the one observed in Madagascar (Fig. 5.4), i.e. very few large cities and lots of smaller cities. In this case, the initial distribution of the population at time $t = 0$ becomes less relevant, and people's preference for particular destinations determines which cities will eventually be the largest ones in the country. Here, we assume that births and deaths are the same, so the only demo-

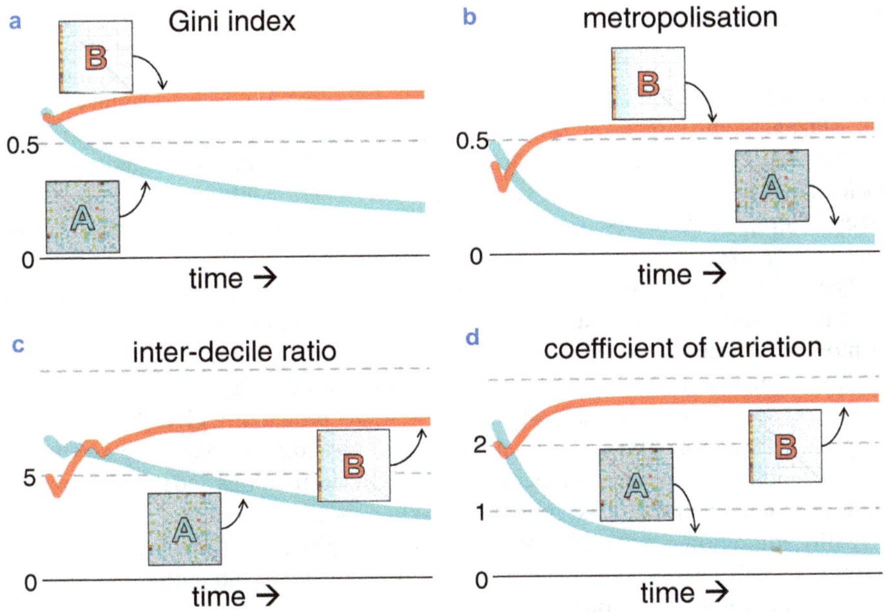

Fig. 5.5 Temporal evolution of the Gini index **a**, the metropolization index **b**, the inter-decile ratio **c**, and the coefficient of variation **d**, according to two migration models. The first is based on proximity between cities A or on proximity and population size of the destination B. The matrices correspond to the tables in Fig. 5.4 on top A and on bottom B

graphic process we observe is migration between cities. Here, if people are slightly more likely to move to a city like Antananarivo, it will eventually become the largest city in the country.

The metrics for the population distribution show substantial differences (Fig. 5.5). Suppose large cities attract more population (second model). In that case, the Gini index and the coefficient of variation are higher, showing the existence of more disparity in the distribution of city population sizes, with just a few large cities hosting most of the population. Also, the metropolization index is higher, indicating that this concentration is observed mainly in a single city. Finally, the coefficient of variation is also high, meaning that the size of cities is very unequal. On the contrary, in the model based on proximity only, the Gini coefficient drops as time passes, meaning that cities evolve to a more similar size. The metropolization index also falls, meaning that no city absorbs most of the population in that system, and the 90-10 inter-decile ratio also drops, showing that the top and bottom cities have a similar size. Finally, a slight variation coefficient means the size distribution is less skewed.

5.7 Conclusion

Human mobility is a broad topic, encompassing daily commuters, residential changes, tourism, visits, and migration. Migration is observed only when the person has moved a sufficiently large distance and for an extended period. Migration profoundly impacts society, including political expressions, use of services, demographic processes, and others. Here, we have discussed only observed migration and its implication on altering city size.

There are many ways to model migration. Our proposed models, where we consider the possible locations as the nodes in a network of cities and the probability of moving between cities as the weighted edges, are just some among many. Our approach relies on a Markov process, which describes the changes over time of an individual's location in the network according to the probability given by the edges' weights. The approach also relies on the assumptions that people move depending only on features of their current location or possible destinations and that their movement does not depend on past migrations. While there is evidence suggesting that, in reality, migration is a complex process driven by the cumulative effect of multiple factors through history, our assumptions help simplify the modelling process and make the interpretation of our models easier.

Essentially, the models are characterized by the transition probabilities of moving between cities, i.e. m_{ij}. The models can be more or less realistic depending on how these probabilities are defined. Sometimes, they can lead to unrealistic city size distributions in the long run, such as an equilibrium distribution where all the cities have the same population. However, most countries and regions display some hierarchy in their urban structure. Hence, a good model should be able to capture small and large cities simultaneously.

We have explored a range of metrics that allow us to compare the distribution of city population sizes in an urban system. These metrics rely on the relative size of the cities but are independent of the number of cities and their absolute size. The size of a city alone does not say much about the role of a city in the urban hierarchy and, therefore, its degree of attractiveness for migrants [45]. For example, China has more than 300 cities with more than one million inhabitants, so a city with one million people would be considered small. However, in Central Africa, a city with one million inhabitants is large enough to be a capital, such as N'Djamena, Libreville, Bangui, and Malabo in the case of Chad, Gabon, Central African Republic, and Equatorial Guinea, respectively [45].

Reasons for moving to one city can be modelled with different techniques. Chapter 6 analyses ways to model the probability Π, depending on city size and distance between cities.

Acknowledgements This research is funded by the Federal Ministry of the Interior of Austria (2022-0.392.231) and by the UK Economic and Social Research Council (project number ES/Y010787/1).

References

1. A. Sîrbu et al. Human migration: the big data perspective, in *International Journal of Data Science and Analytics* (2020), pp. 1–20. https://doi.org/10.1007/s41060-020-00213-5
2. M. Levy, Scale-free human migration and the geography of social networks. Physica A: Statistical Mechanics and its Applications **389**(21), 4913–4917 (2010). https://doi.org/10.1016/j.physa.2010.07.008
3. E. George Ravenstein, The laws of migration, in *Journal of the Statistical Society of London* 48.2 (1885), pp. 167–235. https://doi.org/10.2307/2979333
4. A. Schwartz, Interpreting the effect of distance on migration, in *Journal of Political Economy* 81.5 (1973), pp. 1153–1169. https://doi.org/10.1086/260111
5. D.B. Grigg, E. G. Ravenstein and the laws of migration, in *Journal of Historical Geography* 3.1 (1977), pp. 41–54. ISSN: 0305-7488. https://doi.org/10.1016/0305-7488(77)90143-8
6. M. Schläpfer et al., The universal visitation law of human mobility, in *Nature* 593.7860 (2021), pp. 522–527. ISSN: 1476-4687. https://doi.org/10.1038/s41586-021-03480-9
7. M. Helbling et al., Measuring the effect of climate change on migration flows: Limitations of existing data and analytical frameworks. PLOS Climate **2**(1), 1–10 (2023). https://doi.org/10.1371/journal.pclm.0000078
8. J. Yin et al., Depicting urban boundaries from a mobility network of spatial interactions: a case study of Great Britain with geo-located Twitter data. International Journal of Geographical Information Science **31**(7), 1293–1313 (2017). https://doi.org/10.1080/13658816.2017.1282615
9. X. Yang et al., Understanding the spatial structure of urban commuting using mobile phone location data: A case study of Shenzhen, China, in *Sustainability* 10.5 (2018). ISSN: 2071-1050. https://doi.org/10.3390/su10051435
10. W. Zhu et al., Investigating the complexity of spatial interactions between different administrative units in China using Flickr data, in *Sustainability* 12.22 (2020). ISSN: 2071-1050. https://doi.org/10.3390/su12229778
11. R. Ponce-Lopez, J. Ferreira, Identifying and characterizing popular non-work destinations by clustering cellphone and point-of-interest data, in *Cities* 113 (2021), p. 103158. ISSN: 0264-2751. https://doi.org/10.1016/j.cities.2021.103158
12. Y. Wang et al., Understanding internal migration in the UK before and during the COVID-19 pandemic using Twitter data, in *Urban Informatics* 1 (2022), p. 15. https://doi.org/10.1007/s44212-022-00018-w
13. F. Rowe et al., Urban exodus? Understanding human mobility in Britain during the COVID-19 pandemic using Meta-Facebook data, in Population. Space and Place **29**(1), e2637 (2023). https://doi.org/10.1002/psp.2637
14. Y. Yang et al., A spatiotemporal and graph-based analysis of dockless bike sharing patterns to understand urban flows over the last mile, in *Computers, Environment and Urban Systems* 77 (2019), p. 101361. ISSN: 0198-9715. https://doi.org/10.1016/j.compenvurbsys.2019.101361
15. L. Tang et al., A Clustering Refinement Approach for Revealing Urban Spatial Structure from Smart Card Data, in *Applied Sciences* 10.16 (2020). ISSN: 2076-3417. https://doi.org/10.3390/app10165606
16. N. Sari Aslam et al., Semantic enrichment of secondary activities using smart card data and point of interests: a case study in London, in *Annals of GIS* 27.1 (2021), pp. 29–41. https://doi.org/10.1080/19475683.2020.1783359
17. C. Cabrera-Arnau et al., Inferring urban polycentricity from the variability in human mobility patterns, in *Scientific Reports* 13.1 (2023), p. 5751. https://doi.org/10.1038/s41598-023-33003-7
18. J. Stillwell, P. Boden, A. Dennett, Monitoring who moves where: Information systems for internal and international migration, in *Population Dynamics and Projection Methods* (Springer, 2011), pp. 115–140. https://doi.org/10.1007/978-90-481-8930-4_6
19. R. Prieto-Curiel et al., Gravity and scaling laws of city to city migration, in *PloS one* 13.7 (2018), e0199892. https://doi.org/10.1371/journal.pone.0199892

20. Our World in Data, *Population growth rate with and without migration*. https://ourworldindata.org/grapher/population-growth-rate-with-and-without-migration. [Accessed 01-08-2023]
21. M. Bell et al., Internal migration and development: Comparing migration intensities around the world. Population and Development Review **41**(1), 33–58 (2015). https://doi.org/10.1111/j.1728-4457.2015.00025.x
22. L.M.A. Bettencourt, D. Zünd, Demography and the emergence of universal patterns in urban systems, in *Nature Communications* 11.1 (2020), p. 4584. https://doi.org/10.1038/s41467-020-18205-1
23. V. Verbavatz, M. Barthelemy, The growth equation of cities. Nature **587**(7834), 397–401 (2020). https://doi.org/10.1038/s41586-020-2900-x
24. E.S. Lee, A theory of migration. Demography **3**(1), 47–57 (1966). https://doi.org/10.2307/2060063
25. C. Cabrera-Arnau, S.R. Bishop, The effect of dragon-kings on the estimation of scaling law parameters, in *Scientific Reports* 10.1 (2020), p. 20226. https://doi.org/10.1038/s41598-020-77232-6
26. UNESCO Office in New Delhi and UNICEF India, *National Workshop on Internal Migration and Human Development in India: workshop compendium. Vol. II: workshop papers*. UNESCO and UNICEF, 2012. URL: https://unesdoc.unesco.org/search/N-EXPLORE-bf7bd133-c8f2-4bfa-964b-0492e47d072b
27. L. Christiaensen, Secondary Towns and Poverty Reduction: Refocusing the Urbanization Agenda. The World Bank (2016). https://doi.org/10.1596/1813-9450-7895
28. S. Parnell et al., Introduction: Situating knowledge and action for an urban planet, in *The urban planet: knowledge towards sustainable cities* (2018), pp. 1–16. https://doi.org/10.1017/9781316647554.002
29. D. Hoornweg, K. Pope, Population predictions for the world's largest cities in the 21st century. Environment and Urbanization **29**(1), 195–216 (2017). https://doi.org/10.1177/0956247816663557
30. M. Barthelemy, The statistical physics of cities, in *Nature Reviews Physics* 1.6 (2019), pp. 406–415. ISSN: 2522-5820. https://doi.org/10.1038/s42254-019-0054-2
31. R. Gibrat, *Les In é galit é s É conomiques* (Recueil Sirey, Paris, 1931)
32. A.C. Kelley, L.W. Weiss, Markov processes and economic analysis: the case of migration, in *Econometrica: Journal of Econometric Society* (1969), pp. 280–297. https://doi.org/10.2307/1913537
33. N.W. Henry, R. McGinnis, H.W. Tegtmeyer, A finite model of mobility. The Journal of Mathematical Sociology **1**(1), 107–118 (1971). https://doi.org/10.1080/0022250X.1971.9989790
34. R. Prieto-Curiel et al., Mobility between Colombian cities is predominantly repeat and return migration, in Computers. Environment and Urban Systems **94**, 101774 (2022). https://doi.org/10.1016/j.compenvurbsys.2022.101774
35. A.F. Constant, K.F. Zimmermann, The dynamics of repeat migration: A Markov chain analysis, in *International Migration Review* 46.2 (2012), pp. 362–388. ISSN: 1747-7379. https://doi.org/10.1111/j.1747-7379.2012.00890.x
36. G.A.F. Seber, Multinomial Distribution, in *International Encyclopedia of Statistical Science*. Ed. by Miodrag Lovric. Berlin, Heidelberg: Springer Berlin Heidelberg, 2011, pp. 882–884. ISBN: 978-3-642-04898-2. https://doi.org/10.1007/978-3-642-04898-2_388
37. P. Blanchard, D. Volchenkov, Random Walks and Diffusions on Directed Graphs and Interacting Networks, in *Random Walks and Diffusions on Graphs and Databases: An Introduction* (Berlin, Springer, 2011), pp. 93–106. ISBN: 978-3-642-19592-1. https://doi.org/10.1007/978-3-642-19592-1_7
38. A. Le Maître, *La Métropolitée, ou de l'Etablissement des villes capitales* ··· (B. Boekholt, Amsterdam, French 1682)
39. F. Auerbach, Das Gesetz der Bevölkerungskonzentration, in *Petermanns Geogr. Mitteilungen* 59.74 (1913), pp. 74–76. URL: https://zs.thulb.uni-jena.de/receive/jportal_jparticle_00586390
40. G.K. Zipf, *Human Behavior and the Principle of Least Effort* (Addison-Wesley Press, 1949)

41. Y.M. Ioannides, H.G. Overman, Zipf's law for cities: an empirical examination, in *Regional Science and Urban Economics* 33.2 (2003), pp. 127–137. ISSN: 0166-0462. https://doi.org/10.1016/S0166-0462(02)00006-6
42. W. Pfähler, Relative concentration curve: functional form and measures of non-proportionality. Bulletin of Economic Research **37**(3), 201–211 (1985). https://doi.org/10.1111/j.1467-8586.1985.tb00194.x
43. F.A. Cowell, Chapter 2 Measurement of inequality, in vol. 1. Handbook of Income Distribution (Elsevier, 2000), pp. 87–166. https://doi.org/10.1016/S1574-0056(00)80005-6
44. D. Harre, F. Moriconi-Ebrard, P. Heinrigs, *Urbanisation Dynamics in West Africa 1950–2010* (2015). https://doi.org/10.1787/9789264252233-en
45. R. Prieto-Curiel, P. Heinrigs, I. Heo, Cities and Spatial Interactions in West Africa. Tech. rep. OECD Publishing (2017). https://doi.org/10.1787/24142026

Chapter 6
The Gravity Model for Social Systems

Rafael Prieto-Curiel

Abstract Gravity is one of the most prominent models used across various social areas, including economics, demography, mobility, politics, and other systems where spatial interactions are relevant. The model represents a flexible approach that captures important regularities that might be detected when thousands or millions of people are observed, i.e. bigger origins and bigger destinations increase the intensity of interactions, but it tends to decay with longer distances between them. Therefore, the number of interactions between two locations is frequently modelled with an equation that resembles Newton's Law of Gravitation (also known as the Law of Gravity). Here, I explore different aspects, i.e. essential components, techniques for estimating the parameters, interpretation of results, and settings where gravity is a helpful tool, including mobility and migration.

6.1 Introduction

Let us start with a real case of migration. Consider Houston, Dallas and El Paso, three cities in Texas, USA. In 2018, more than 18,000 people moved from Houston to Dallas, but only 2,000 moved from Houston to El Paso.[1] Why is Dallas a more common destination for people from Houston than El Paso? There are too many reasons why people move and choose, among many possible destinations, one city as their new home. It is impossible to model all the reasons behind moving at an individual level. We can detect a collective pattern when we observe thousands or perhaps millions of people deciding whether or not to move and where to move. In particular, Dallas is only 384 km away from Houston, whereas El Paso is more than 1,000 km away. Dallas has nearly nine times more population than El Paso. If a person moves for work, to study, or to meet a family member, it is more likely that

[1] Data from https://www.census.gov/data/tables/2018/demo/geographic-mobility/metro-to-metro-migration.html Last accessed in June 2024.

R. Prieto-Curiel (✉)
Complexity Science Hub, Vienna, Austria
e-mail: prieto-curiel@csh.ac.at

D. Rybski (ed.), *Compendium of Urban Complexity*, Understanding Complex Systems,
https://doi.org/10.1007/978-3-031-82666-5_6

the person will find a job, an offer at a school, or a family member in Dallas than in El Paso (Fig. 6.1).

The size of the destination and the physical distance are the main factors that help us model a social process in which geography plays a role. Perhaps at an individual level, people do not frequently think of the city size they are moving to, but rather, reasons that attract people, as many as they can be, are more commonly found in larger cities. But the size of the destination is not the only reason. From Houston, for example, nearly 5,000 people move each year to Chicago, a huge city (even larger than Dallas) in the Northern part of the US. The distance between Houston and Chicago is nearly 1,800 km, so a person would have to drive more than 16 hours to return home (as opposed to Dallas, which is only 3 hours away). The likelihood of moving to a new destination decays with distance. Finally, suppose instead of considering people who moved *from* Houston, we consider people who moved *to* Houston. In that case, only 5,000 people moved from Beaumont (a medium-sized city one hour away from Houston) against the 16,000 people who moved from Dallas. As with the destination, we expect more people to move to Houston from larger cities simply by frequency.

Whether it is intracity migration, the number of visitors to a tourist destination, the number of journeys inside a city or the number of customers of a restaurant at the beach, gravity is one of the most powerful yet simple models to explore [1]. Whether we apply it for mobility, migration, trade, or others, the gravity model for human constructs should be considered a general and very flexible model, which captures the impact of size at the origin and destination and their distance [2]. It is a spatial model used for decades, perhaps under different names but capturing similar principles [1, 3–5]. The gravity model usually depends on the size of the origin, usually captured by its population. The flow of people, specific products, or the number of migrants is expressed as an increasing function of the size of origin,

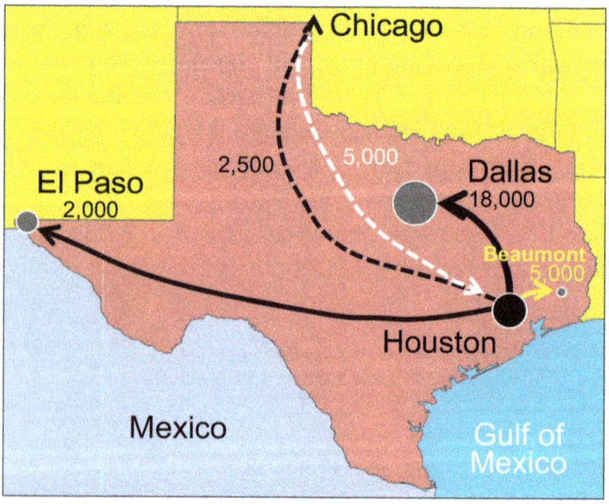

Fig. 6.1 Migrants to and from Houston, Texas, in 2018. The width of the arrows represents the intensity of the flow

assuming that a higher flow is expected with more people. The gravity model captures the *attractivity* of the destination and, in general, utilizes population size as a proxy. Howevcr, the size of the origin might be captured by different means. For example, we could use GDP if we are interested in trade or the number of hotel rooms for an application in tourism [6]. In terms of the distance, the actual physical distance between origin and destination is frequently used, but the model allows other forms of pairwise relationships that might capture proximity, such as travel time or other aspects, including frictions or barriers [7–9]. The gravity model, for example, has been used to model trade between countries, considering the population (or income) size and the physical or cultural distances between two distinct locations [3, 7, 10].

Let $F_{o,d}$ be the number of "visitors" from o to d during some period. That can be the number of customers from the neighbourhood o to restaurant d or the number of people who moved from Houston to Dallas, so $F_{Houston,Dallas} = 18,000$. Then, we aim to construct a model such that $F_{o,d}$ increases with the size of the origin P_o, also increases with the size of the destination P_d and decreases with the distance between them, $D_{o,d}$. Many equations satisfy those three principles. Commonly, one such expression is used, as it has many desirable properties, including parameters that we know how to interpret and ways to estimate them [11]. Such an equation is frequently expressed as

$$F_{o,d} = \kappa \frac{P_o^{\mu} P_d^{\nu}}{D_{o,d}^{\gamma}}, \tag{6.1}$$

where κ, μ, ν, and γ are non-negative model parameters that we frequently estimate with data or make some "educated guess" based on some knowledge of the process. μ and ν capture non-linear impacts of the size of origin and destination; γ captures the impact of distance, and finally, κ is a constant which helps adjust figures and units. For some values of κ, μ, ν and γ, the Eq. (6.1) may be interpreted as the expected frequency at which people from the n origins visit the m destinations.

The expression, with $\mu = \nu = 1$, and with $\gamma = 2$, was conceived by Isaac Newton in 1687 for modelling the attraction between two objects, and it is usually called "the law of universal gravitation". Due to its resemblance to the law of universal gravitation, the social model is called the *gravity model*. The gravity model in social settings has many expressions, one of which mimics Newton's law of universal gravitation. However, many expressions capture similar aspects, for instance, using exponentials instead of functions like P_o^{μ} or P_d^{ν}.

Newton noticed that to transform the product of two masses divided by the square of their distance from the right-hand side (with units kg^2/m^2) to a force on the left-hand side (with units kgm/s^2), he needed a constant with funky units, known as the *gravitational constant* (approximately $6.674 \times 10^{-11} m^3/kgs^2$. The Cavendish experiment measured the gravitational constant more than 100 years after the law of universal gravitation was published. In our expression, κ acts as such a constant, which helps transform between units. The gravity model is frequently used to model the flux for a given period, usually one year in the case of migration and one day or week in the case of visitors. In the model, the value of κ depends on the interval of time considered. If we double the time interval, we should also double the values of κ to compensate.

The gravity model for human constructs should be thought of as a general and very flexible model, which captures the impact of size and distance and might capture other aspects, such as frictions or barriers [3, 7–9]. Different expressions of the gravity model are frequently used to inform relevant aspects of migration, transport, mobility or trade, so they are a valuable tool for policymakers [12, 13].

The model should not be conceived as Newton's gravity equation with one or two parameters to fit. Migration between two cities does not double if the population of the destination doubles, and migration between two cities is not four times larger with half the distance. Nothing forces humans to behave in such strict manners. Two particles are attracted equally under similar circumstances, but no two individuals are so identical to ensure they will visit or move to the same destination under the same circumstances. Thus, any physical model used in a social setting must be carefully analyzed, including all assumptions needed to use a specific expression. Only with one particular expression, and if we fix the parameters (with $\mu = \nu = 1$ and $\gamma = 2$), we obtain Newton's universal gravitation law.

The gravity model captures the *attractivity* of the destination and utilizes size as a proxy. However, other metrics may be used, for instance, the number of beds in a tourist destination or job offers. It depends on the distance, but it can be physical distance, road distance, monetary costs, and time travel. The model has some parameters (perhaps too many) and at least four. The gravity model is very flexible, and many insights may be obtained from it [1]. With the correct data, it is possible to detect if physical distance has a higher impact among young people or the elderly or to check if people prefer moving to locations with similar weather. Using a gravity-based model, it was detected that people prefer moving to cities of similar size as their origin, and rarely does a person from a big city move to a small town or the countryside, and vice-versa [14]. If a company is deciding where in the city to open a new store, they can estimate how many people would visit it from different neighbourhoods. Also, using some of the principles and techniques from the gravity model, they can estimate how many customers they could lose at their other locations. The gravity model depends on data and might rely on some assumptions that must be carefully considered. There are many techniques to estimate its parameters, ranging from simple linear regression to minimizing the differences between what is observed and modelled.

6.2 The Building Blocks of Gravity

Whether we are using the gravity model to forecast or detect the impact of some environmental or social factors, the first step is to define the four components clearly: origins, destinations, distances and fluxes.

- *Origins*–define if the model is based on cities, metropolitan areas, countries, neighbourhoods, or similar spatial units and clearly define what we mean about size. Frequently, the population size of each origin is the most suitable metric.
- *Destinations*–perhaps the easy choice here is to mimic origins, so consider flows between cities or trade between countries and use their corresponding size. Perhaps the model considers tourist destinations or shopping malls, and the corresponding size metric is the number of hotel rooms or the surface of distinct commercial areas. Depending on the considered indicator, the destinations and the corresponding way they are measured might be modelled in different ways. It quantifies how attractive each destination is.
- *Distances*–depending on the objective, the distance might be commuting time, actual physical distance, costs or other units. Often, the objective of gravity models is to detect the impact of barriers, like an international border, certain interventions, like a new train station, or some social factors, such as the impact that cultural or language differences impose on migration.
- *Flux*–some data which captures the phenomenon of the model is needed. Suppose the objective is to capture tourism, for instance. In that case, some data that suits the origins and destinations is needed, such as the number of visitors from different origins to each destination during a certain period. A frequent issue with this type of data is having a very short period for a process that takes longer times or dealing with missing or incomplete observations.

For the notation, usually, for origin o with size P_o and destination d with size or attractiveness P_d, the flux (or number of migrants, say) between o and d is written as $F_{o,d}$ and the distance is $D_{o,d}$. When the flux is estimated, usually we write $\hat{F}_{o,d}$. Also, it is common to think of the total outflow of some origin (the sum of $F_{o,d}$ over all possible destinations), and it is usually expressed as $\sum_d F_{o,d} = F_{o,\bullet}$. Similarly, the total inflow is the sum over all possible origins, and it is expressed as $\sum_o F_{o,d} = F_{\bullet,d}$.

6.2.1 Assumptions

All models rely on data and assumptions. In the case of the gravity model, it is worth considering first if a linear impact of the origin and destination sizes P_o and P_d is reasonable. Many things have been detected to scale non-linearly with city size, for instance, GDP [15] (Chap. 2). That means that if city B is ten times larger than city A, then the GDP of city B is substantially more than ten times the GDP of city A, which means that city B has a higher GDP per person than city A. Migration between cities is affected by city size, meaning that assuming a linear model might be too restrictive [14]. However, assuming a linear impact of the origin size might be reasonable for other aspects. For example, if we deal with some trade data, thinking that a city with twice the number of people consumes two times the amount of food or twice the number of beers might be reasonable. Still, it is worth checking if some non-linear departures with size are observed, and often, the linearity assumption is too restrictive.

A second aspect to consider is symmetry. The gravity model is often applied to a set of units, for example, migration between cities or trade between countries. Therefore, the gravity model frequently considers some units (say, a city) as the origin and destination of some flux. Then, the flux and the distances can be structured on a square matrix, where rows and columns correspond to separate units. Notice that distances are symmetric (the matrix $D_{o,d}$), and so the right-hand side of Eq. (6.1) gives the same values if $\mu = \nu$. Unless the parameters μ and ν in Eq. (6.1) have different values, symmetry is somehow being assumed and forced into the model. Thus, it is assumed that outflows and inflows are equal. Unless there is a solid reason to think that the process is symmetric, it is preferable to let μ and ν as different parameters (and perhaps obtain symmetry as a result of the parameter estimation, if $\hat{\mu} \approx \hat{\nu}$).

6.2.2 Measuring Units

One issue with the gravity equation is related to the units. The product of the population of two cities on Eq. (6.1) could have too many digits, and divided by some distance in km, it still has too many digits. Therefore, often, to get an estimated flow (which frequently is not that big), the estimated value of $\hat{\kappa}$ in Eq. (6.1) has to be very small to compensate. Dealing simultaneously with huge and tiny numbers may lead to numerical errors, so the best thing to do is to think first of the units. Depending on the case, expressing population size in millions and distance in hundreds of km might prevent dealing with large and small units simultaneously.

If the distance between two observations is very small, then the impact that distance has on Eq. (6.1) might not be correct and might lead to an unbounded estimate. In such cases, perhaps it is worth considering those two units as the same and merging the number of visitors from them.

Something to consider concerning the gravity model is that it requires large amounts of data, including population sizes, distances and the flow between every pair of cities. For instance, the gravity model for the $N = 118$ cities with more than 100,000 inhabitants in Nigeria requires capturing 118 population sizes, measuring 7,000 distances $D_{o,j}$ and knowing roughly $N(N-1) \approx 14,000$ flows $F_{o,d}$.

6.3 Applying the Gravity Model

Consider, for example, a $n \times n$ square grid where one person occupies each square. On top of that grid are m destinations (say, stores or cities) with some level of attractiveness P_k for each destination (with $k = 1, 2, \ldots, m$). Person i is attracted to destination k according to

$$A_{i,k} = \frac{P_k^{\nu}}{D_{i,k}^{\gamma}},$$

where $D_{i,k}$ is the Euclidean distance between the square i and the destination k and $\nu > 0$ and $\gamma > 0$ are some parameters. Here, each square has some "gravity" to each destination. For some values of ν and γ, it is possible to detect, for each person, what is their most attractive destination. The most attractive destination for most people is frequently the nearest one, but it depends strongly on attractiveness (Fig. 6.2a). For example, if we consider cities in some countries instead of destinations, their attractiveness could be their size or the number of jobs available. If each person goes to their most attractive destination, the result is partitioning the plane into regions (Fig. 6.2b).

With a gravity model, it is relevant to consider that choosing the most attractive destination is always possible, no matter how far or attractive the destinations are. Therefore, if we consider that the destinations are shops where people might move, we might assume everyone picks a shop. However, this is not necessarily

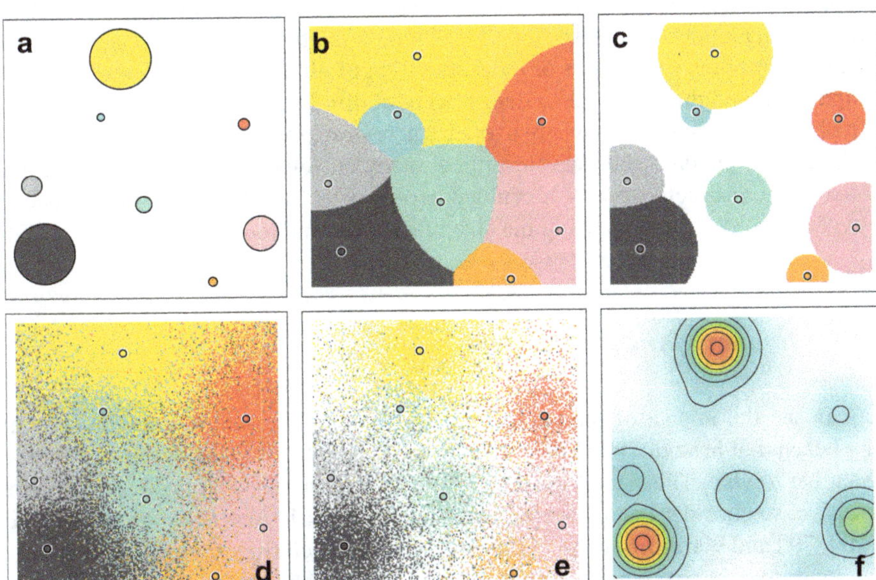

Fig. 6.2 **a** Different ways in which the attractiveness to some place can be observed, where the location of $m = 8$ destinations is shown by the size of the disc. **b** Detection of the most attractive destination for each square in the grid. **c** Detection of the most attractive destinations, where the attractiveness is larger than some positive threshold, so the attractiveness below some threshold is not coloured. **d** Selection of a destination based on the probability of being selected is proportional to the attractiveness of each destination. **e** Selection of a destination based on the probability of being selected is proportional to the attractiveness of each destination, considering also the probability of not moving. **f** Total gravity is computed as the sum of the attractiveness of each destination experienced by each square in the grid **f**

the case. People far away from all destinations might choose not to go to any shop (Fig. 6.2c). An option here is to consider the attractiveness of any destination above a certain threshold. That is, take some value of $\omega > 0$ and consider only values of the attractiveness where $A_{i,k} > \omega$.

The gravity model returns the attractiveness from each location to every destination. The level of attractiveness of some destinations can also be used to consider the probability that a person will select them. For example, let's ask a person to choose among their m options. The probability that they choose destination k is given by $\pi_{i,k} = A_{i,k}/\sum_k A_{i,k}$, so the sum of the probability of selecting all destinations is one. This gives us a stochastic model, where the probability of moving to some destination is a realization of a multinomial distribution (Fig. 6.2d). The term "multinomial" here refers to a probability distribution that can be thought of as if the person rolls an "unfair" dice with k sides and picks the corresponding destination (unfair since the probability of each side will not be the same). Again, if a person is too far from any destination, they might choose not to move. The model can be easily adjusted if we consider an additional "destination" that represents the option of not moving (Fig. 6.2e).

For each square in the grid, we can consider the maximum attractiveness they receive from all possible destinations or the total attractiveness (Fig. 6.2f) computed as the maximum attractiveness or the summation of all destinations. These types of computations might be useful to detect accessibility [16]. Also, these calculations help plan strategies to detect where to open a new store, for example, by adding a new destination to the data and measuring how many customers they expect. A similar model was constructed nearly a century ago to detect the "point of indifference" between two retail centres, using the size of each destination and the distance [17].

6.3.1 *Impact of the Parameters*

We can see the impact of the parameters with a simple example. Imagine a 1 km long beach that has only three bars. Every day, n people go to the beach and decide which bar to visit. The size of the dance floor of each bar is what makes them more attractive, so consider that their sizes are $P_1 = 5$, $P_2 = 10$ and $P_3 = 20\ \text{m}^2$, located at 100, 700 and 900 m from one of the edges of the beach (Fig. 6.3). The probability that the person i goes to bar k, written as $\pi_{i,k}$, is proportional to

$$\pi_{i,k} \propto \frac{P_k^{\nu}}{D_{i,k}^{\gamma}}, \tag{6.2}$$

where $D_{i,k}$ is the distance between the person i and the bar k. The symbol $\propto$ means "proportional", and it is frequently used to simplify the notation. Remember that we deal with probabilities, so $\sum_{k=1,2,3} \pi_{i,k} = 1$. Thus, the bar that each person could

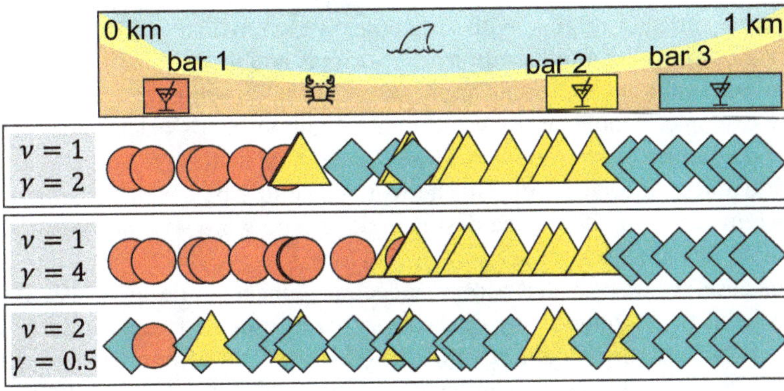

Fig. 6.3 Three bars located on a 1 km beach. Bar 1 is located 100 m away from one beach edge, Bar 2 at 700 m and Bar 3 at 900 m. The corresponding size of the dance floor at each bar is 5, 10 and 20 m^2. At the beach, there are 25 people located in random positions. Each person chooses a bar based on their position and the size of the dance floor. They prefer bars with a bigger dance floor, but they would also rather walk short distances at the beach. Each customer is represented by a red circle when they choose Bar 1, a yellow triangle when they choose Bar 2, or a blue square when they choose Bar 3. Different values of ν and γ in Eq. 6.1 change how bars are picked. For instance, for large values of γ, the size becomes less relevant, and it's "all about location". For larger values of ν, the size of the dance floor becomes more relevant, so bigger bars are more popular

choose can be modelled with a multinomial distribution, and the bar they go to is a realization of that distribution.

Simulating how a customer picks a bar at the beach is possible. Bars are visited more frequently by nearby customers. However, for higher values of ν, people care more for the size of the dance floor, whereas higher values of γ mean that distances are more relevant. This version of the gravity model is known as the *Huff model*, formulated in 1963 by David Huff [18, 19]. It is frequently used in economics, retail research, and urban planning to capture "competition" between distinct destinations, such as bars.

If bars are too small or far away, people might not visit them! In the Huff model, every customer has to choose a bar, regardless of how far away or unattractive they might find them. But what if $P_k^{\nu}/D_{i,k}^{\gamma}$ is too small for all bars? It is possible to extend the Huff model so that not everyone has to pick a bar.

6.3.2 *Gravity Model on a Network*

Thinking of the gravity model on a grid is the simplest one since each person is represented by a square in the grid (so each square is one person). What if, instead of one person on each square, we imagine there are P_i people on each square? For example, if each square was a neighbourhood in a city and the m destinations were shops where people would buy groceries. Instead of an entire grid, we can imagine only a few squares (or origins) with P_i people on each, with $i = 1, 2, \ldots, n$. So, we consider n origins with m destinations. The origins and the destinations do not need to be the same (although, in some applications, it makes sense to consider the same set). For example, people across n different neighbourhoods (each neighbourhood with P_i population) might shop in m stores (each store with some attribute S_k that is the attractiveness of each store). This network can either be a bipartite with n nodes as the origins and m nodes as the destinations or a network with n nodes if origins and destinations are the same set.

In this case, we assume that people of each origin act similarly, so they have the same probability of going to any of the m stores or moving to any of the m cities. Instead of looking at probabilities in Eq. (6.2), we look at frequencies when we multiply both sides with the population of the origin and obtain $F_{o,d} = \kappa P_o^{\mu} P_d^{\nu} / D_{o,d}^{\gamma}$, i.e. Eq. (6.1).

Suppose we have n origins (that represent neighbourhoods in a city) with m destinations (for example, green areas in the city). For each origin, we consider its population P_o, and for each destination, we consider its level of attractiveness, P_d. We then measure the distance between each origin and destination, $D_{o,d}$, which might be the Euclidean distance, but also the road distance, commuting times or other units that reflect proximity [20]. For example, some distances could be zero when the origins and destinations are the same. However, the gravity model is designed to capture the impact of *long* distances. Therefore, avoiding distances equal to zero (or even too close to zero) is better.

6.4 Using the Gravity Model

Although Eq. (6.1) may be used to estimate the flow of people or trade between distinct locations, the model is frequently used to measure the impact of distance or political barriers. Based on some data (that is, having values of $F_{o,d}$ for different origins and destinations and having also defined the size of the origin, the attraction of the destinations and the distance between them), a common challenge is to estimate the values of the parameters. The gravity Eq. (6.1) has at least two parameters, κ and γ if a linear and symmetric impact of city size was assumed (so that $\hat{\mu} = \hat{\nu} = 1$ as an assumption) or three or four parameters, depending on how it was conceived.

Although rarely observed, one ideal scenario is when we have a complete dataset (that is, no $F_{o,d}$ is missing) and all flux values are well above zero. Then, we can take the logarithm on both sides of Eq. (6.1), and the result is

$$\log F_{o,d} = \log \kappa + \mu \log P_o + \nu \log P_d + \gamma \log D_{o,d}\,. \tag{6.3}$$

The expression is a linear equation concerning the logarithm of the populations and distances. The sign in front of $\gamma \log D_{o,d}$ should be negative, but we can also write it with a positive sign and keep in mind that γ should be negative. Also, $\log \kappa$ can be considered a new constant number. Then, the coefficients $\kappa' = \log \kappa$, μ, ν and

γ need to be estimated. Assuming that our model gives an unbiased flow estimate, a regression gives us the four parameters. If we further assume that everything else that is not captured only by population size and distance follows a normal distribution with constant variance (in Eq. (6.3), with logarithm on both sides). In that case, we also get estimates of the variance of each parameter so that we can also obtain statistical significance. Thus, parameters may be obtained with a regression.

Almost always, there will be pairs of cities for which the number of visitors is zero, meaning that there will be some $F_{o,d} = 0$ for which the logarithm is not defined. Adding some small constant ϵ to avoid zeros might create substantially different estimates, depending on how small or large is that arbitrary ϵ, so it might create more issues and uncertainty [8]. A better option might be to use a Poisson regression, which is appropriate when the dependent variable is a count, such as the number of people, transactions, or others. In a Poisson regression, the flow $F_{o,d}$ is assumed to follow a Poisson distribution with rate $\exp(\kappa + \mu P_o + \nu P_d + \gamma D_{o,d})$ and the estimation works similarly with a regression.

Here is where other elements might be added to the model. For example, the unemployment rate at the city of origin and destination might help capture its impact on pushing and pulling people. We could also consider a binary variable $I_{o,d}$ that has values of one if two cities share some attribute (say, they are in the same state or have similar weather) and zero otherwise. See [3, 9, 21] for ways in which parameters might be estimated, and see [8] for the challenges that the estimation poses.

Besides regression techniques, the gravity model can also be considered an optimization problem. For a set of values $(\kappa, \mu, \nu, \gamma)$, the right-hand side of Eq. (6.1) produces some estimate of the flow $\hat{F}_{o,d}$ and some error $F_{o,d} - \hat{F}_{o,d}$. Adding all the square values of the errors gives a function of the four parameters, which then should be minimized.

6.4.1 Parameter Interpretation

Even before estimating the parameters, it is worth thinking about the expected values of each parameter and letting the regression or the optimization only confirm our impressions. The coefficients μ and ν affect city size. Suppose the smallest city among our observations has $P_s = 100,000$ inhabitants and the largest has $P_l = 10$ million so that $P_l = 100P_s$. With a value of $\hat{\mu} = 1$, the model says that the probability of moving is the same for the two cities. That is, the flow is proportional to city size, so we expect 100 times more flow from the large city simply because it is 100 times larger. However, with $\hat{\mu} = 1.15$, we already get that the flow for the large city is $P_l^{1.15} = (100P_s)^{1.15} \approx 200(P_s^{1.15})$ so that the flow with respect to population is twice as intense as it is for the small city. Things might explode quickly for these types of coefficients. With $\hat{\mu} = 1.5$, the process is ten times more intense (or the frequency of migration is ten times larger for the biggest city), and with $\hat{\mu} = 2$, it is 100 times more intense. In the case of $\hat{\mu} = 2$ and if 1 % of the population from the small city moves, then 100 % of the population of the large city moved, so the value is already outside a reasonable range. We get a similar effect for values smaller than one, where values below $\hat{\mu} = 0.05$ should not be observed. And the same occurs with $\hat{\nu}$. Values close to 1 are reasonable, and above 2 or below 0.05 are not. For values of $\hat{\mu} > 1$, the process is more intense in large cities, and for values of $\hat{\mu} < 1$, the process is more intense in small cities. If uncertain, consider what happens on both sides of Eq. (6.1) when you divide by the size of the origin (or the destination). The left-hand side gives the fraction of visitors, and the right-hand provides some function of the sizes and distances.

The relevant null hypothesis to test concerning μ and ν is whether their estimated values might equal one. If an interval for $\hat{\mu} \in [a, b]$ contains 1, there is not enough evidence to say that visitors are more intense in larger or smaller cities.

For distance (whether physical, road distance, commuting time or other distance measure), a value of $\hat{\gamma}$ close to zero means that it is not very significant, and higher values indicate that the observed process is highly sensitive to distance. With $\hat{\gamma} = 1$, the estimated flow between two cities decreases by half if the distance between two cities doubles. With $\hat{\gamma} = 1/2$, if the distance doubles, the estimated flow is reduced only by 30%.

In general, a regression (or an optimization problem, minimizing the mean square error) is highly sensitive to extreme values, so if the data considers two very close or distant cities, those values will greatly impact the parameters. The same happens with the size of origins and destinations. If the model has a huge or tiny city, many results will be based on the corresponding flow. And the same might occur with migration or the number of visitors. If, for some reason, there is a very high flow between two cities (for instance, if a large company decides to move its offices one year), then that process will have an enormous impact on the estimation process.

A useful tip that works here but is also valuable in other social areas is plotting the data and checking if the model and the assumptions make sense (Fig. 6.4). For the gravity model, four very useful plots are:

- size of the origin P_o in the horizontal axis and the total outflow $F_{o,\bullet}$ in the vertical axis,
- size of the destination P_d in the horizontal axis and the total inflow $F_{\bullet,d}$ in the vertical axis,
- for one observation (perhaps a large city or a small one), take the distance to other observations in the horizontal axis and the outflow from it in the vertical axis, and
- for the same observations, consider the inflow to it in the vertical axis.

The idea with these four plots is to observe a higher total inflow and outflow from bigger cities. We expect longer distances to discourage interactions with the third and fourth plots so there are fewer migrants in both directions. Here, we observe the impacts of the distance between cities and their size (Fig. 6.4, bottom). There is a general trend where longer distances discourage migration, and more populated cities (so bigger discs) attract and push more people from and to Houston. Any parameter fitting should come after we are sure the data is correct and the model's assumptions correspond with the observed patterns.

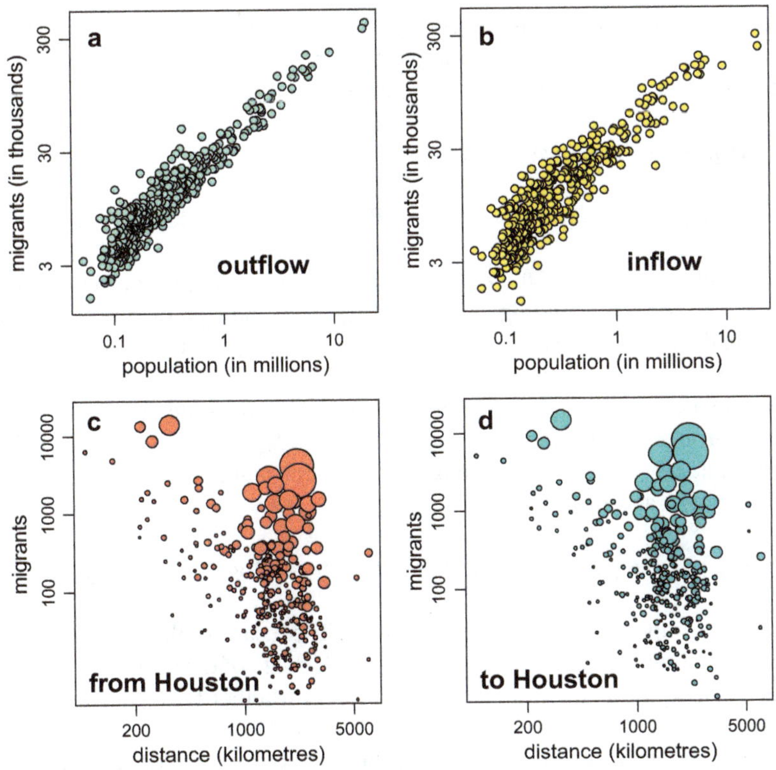

Fig. 6.4 **a** Number of outflow and **b** inflow migrants from one year to the next one between US metropolitan areas in the vertical axis and the corresponding population in the horizontal axis. **c** Number of migrants from and **d** to Houston (bottom) from different metropolitan areas (vertical axis) and the Euclidean distance between Houston and other areas (horizontal axis). The size of the disc is proportional to the population of the city. All panels are on a logarithmic scale on both axis

6.4.2 City to City Migration

We can use the gravity model to verify some aspects of trade between countries or to model migration. For instance, when people move between cities in the US, do they prefer to stay in the same state? The American Community Survey captures metro area-to-metro area migration flows, so it helps us answer this question. Let $F_{o,d}$ be the number of people that moved from some city o to some destination d in a year, and let P_o and P_d be the corresponding populations (in millions). Also, let $D_{o,d}$ be the Euclidean distance between cities (in km). Taking the logarithm on both sides of Eq. (6.1), we obtain that

$$\log F_{o,d} = \log \kappa + \mu \log P_o + \nu \log P_d + \gamma \log D_{o,d} . \tag{6.4}$$

Theoretically, we could adjust a regression and obtain values of κ, μ, ν, and γ. However, is it possible to take the logarithm on both sides? Not really. Of all the possible pairs of cities in the US, there are zero migrants in 67 % of them, so the logarithm on both sides of the equation is not well defined. Thus, it is better to use a Poisson regression here. Instead of Eq. (6.4), we write

$$E[F_{o,d}] = \exp\left(\log \kappa + \mu \log P_o + \nu \log P_d + \gamma \log D_{o,d}\right), \tag{6.5}$$

where the left-hand side of the equation is the expected number of migrants from o to d, given the corresponding sizes and distances. The sign of μ and ν should be positive, and we expect that distance discourages migration, so the sign of γ should be negative.

Since we are wondering whether being part of the same state impacts the intracity migration propensity, we define the binary variable $S_{o,d} = 1$ if both cities belong to the same state and zero if they belong to a different state. We can add that variable to the regression and express it as

$$E[F_{o,d}] = \exp\left(\log \kappa + \mu \log P_o + \nu \log P_d + \gamma \log D_{o,d} + \eta S_{o,d}\right), \tag{6.6}$$

where η is the impact of being in the same state. The results of the Poisson regression models are in Table 6.1. Model 1 only considers the size of the origin, the destination and the distance, whereas Model 2 also considers whether two cities belong to the same state.

It is easy to interpret the results in Table 6.1. For example, for Model 1, we take two cities (their population in millions and the distance between them in km), and we estimate that the number of migrants between them is given by $E_{o,d} = \exp(12.314) \times P_o^{0.870} \times P_d^{0.814} \times D_{o,d}^{-1.086}$. Thus, we confirm that with a bigger origin and a bigger destination, we expect more migrants. The flow, however, is not linear, not symmetric and more intense in smaller cities (since both $\hat{\mu} < 1$ and $\hat{\nu} < 1$). Also (because of the negative sign), longer distances mean less migration. The impact of

Table 6.1 Results of the Poisson regression models. The dependent variable is the expected number of migrants between the two cities. Model 1 gives the values of the parameters of Eq. (6.5). Model 2 also considers the impact of two cities being part of the same state in the US, from Eq. (6.6)

	Model 1	Model 2
(Intercept)	12.314 (0.002)***	10.036 (0.003)***
$\log P_o$	0.870 (0.000)***	0.878 (0.000)***
$\log P_d$	0.814 (0.000)***	0.821 (0.000)***
$\log D_{o,d}$	−1.086 (0.000)***	−0.777 (0.000)***
same_state		1.348 (0.001)***
AIC	11, 833, 941	9, 979, 479
Num. obs.	147, 840	147, 840

$^{***}p < 0.001$; $^{**}p < 0.01$; $^{*}p < 0.05$

distance is quite significant. If, for example, the distance between two cities doubles, then the size of the destination needs to increase by 2.5 times to compensate (so our estimate remains constant). The distance coefficient is -1.086, meaning the impact is roughly linear, not quadratic. Nothing forces humans to behave like planets attracting each other.

For Model 2, we added a binary variable to check whether two cities in the same state have higher migration flows. The Akaike Information Criterion (AIC) evaluates how well a model fits the data. It is computed as $AIC = 2k - 2 \log L$, where k is the number of variables used and L is the likelihood estimate. The AIC balances the model performance and the number of parameters used. Models with a smaller AIC are better, so adding more variables (larger k) is preferred only if it increases the likelihood sufficiently. We see that Model 2 is an improvement over Model 1.

The results show that adding the binary variable $S_{o,d}$ to our model made it more accurate. Thus, there is an impact of two cities being from the same state and due to the sign of the variable, being from the same state significantly increases the migration between cities. However, the coefficient of $\log distance$ in Model 2 is smaller than in Model 1, reflecting perhaps that cities from the same state tend to be nearby.

How precise is the gravity model? One way to verify its precision is by plotting the observed flow (so the number of migrants between every pair of cities in our case) against the predicted number (Fig. 6.5). Since the size of cities covers a wide range

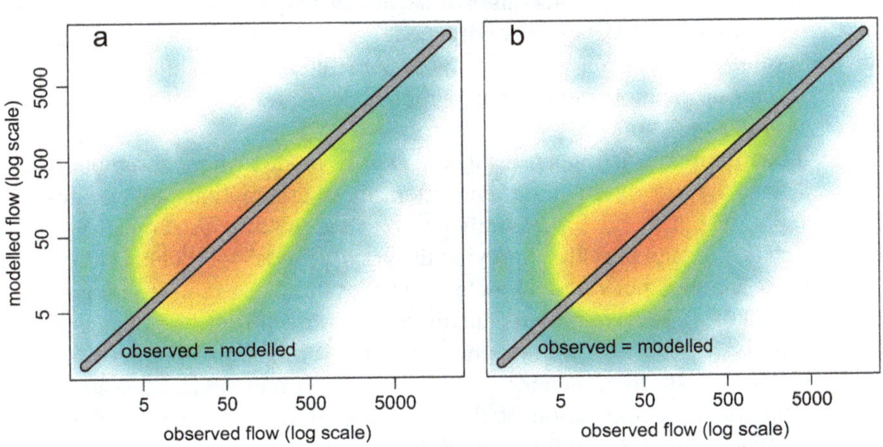

Fig. 6.5 Observed number of the metro to metro migrants in the US in a year between 2015 and 2019 (horizontal axis) and modelled number (vertical axis). Model 1 **a** only considers the impact of the origin and destination size, and Model 2 **b** considers whether two cities belong to the same state. This type of figure is a smooth scatterplot, which is useful for detecting patterns when we have thousands of data points. The part of the figure with more points is red, with fewer points are yellow, then blue, and a white region means no points. The dark diagonal line marks when the observed values match the predicted values for any two cities. The section above the line indicates the region where the model overestimates migration, and below the line, migration is underestimated. Both panels are in logarithmic scale in both axis

and the intracity migration also has a wide range of values, using a logarithmic scale on both axes is common. Values closer to the diagonal mean the predicted values are close to the observed values. For a small number of visitors or migrants, the gravity model tends to give values that may be perceived as far from the observed figures, but it means that the gravity model shows values that are maybe a few dozen away. When there are many migrants between two cities, the gravity model seems more accurate. And the gain of adding the binary variable $S_{o,d}$ is that the predicted values, particularly the large ones, are closer to the observed values.

6.5 Conclusion

Predicting whether a person will move to some city, travel to some neighbourhood, or visit a park or a shop is very challenging at an individual level. But when we observe hundreds or maybe thousands of people, we can detect patterns. Based mainly on the origin's size, how attractive the destination is, and the distance between them, the gravity model is designed to capture most aspects of that collective pattern.

The gravity model is frequently used when data is scarce to obtain some rough estimate of the flow between two locations, simply applying Eq. (6.1) or some variation directly to the data. For each person, we are considering the process of choosing among many options as a multinomial variable. Also, repeating the same simulations makes it possible to obtain intervals, so the model is quite powerful in detecting departures. With enough data, the gravity model may also be used to fit a regression and fetch values of the parameters. With those parameters, we can infer what attracts people to some destination, including geographic, demographic, economic, climatic, and environmental factors [22].

Some caution is needed when using the gravity model. Many assumptions are involved with the gravity equation, and often, these assumptions are not desired. For example, in Eq. (6.1), if we assume that $\mu = \nu$, we impose symmetry on the flow, which is often not the case. Applying the gravity model without considering all assumptions and parameters could yield wrong results. For example, applying the gravity model using some temporal data might give some misleading forecasts [23]. The model is not bounded, and in some scenarios, it might also predict that more people will leave the origin than there are [24]. Also, two numerical considerations are relevant for the computation of the gravity model. Some of the values may be simultaneously too large and too small. For example, when two populations are multiplied in Eq. (6.1). Avoiding very small and very large units in the same equation gives better numerical estimates.

One of the most important things to remember about the gravity model is that some physical models might help us structure a social process systematically and concisely. They work great to form an analogy between known processes and their equations and social behaviours. However, humans do not behave like planets in the solar system or like drops of water, so no model will perfectly capture all aspects of migration.

Acknowledgements This research is funded by the Federal Ministry of the Interior of Austria (2022-0.392.231).

References

1. A.T. Philbrik, A short history of the development of the gravity model, in *Australian Road Research* 5.4 (1973), pp. 40–54. https://trid.trb.org/View/140675
2. W.-S. Jung, F. Wang, H. Eugene Stanley, Gravity model in the Korean highway, in *EPL (Europhysics Letters)* 81.4 (2008), p. 48005. https://doi.org/10.1209/0295-5075/81/48005
3. J.J. Lewer, H. Van den Berg, A gravity model of immigration. Economics letters **99**(1), 164–167 (2008). https://doi.org/10.1016/j.econlet.2007.06.019
4. E. George Ravenstein, The laws of migration, in *Journal of the Statistical Society of London* 48.2 (1885), pp. 167–235. https://doi.org/10.2307/2979333
5. S.A. Stouffer, Intervening opportunities: a theory relating mobility and distance. American Sociological Review **5**(6), 845–867 (1940). https://doi.org/10.2307/2084520
6. M. Batty, R. Milton, A new framework for very large-scale urban modelling, in *Urban Studies* (2021), p. 0042098020982252. https://doi.org/10.1177/0042098020982252
7. J.E. Anderson, The gravity model. Annu. Rev. Econ. **3**(1), 133–160 (2010). https://doi.org/10.1146/annurev-economics-111809-125114
8. M. Burger, F. Van Oort, G.-J. Linders, On the specification of the gravity model of trade: zeros, excess zeros and zero-inflated estimation. Spatial Economic Analysis **4**(2), 167–190 (2009). https://doi.org/10.1080/17421770902834327
9. D. Karemera, V. Iwuagwu Oguledo, B. Davis, A gravity model analysis of international migration to North America, in *Applied Economics* 32.13 (2000), pp. 1745–1755. https://doi.org/10.1080/000368400421093
10. H. Barbosa et al., Human mobility: Models and applications. Physics Reports **734**, 1–74 (2018). https://doi.org/10.1016/j.physrep.2018.01.001
11. A.-L. Woelwer et al., *Gravity: Estimation Methods for Gravity Models.* R package version 0.9.9. 2020. https://CRAN.R-project.org/package=gravity
12. D. Gnisci, West African mobility and migration policies of OECD countries, in Paris, Organisation for Economic Cooperation and Development. Sahel and West Africa Club (2008). https://doi.org/10.1787/2074353x
13. WFP, *At the root of exodus: Food security, conflict and international migration.* (World Food Programme, 2017)
14. R. Prieto-Curiel et al., Gravity and scaling laws of city to city migration, in *PloS one* 13.7 (2018), e0199892. https://doi.org/10.1371/journal.pone.0199892
15. L.M.A. Bettencourt et al., Growth, innovation, scaling, and the pace of life in cities. Proceedings of the National Academy of Sciences **104**(17), 7301–7306 (2007). https://doi.org/10.1073/pnas.0610172104
16. M.F. Guagliardo, Spatial accessibility of primary care: concepts, methods and challenges. International Journal of Health Geographics **3**(1), 1–13 (2004). https://doi.org/10.1186/1476-072X-3-3
17. W. John Reilly, *The law of retail gravitation* (Knickerbocker Press, New York, USA, 1931)
18. D.L. Huff, A probabilistic analysis of shopping center trade areas. Land Economics **39**(1), 81–90 (1963). https://doi.org/10.2307/3144521
19. J. Luo, Integrating the Huff model and floating catchment area methods to analyze spatial access to healthcare services. Transactions in GIS **18**(3), 436–448 (2014). https://doi.org/10.1111/tgis.12096
20. R. Prieto-Curiel et al., Detecting cities with high intermediacy in the African urban network, in Computers. Environment and Urban Systems **98**, 101869 (2022). https://doi.org/10.1016/j.compenvurbsys.2022.101869

21. J. Westerlund, F. Wilhelmsson, Estimating the gravity model without gravity using panel data. Applied Economics **43**(6), 641–649 (2011). https://doi.org/10.1080/00036840802599784
22. A.J. Garcia et al., Modeling internal migration flows in Sub-Saharan Africa using census microdata. Migration Studies **3**(1), 89–110 (2015). https://doi.org/10.1093/migration/mnu036
23. R.M. Beyer, J. Schewe, H. Lotze-Campen, Gravity models do not explain, and cannot predict, international migration dynamics. Humanities and Social Sciences Communications **9**(1), 1–10 (2022). https://doi.org/10.1057/s41599-022-01067-x
24. F. Simini et al., A universal model for mobility and migration patterns. Nature **484**(7392), 96–100 (2012). https://doi.org/10.1038/nature10856

Chapter 7
Segregation in Cities

Laetitia Gauvin and Julien Randon-Furling

Abstract This chapter focuses on segregation phenomena in urban contexts–that is, situations in which different population groups do not enjoy the same access to what a city has to offer. While there are "macroscopically"-enforced types of segregation, such as what existed under the apartheid regime in South Africa, segregation also appears through *emergence* in the complex system formed by the web of interactions that connects individuals' decisions on where they choose to live, work and play within the city. Here, we provide a brief introduction to various aspects of segregation: its multiple definitions and dimensions, available data sources (from traditional surveys and censuses to digital human traces), indices and other types of measure, and finally, a selection of approaches to study and analyze segregation.

7.1 Introduction

In 2015, the United Nations adopted a blueprint titled the 2030 Agenda for Sustainable Development. It contains 17 key Sustainable Development Goals [1], the 11th of which aims at making "cities and human settlements more inclusive, safe, resilient and sustainable". Indeed, while urbanization has become faster in the past decades, with the global urban population overtaking the rural one in 2008 and its share set to hit 2/3 by 2050 [2]), this has not always led to more inclusive cities–often this has even been accompanied by the development of slums, with over 1 billion people living in slums in 2018 according to UN figures [2]. Albeit extreme, slums

L. Gauvin (✉)
IRD, UMR 215 PRODIG, Aubervilliers cedex, France
e-mail: laetitia.gauvin@ird.fr

ISI Foundation, Turin, Italy

J. Randon-Furling
Centre Borelli, ENS Paris-Saclay, Gif-sur-Yvette, France
e-mail: j.randon-furling@cantab.net

MSDA, UM6P, Ben Guerir, Morocco

D. Rybski (ed.), *Compendium of Urban Complexity*, Understanding Complex Systems,
https://doi.org/10.1007/978-3-031-82666-5_7

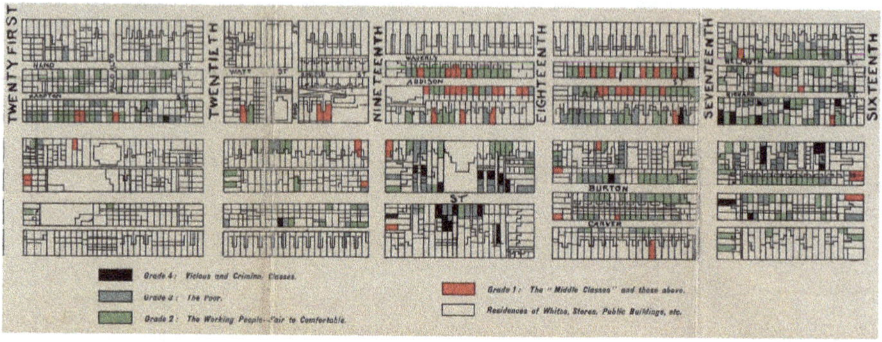

Fig. 7.1 Excerpt from Du Bois's social study [3] of Philadelphia, showing the social and economic segregation of African-American inhabitants in Philadelphia at the end of the 19th century

are one example of urban segregation: a phenomenon resulting in different population groups unequally sharing or using urban spaces and services. First studies on the phenomenon of segregation date back at least to the early 20th century with the Chicago School of Sociology (the first school of urban sociology), and even further, if one recalls the pioneering work of African-American sociologist W. E. B. Du Bois in Philadelphia [3] (see Fig. 7.1).

In 1925, E. Burgess and R. Park, two of the founders of the Chicago School of Sociology, introduced the concept of urban ecology to describe the organization of the city [4]. They proposed a concentric scheme describing urban growth, where concentric zones are socially differentiated (a central zone inhabited by workers, followed by a residential area and the suburbs). This geographical representation formed a global interpretation of the phenomenon of specialization of urban spaces.

Closer to us in time, American economist T.C. Schelling, through two papers published around 1970 [5, 6] and a book in 1978 [7], also devoted his attention to the study of the distribution of social groups in cities, but at a local level. His analysis of segregation, however, is not based only on the geographical factor as in Park & Burgess's approach. According to Schelling, segregation results from three mechanisms: organized discrimination (for example racial segregation laws), social differentiation, and/or the somewhat surprising collective result of individual decisions even when these decisions lean another way.[1] The latter point has been instrumental in making Schelling's dynamic model of segregation an exemplar of Complex Systems approaches in city science: it shows via a simple agent-based model how a collective, macroscopic property (segregation) may emerge from a host of individual, microscopic interactions. If this result does not necessarily explain segregation in real cities (nor does it claim to do so), it is most valuable because it complexifies our vision and understanding of segregation phenomena. Segregation

[1] The third point was contemporaneously investigated within a more general model of social interactions by Japanese American psychologist and sociologist J.M. Sakoda [8], who introduced an early version of this model in his 1949 PhD Dissertation at the University of California [9].

may manifest itself through what is called *emergence*: a characterizing property of Complex Systems, wherein *the whole is more than the sum of its parts* and the collective, macroscopic behavior differs from what may be inferred at the individual, microscopic level.

Since Schelling's work, many developments have been proposed and explored. Ways of looking at segregation, particularly at the diversity of its natures and meanings according to scale (local, intermediate, citywide) and space (geographical and/or social space), have evolved. We do not aim to provide a comprehensive review of the topic but instead offer a brief introduction to a student or any reader interested in, but new to, the subject. We will, in particular, discuss definitions, data sources, measures, and indices, as well as several approaches for both modeling and analysis–a Complex Systems perspective is the one we will favor, for this compendium on urban complexity, but not exclusively.

7.2 Definition(s) of Urban Segregation

To segregate literally means the act of separating a herd, disassembling a group. Thus, in the context of city science, urban segregation may be defined as a phenomenon (or the result thereof) by which the population living in a city comes to be split into groups that do not share the same characteristics with respect to at least one aspect of urban life.

One such example is when the population is composed of various pre-identified groups (*e.g.* religious, ethnic, gender, age, or income groups) who do not enjoy the same access to various places in the city. This may be because of laws enforcing institutional segregation: the apartheid in South Africa until 1991, the Jim Crow laws in the USA until 1964, and the 1516 decree instituting the Jewish Ghetto in Venice. But this may also result from more insidious causes, as an emergent property of the complex system that a city constitutes (see Sect. 7.5).

When focusing on access to various places in the city, the type of segregation we are looking at is *spatial segregation*. This includes "static", residential segregation (when the neighborhoods where people live depend on the groups they belong to), and "dynamic" segregation (when mobility and individual trajectories in the urban area depend on the groups). However, the urban landscape is not just space, at least not just physical, geographical space. One could even argue that the city is primarily a social space and that physical proximity may not always be a proxy for social contact. Therefore, beyond spatial segregation, it is important to also define and investigate *social segregation*: do patterns of behaviors and interactions depend on the groups to which individuals belong? Occupational segregation, for instance, is often observed between genders: women, men, and other gender groups tend to work in different sectors.

As we have just explored, the term "segregation" is not singular in meaning. Moreover, there can be multiple aspects even within a specific type, such as spatial segregation. In line with this idea, Massey and Denton's seminal 1988 paper [10] introduced the concept that segregation actually consists of five distinct dimensions:

- evenness,
- exposure,
- clustering,
- centralization,
- concentration.

Given a population composed of pre-identified groups, evenness corresponds to how uniform the spatial distribution of each group is; exposure relates to the probability for a member of a group to come across members of each of the other groups; as for centralization, it incorporates the notion that different areas in a city are not equivalent: some are more sought after than others (*e.g.* central areas in many European cities), and therefore if a group tends to be over-represented in such areas, this is something more than simple unevenness. As argued later by Reardon & O'Sullivan [11], clustering (the tendency of members of a group to cluster together) is another way to describe evenness, and concentration (how much space is occupied by each group) also relates directly to evenness. Hence, foregoing differences in the intrinsic quality of various areas, one may retain two main dimensions of segregation: evenness and exposure. These two concepts do not rely on physical space between individuals; rather, they can be defined on any relational structure connecting them. This highlights that space is just one of many ways individuals relate to each other-social interactions, whether in physical or virtual spaces, represent another. While spatial segregation may arise from the former and social segregation from the latter, in reality, these forms often overlap. However, they may not be equally observable with the same type of data, as the nature of the data available will determine which form of segregation can be measured and analyzed.

7.3 Data Sources

Reliable data are essential to measure the extent to which given groups are segregated. Their analysis is a step towards a better understanding of segregation and might hint at possible refining of integration policies. Census data have been extensively used to identify patterns of segregation in cities. They can provide insight into several forms of segregation, such as segregation by age, gender, education level, ethnicity, or socioeconomic level [12]. They usually cover a representative population sample and are provided every few years; in very few countries, such as the Netherlands, they represent the whole population [13]. While census data helps to apprehend the level of spatial segregation, social interactions cannot be captured through such data. Alternatively or in complement, surveys are a possible tool to track the patterns of social segregation. They have the advantage of being specially designed for a given scope and context, but they usually cover a small part of the population at one point in time. However, with the ongoing urbanization, more frequent and granular information across the city is needed to explore the evolution of the social mix across different neighborhoods. Data collection of human digital traces enables the exploration of the micro-dynamics of social interaction and/or of the characteristics of the residential settlements at an unprecedented spatial and temporal scale. For instance, mobile phone companies' data from anonymous users have been used to investigate the presence of segregated patterns for instance in cities in South Asia [14], in Turkey [15] and in Serbia [16].

Here we describe the main data sources considered to study segregation. Each type of data might serve a different purpose: to assign a group to individuals (economic status, gender, ethnicity) and to establish the spatial location of the residence or of the activity types (mobility, place visits) of each individual or group of individuals. There exist essentially two main classes of data: traditional census data and emerging data from digital sources.

7.3.1 Census and Survey Data

Census data provide information about the inhabitants of a country. Usually, they include demographic attributes such as age or gender, economic status (employment status, income), place of residence, and in a few countries, e.g. the United States,[2] ethnicity might be provided. Many countries release national census data with a time interval of several years. The Demographic and Health Surveys (DHS) Program[3] collects data about population well-being in over 90 low- and middle-income countries with different spatial resolutions. They include information about the living conditions of households (assets ownership which also enables the calculation of the wealth index of these households) as well as their place of residence. These data can be used to study segregation of deprivation [17]. In general, census and survey data are relevant to identify patterns of residential segregation, they can be combined with other data sources to look at other types of segregation.

7.3.2 Location Data

Location data are useful for investigating whether individuals from different groups visit different kinds of places. Social media might be possible data sources as some users provide their geolocalization while posting content. Enriching these data with

[2] https://www.census.gov/ last accessed October 14, 2024.

[3] https://dhsprogram.com/data/ last accessed October 14, 2024.

other information (social media content, census data...) allows us to characterize different groups in terms of residential location and mobility in the city. Mobile phone data such as Call Detail Records offer proxy of location data and have been largely used to study socio-spatial segregation by gender, ethnicity, or income [14, 18, 19]. Such data are usually given as a tuple (timestamps, cell tower, duration of the call) together with the positions of the cell tower so that every time a mobile phone user makes a call, an approximate location can be assigned. Looking for the location where most of the activity is recorded during the night hours provides an estimation of the residential location with varying levels of accuracy [20]. Global Positioning System (GPS) coordinates, also coming from mobile phone data, offer an even more detailed picture of the places visited in the city by different groups of users [21]. These data are typically collected from anonymized users who have opted-in to provide access to their location data anonymously. Location accuracy is determined by the device and is variable. Temporal sampling of de-identified users' location is also variable and dependent on app/OS characteristics and user behavioral patterns. Typically these data are listed as event of the form (anonymous hashed user id, time, latitude, longitude).

7.3.3 Social Interaction Data

Call Detail Records, previously introduced, opens up the open up the possibility to study not only residential segregation but also social segregation by analyzing calling patterns. Social network data is another data source that can be used to assess the extent to which social interactions are segregated in a city. Usually provided as a list of pairs of anonymous identifiers and their localization (when available), such data make it possible to investigate the relationship between online fragmentation and physical segregation [22, 23].

One way to extract meaningful information from data is to compute segregation indices. Hereafter, we describe some of the most widely known indices, as well as a few recent additions to the family of mathematical measures of segregation.

7.4 Measure and Indices

There is a multi-decade history and a vast literature of mathematical indices designed to measure segregation [11, 24–38]. We will not review all of them but we shall discuss a few ways to measure segregation according to the type of data and therefore to the type of segregation.

7.4.1 Spatial Segregation Indices

A first set of indices are designed to work with census or census-like data. Typically, such data provide population composition (in terms of pre-defined groups) in each of a number N of areal units that cover the urban area of interest. These areal units may be census tracts or blocks or other units designed for statistical purposes (*e.g.* IRIS in France). By their nature, such data sources allow to measure spatial segregation rather than social segregation–in the most standard case, one only measures static, residential segregation, although surveys also give access to some form of dynamics (*e.g.* the way spatial segregation changes around the clock [39], according to who is in the various areal units at given times, typically daytime *vs.* nighttime).

Dissimilarity

One of the basic ideas consists of measuring how each unit differs from (or is *dissimilar* from) the total urban area. For instance, suppose there are two pre-defined groups in the population, say A and B. If these were uniformly distributed across the urban area, then in each areal unit i, the number a_i of individuals belonging to group A would represent the same fraction of that group as the b_i individuals of group B with respect to their group. In other words, by writing simply A (resp. B) for the total number of individuals from that group, one would have that $a_i/A - b_i/B = 0$ for every i, hence summing $|a_i/A - b_i/B|$ over all $1 \leq i \leq N$ would also give 0. In the other extreme case, when each areal unit comprises individuals from only one of the two groups, then

$$\sum_{i=1}^{N} \left| \frac{a_i}{A} - \frac{b_i}{B} \right| = \sum_{\substack{1 \leq i \leq N \\ b_i = 0}} \frac{a_i}{A} + \sum_{\substack{1 \leq i \leq N \\ a_i = 0}} \frac{b_i}{B} = 2 .$$

Whence the definition of the so-called *Dissimilarity Index*,

$$D = \frac{1}{2} \sum_{i=1}^{N} \left| \frac{a_i}{A} - \frac{b_i}{B} \right| ,$$

that ranges from 0 (uniform distribution of both groups A and B across the areal units) to 1 (only one group represented in each unit).
To date, this is one of the most widely used indices to measure segregation. Yet, it suffers from three strong limitations:

- It is defined only for populations composed of *two groups*. This can be easily remedied, as seen in the next paragraph.
- It is *partially aspatial*. The spatiality of the population's distribution plays a role only insofar as it determines the composition of each areal unit: individuals belong

to the same unit according to their spatial proximity. However, the spatial arrangement of areal units is not taken into account: D simply treats them as boxes without any spatial relationship between them.

- It runs into the *modifiable areal unit problem* (MAUP) [40, 41]: D is hugely affected by the design of the areal units, *e.g.* if A and B have equal numbers of individuals across the city and if the initial areal units form a checkerboard of mono-group units (so that $D = 1$), then redesigning the units by merging pairs of adjacent units will lead to a radical change in D, from 1 to 0–even though the actual spatial distribution of the population has not changed.

Each of the three points mentioned above has been addressed in the literature, and we will give examples of indices that circumvent these limitations.

Multigroup indices

Many of the indices proposed to measure segregation for populations composed of more than two pre-defined groups are derived from one form or another of entropy measure[4], see also Chap. 12. To examine some of them here, let us introduce the entropy of a multigroup distribution. Suppose there are M pre-defined groups in a population and write p_m for the fraction of individuals belonging to group m; then the entropy E of this distribution is defined as

$$E = -\sum_{m=1}^{M} p_m \ln p_m .$$

Now, a natural way to measure dissimilarity in a multigroup context is to compare the entropy of the population distribution within each areal unit with the entropy of the distribution in the whole population. Specifically, if the population is spread across N units, writing T for the total number of individuals in the population, t_i for the number of individuals in unit i and p_{im} for the fraction of individuals in unit i who belong to group m, then one may define the local entropy $E_i = -\sum_{m=1}^{M} p_{im} \ln p_{im}$ and compute the weighted average of the differences between the global entropy E and the local entropies E_i:

$$H = \frac{1}{E} \sum_{i=1}^{N} \frac{t_i}{T} (E - E_i) ,$$

where the division by E provides a natural scale ($H = 0$ if $E_i = E$ for every i, i.e. if every unit has the same multigroup distribution entropy as the whole, and $H = 1$

[4] Entropy-based measures may be derived from a more general framework based on information theory [42, 43].

if $E_i = 0$ for every i, i.e. if every unit consists of only one group).[5] This quantity H is known as Theil's index [44] and may also be seen as the weighted average of the Kullback-Leibler (KL) divergence between the global multigroup distribution and the local multigroup distribution–indeed:

$$\begin{aligned} H &= \frac{1}{E} \sum_{i=1}^{N} \frac{t_i}{T} (E - E_i) \\ &= \frac{1}{E} \left(- \sum_{m=1}^{M} p_m \ln p_m + \sum_{i=1}^{N} \frac{t_i}{T} \sum_{m=1}^{M} p_{im} \ln p_{im} \right) \\ &= \frac{1}{E} \left(- \sum_{m=1}^{M} \sum_{i=1}^{N} \frac{p_{im} t_i}{T} \ln p_m + \sum_{i=1}^{N} \frac{t_i}{T} \sum_{m=1}^{M} p_{im} \ln p_{im} \right) \\ &= \frac{1}{E} \sum_{i=1}^{N} \frac{t_i}{T} \sum_{m=1}^{M} p_{im} \ln \frac{p_{im}}{p_m} \\ &= \frac{1}{E} \sum_{i=1}^{N} \frac{t_i}{T} D_{\mathrm{KL}} (\{p_{im}\} \,||\, \{p_m\}) \ , \end{aligned}$$

where $D_{\mathrm{KL}} (\{p_{im}\} \,||\, \{p_m\}) = \sum_{m=1}^{M} p_{im} \ln (p_{im}/p_m)$ is by definition the Kullback-Leibler divergence (also called relative entropy) from the global distribution to the local distribution in unit i [45].

The latter formulation shows that H is non-negative but also unbounded (since KL divergence is not bounded). It is equal to 0 if the multigroup distribution in each areal unit is the same as in the whole population. It equals 1 for a two-group population distributed in a checkerboard pattern.

An illustration of several segregation indices is shown in Fig. 7.2. *Spatial versions.*

Multigroup indices, and in particular the type of entropy-based indices introduced in the previous paragraph, still fall short of taking into account spatial relationships between the statistical units.

One way of incorporating space is to use weighted spatial averages of population counts rather than single unit counts [11, 33]. This is a common approach in spatial statistics: *e.g.* replacing the number t_{im} of individuals in unit i belonging to group m by a local average:

$$\hat{t}_{im} = \sum_{j=1}^{N} k \left(\mathbf{u}_i, \mathbf{u}_j \right) t_{jm} \ ,$$

where $\mathbf{u}_i$ is the spatial location of (the centroid of) unit i and k is a kernel function, typically a decreasing function of the distance $||\mathbf{u}_j - \mathbf{u}_i||$. A standard choice of kernel is the Gaussian one,

[5] If one computes just the weighted average of the local entropies, one obtains the Mutual Information index of segregation, that is, $\mathrm{MI} = \frac{1}{T} \sum_{i=1}^{N} t_i E_i$.

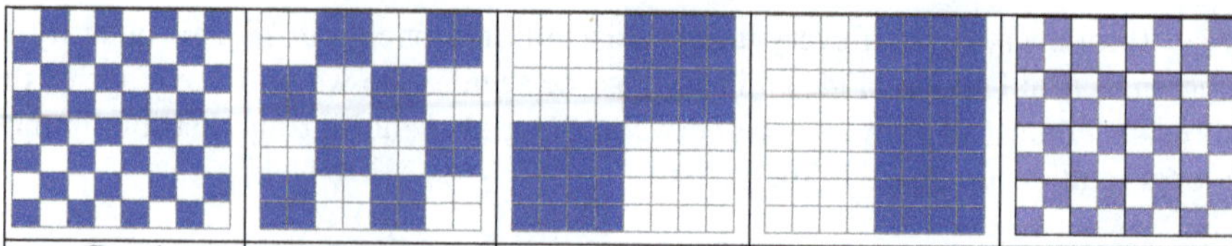

$D = 1$	$D = 1$	$D = 1$	$D = 1$	$D = 0$
$H = 1$	$H = 1$	$H = 1$	$H = 1$	$H = 0$
$G = 1$	$G = 1$	$G = 1$	$G = 1$	$G = 0$
$SD = 0$	$SD = 1/2$	$SD = 3/4$	$SD = 7/8$	$SD = 0$

Fig. 7.2 Various configurations of a two-group population arranged in single group boxes on a grid, to serve as exemplars for the computation of three standard segregation indices, Dissimilarity (D), Theil (H) and Gini (G), plus a spatial version (SD) of the D index (in SD, the population count of each unit is replaced by a weighted average between the unit, with weight $1/2$, and its four nearest neighboring units with weight $1/8$ each). The fifth configuration illustrates the MAUP: it is the same as the first configuration, but the areal units used to compute the indices are larger (shown in black lines, and the smaller units have been subdued for visual clarity)

$$k\left(\mathbf{u}_i, \mathbf{u}_j\right) \propto \exp\left[-\frac{||\mathbf{u}_j - \mathbf{u}_i||^2}{2\sigma^2}\right],$$

where σ is sometimes called the bandwidth of the kernel–it is the typical distance beyond which units are quasi-neglected in the weighted average.

Note that, using a Dirac delta kernel is equivalent to using the aspatial versions of the indices.

Multiscale measures

While traditional spatial versions of standard segregation indices address, in part, the MAUP, they often include some arbitrary choice of a scale. Think, for instance, of the bandwidth parameter in a Gaussian kernel. Another approach that has been mainly investigated over the past two decades seeks to integrate information over multiple scales [34, 35, 37, 38]. This may be done by looking at sequences of intermediate scales rather than simply the finest or coarsest one [35, 37]. Eventually, one may take into account all possible scales and extract information from the sequence of values obtained for any chosen aspatial index computed at every scale around any starting point [38, 46, 47].

7.4.2 *Social Segregation Indices*

Spatial segregation indices are often used as a proxy to measure social segregation, but they may misrepresent the reality of social segregation. Indeed, people from different groups may inhabit or work in the same spatial area (*e.g.* a few blocks or a neighborhood) without actually coming into contact with each other, let alone develop social bonds, exchange views, become friends, etc. Two axes along which one may work to address this problem are (i) using spatial data at the individual level and/or (ii) working directly with social interaction data.

Personalized (multiscale) indices

As an example of how individual-level spatial data can be used to approach social segregation, consider a personalized version of the multiscale index mentioned at the end of the previous section. That is, instead of using spatial proximity for the aggregation procedure, one may use an aggregation path derived from the sequence of locations actually visited by an individual. In addition, if these locations, rather than blocks or tracts, are places like bars or shops, so much the better–this way, space is more likely to serve as a good proxy for social interaction.

Network measures

Data permitting, one may directly have access to social interaction networks, i.e., social interactions modeled in the form of a graph defined by its adjacency matrix: a set of nodes and edges (potentially weighted and directed edges). This allows one to use the toolbox of network analysis to measure segregation. For instance, one can compare the network's actual modular structure to the known underlying groups of the networks (gender, ethnicity, etc). Detecting the modular structure of a network is usually done by maximizing the modularity

$$Q = \frac{1}{2l} \sum_{i,j} \left(a_{ij} - \frac{k_i k_j}{2l} \right) \delta(c_i, c_j),$$

where i and j are nodes of the networks and l is the number of links. k_i is the degree of node i, i.e., the number of nodes to which it is connected. The element a_{ij} is a binary variable that takes the value 1 if nodes i and j are connected. c_i indicates which community node i belongs to. Modularity quantifies to which extent nodes of the same community are more connected with respect to the random case given their degree. The assortativity index r as defined in [48] quantifies to which extent nodes of the same type are connected:

$$r = \frac{\sum_i e_{ii} - \sum_i a_i^2}{1 - \sum_i a_i^2} = \frac{\text{Trace}(e) - ||e^2||}{1 - ||e^2||},$$

where e_{ii} is the observed fraction of links between nodes of type i and a_i is the expected fraction of links between nodes of type i given the number of nodes of that type and the total number of links.

Let us also mention here an approach similar in spirit to the multiscale one mentioned before for spatial segregation. Suppose nodes on a network correspond to individuals who each belong to one of several predefined groups. By looking at the average time it takes for a random walk to discover the different groups and/or to visit nodes from these groups more or less in the same proportions as the distribution in the whole population (this can be measured again using the KL divergence), then one has a measure of how segregated the network is [49–51].

7.4.3 Libraries and Packages

Several libraries exist to compute some of the indices mentioned above easily. In R, the package *segregation*[6] provides an easy way to calculate segregation measures based on the Mutual Information Index and Theil's Entropy Index. In python, there is also such a library under pysal,[7] where many indices are implemented, both spatial and non spatial, including the most common ones (dissimilarity index, Gini index, entropy, isolation, exposure).

7.5 Complex Systems Approaches to Segregation

Different approaches exist in Complex Systems to study urban segregation. We are listing the main ones.

7.5.1 Agent-Based Modeling

Agent-based models[8] are useful tools for studying the emergence of segregation by simulating interactions between agents, which can represent individuals or groups. In these models, agents independently assess their situations and make decisions based on that evaluation. The origins of multi-agent systems can be traced back to the 1940s, when mathematician and physicist J. von Neumann introduced the concept of cellular automata [53]. Later, J.H. Conway developed the "Game of Life," a simulation depicting the birth and death of "cells" according to predefined rules [54]. Before these games, one of the earliest multi-agent simulations was created by economist T.C. Schelling. In his 1971 paper, "Dynamic Models of Segregation" [6], Schelling presented a model in which a segregated structure emerges, even though individual agents do not actively seek segregation. These early simulations were performed manually, using pawns on a chessboard. With the rise of computers, multi-agent models, particularly Schelling's segregation model, gained widespread attention. His model,

[6] https://github.com/cran/segregation last accessed October 14, 2024.

[7] https://github.com/pysal/segregation last accessed October 14, 2024.

[8] This section is partially adapted from [52], originally in French.

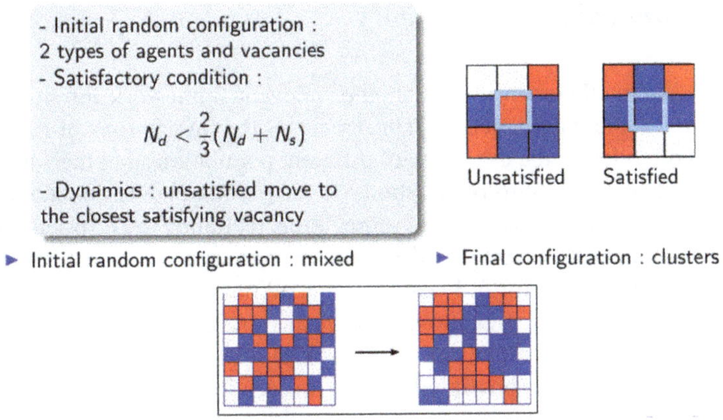

Fig. 7.3 Schelling's original model illustration. Here we show the original tolerance threshold value. N_d and N_s respectively stand for different and similar neighbors. On the right of the figure we show the two possible configurations based on the number of different neighbors of the central agent. Finally, at the bottom of the figure, we show an example of an initial and a final configuration

further expanded in the book *Micromotives and Macrobehavior* [7], remains one of the first significant attempts at multi-agent simulation.

The simulation proceeds as follows: initially, two types of agents are randomly placed on the squares of a chessboard. Some squares are left vacant to allow for potential moves. The types might represent agents from different demographic or socio-economic groups.

Agents derive satisfaction–equivalent to a utility of 1–when a sufficient number of neighboring agents belong to the same group. Specifically, an agent is considered satisfied if at least one-third of the eight surrounding squares (excluding edges) are occupied by agents from the same group, setting the tolerance threshold at one-third. Starting from a random configuration, some agents may not be satisfied (with a utility of 0). To resolve this, unsatisfied agents move to the nearest available empty spaces where they can be surrounded by a sufficient number of similar agents. This relocation process is iterative: moving unsatisfied agents may lead to new unsatisfied agents, triggering further moves.

The cycle repeats until all agents are satisfied, or no further movement can improve satisfaction. The simulation ultimately reveals the formation of clusters of similar agents, leading to segregation (see Fig. 7.3 for a visualization). Multiple variants of the Schelling segregation models were then considered in Complex Systems. Agent-based models such as the Schelling segregation model can be fully implemented using languages such as C or Python, or they can also be studied using applications already built but less flexible such as Netlogo.[9]

[9] https://ccl.northwestern.edu/netlogo/models/Segregation last accessed October 14, 2024.

7.5.2 Dynamical Systems Theory

Dynamical systems theory provides another relevant framework for studying self-organization, i.e. the emergence of patterns out of the interactions of multiple elements. Indeed modeling the evolution of different populations and their interactions can be achieved using systems of equations. In [55], a model of residential neighborhoods demonstrates that weak forms of segregation, including mixed populations, can emerge. Additionally, describing well-known segregation models like Schelling's as dynamical systems allows for fully categorizing all possible equilibrium states [56].

In the context of urban segregation, dynamical systems theory offers an analytical tool for investigating city organization in a simplified form and exploring various scenarios to understand the possible mechanisms driving segregation.

7.5.3 Statistical Physics

Methods from statistical physics are commonly adopted in Complex Systems. The analogy between the tolerance of agents with respect to different neighbors and the temperature of a spin model proved fruitful in describing the dynamics of segregation [57]. Also physical quantities are helpful to this matter as they can be seen as control parameters and taken to identify different phases. Segregation phenomena can also be seen as a surface tension problem. Replacing the utility of an agent (i.e., the function on which they base their decision) with an energy, as in physics, can contribute to explaining what are possible mechanisms for the emergence of segregation [58].

7.5.4 Network Analysis

As alluded to earlier when discussing network-based segregation measures, thinking about segregation as resulting from interactions naturally leads to representing individuals and their neighbors as networks. This approach not only allows segregation to be quantified using tools from network science but also provides a framework for identifying the underlying mechanisms driving segregation in urban environments. Indeed, network models have been primarily used to study segregation [59–61], they can help understand whether there exist networks' properties that may lead to different levels of segregation—for instance, size effects have been found to play a role. Generally, network models enable one to investigate what mechanisms can lead to the emergence of segregation. Moreover, networks are not only appropriate tools to represent interactions, but they can also include a spatial dimension that is particularly relevant to studying urban segregation. Indeed, spatial networks have been considered in several studies. Networks are also natural representations of large-scale communication data like social media interactions or calls between individuals that can be used as a proxy to quantify the level of social and spatial mixity [14, 62].

7.5.5 Statistical Modeling and Machine Learning

Statistical modeling consists of suggesting a possible stochastic generation mechanism for data to test some hypothesis or to make some prediction. In the case of the study of urban segregation, this type of modeling can provide a way to measure segregation or make statistical inferences about the underlying patterns of segregation [63]. These models also allow one to explore the relationships between different types of segregation, such as work and residential segregation [64].

Statistical modeling and machine learning are closely related in that in both approaches, one attempts to establish the link between independent variables (resp. features) and dependent variables (resp. targets). Machine learning techniques, such as neural networks or spectral methods, prove to be efficient tools for the exploration of segregation patterns in socio-economic and geographical data [42, 65] or to look for the association between segregation and other phenomena, such as the spread of diseases, for instance [66].

7.6 Conclusions

In this chapter, we provided an overview of the various types of segregation and gave some insight into how the phenomenon was investigated in the field of Complex Systems. We have seen that approaches coming from network science, among others, are particularly appropriate to study social segregation in addition to spatial segregation. Moreover, digital traces increasingly used in Complex Systems studies allow the capture of segregation dynamics with high spatial resolution for different layers of the population, providing a more detailed picture of segregation phenomena. For instance, in recent years, large-scale GPS data have been used to estimate a measure of experienced segregation [67].

While the increasing availability of new digital traces and the development of tools in Complex Systems represent an unprecedented opportunity to advance the study of urban segregation, it also comes with some limitations and challenges. Indeed, since data collected passively might not be representative of the population, the development of methods to compensate for possible misrepresentation is essential to have a faithful representation of the patterns of segregation. Similarly, it is crucial to understand how approaches used in the Global North can be replicated in the Global South, where data are usually scarcer. Another aspect relates to privacy and the choice of the proper aggregated data that preserves privacy and, at the same time, gives a meaningful insight into the segregation phenomenon.

Finally, we would like to provide some recommendations when initiating a study about urban segregation. While Complex Systems have provided insightful studies on segregation in cities, it is necessary to combine these approaches with those from other fields, such as those from social science, to maintain a critical view of the findings. Moreover, experienced segregation is a complex phenomenon with multiple dimensions, it should be studied in the light of other aspects of everyday life, such as urban mobility, a topic developed in Chap. 4. Also, segregation is not a static phenomenon and needs to be re-evaluated in a timely fashion—taking advantage, for instance, of the continuous flow of digital traces.

References

1. *The 17 Goals–Sustainable Development*. https://sdgs.un.org/goals. Accessed: 2023-01-20. https://doi.org/10.1093/oso/9780192848758.002.0009
2. *UN Goal 11*. https://sdgs.un.org/goals/goal11. Accessed: 2023-01-20. https://doi.org/10.4324/9781315414218-4
3. W.E.B. du Bois, The Philadelphia Negro,–A social study (1899). Accessed 2024-02-05, via the Open Collections Program at Harvard University. Publications of the University of Pennsylvania (2024). https://doi.org/10.2307/1834717
4. R.E. Park, E.W. Burgess, The city. University of Chicago Press (1925). https://doi.org/10.1068/b4101ed
5. T.C. Schelling, Models of segregation. The American economic review **59**(2), 488–493 (1969). https://doi.org/10.1007/978-0-387-09789-3_7
6. T.C. Schelling, Dynamic models of segregation. Journal of mathematical sociology **1**(2), 143–186 (1971). https://doi.org/10.2307/j.ctv1dv0tg9.21
7. T.C. Schelling, *Micromotives and Macrobehavior*. (WW Norton & Company, 2006). https://doi.org/10.2307/134812
8. J.M. Sakoda, The checkerboard model of social interaction. The Journal of Mathematical Sociology **1**(1), 119–132 (1971). https://doi.org/10.1080/0022250x.1971.9989791
9. J.M. Sakoda, Minidoka: An Analysis of Changing Patterns of Social Interaction. PhD thesis. University of California, Berkeley, 1949. https://doi.org/10.1300/j079v02n02_05
10. D.S. Massey, N.A. Denton, The dimensions of residential segregation. Social forces **67**(2), 281–315 (1988). https://doi.org/10.2307/2579183
11. S.F. Reardon, D. O'Sullivan, Measures of spatial segregation. Sociological methodology **34**(1), 121–162 (2004). https://doi.org/10.1111/j.0081-1750.2004.00150.x
12. S. Musterd et al., Socioeconomic segregation in European capital cities. Increasing separation between poor and rich, in Urban Geography **38**(7), 1062–1083 (2017). https://doi.org/10.1080/02723638.2016.1228371
13. J. van der Laan et al., A whole population network and its application for the social sciences. European sociological review (2022). https://doi.org/10.1093/esr/jcac026
14. J. Blumenstock, L. Fratamico, Social and spatial ethnic segregation: a framework for analyzing segregation with large-scale spatial network data, in *Proceedings of the 4th Annual Symposium on Computing for Development*. 2013, pp. 1–10. https://doi.org/10.1145/2537052.2537061
15. D. Rhoads et al., Measuring and mitigating behavioural segregation using Call Detail Records, in *EPJ Data Science* 9.1 (2020), p. 5. https://doi.org/10.1140/epjds/s13688-020-00222-1
16. N. Grujić, Evidence of urban segregation from mobile phone data: A case study of Novi Sad, in *27th Telecommunications Forum (TELFOR)*. IEEE. **2019**, 1–4 (2019). https://doi.org/10.1109/telfor48224.2019.8971290

17. F. Temporin, A multilevel structural equation modelling approach to study segregation of deprivation: an application to Bolivia. Quality & Quantity **53**(3), 1657–1674 (2019). https://doi.org/10.1007/s11135-018-00832-y
18. R. Goel, R. Sharma, A. Aasa, Understanding gender segregation through call data records: an Estonian case study, in *Plos one* 16.3 (2021), e0248212. https://doi.org/10.1371/journal.pone.0248212
19. B. Moya-Gómez et al., Exploring night and day socio-spatial segregation based on mobile phone data: The case of Medellin (Colombia), in Computers. Environment and Urban Systems **89**, 101675 (2021). https://doi.org/10.1016/j.compenvurbsys.2021.101675
20. L. Pappalardo et al., Evaluation of home detection algorithms on mobile phone data using individual-level ground truth, in *EPJ data science* 10.1 (2021), p. 29. https://doi.org/10.1140/epjds/s13688-021-00284-9
21. E. Moro et al., Mobility patterns are associated with experienced income segregation in large US cities, in *Nature communications* 12.1 (2021), p. 4633. https://doi.org/10.1038/s41467-021-24899-8
22. A.J. Morales et al., Segregation and polarization in urban areas, in *Royal Society Open Science* 6.10 (2019), p. 190573. https://doi.org/10.1098/rsos.190573
23. G. Tóth et al., Inequality is rising where social network segregation interacts with urban topology. Nature communications **12**(1), 1–9 (2021). https://doi.org/10.1038/s41467-021-21465-0
24. D.O. Cowgill, M.S. Cowgill, An index of segregation based on block statistics. American Sociological Review **16**(6), 825–831 (1951). https://doi.org/10.2307/2087511
25. W. Bell, A probability model for the measurement of ecological segregation. Social Forces **32**(4), 357–364 (1954). https://doi.org/10.2307/2574118
26. O. Dudley Duncan, B. Duncan, A methodological analysis of segregation indexes, in *American sociological review* 20.2 (1955), pp. 210–217. https://doi.org/10.2307/2088328
27. J.M. Sakoda, A generalized index of dissimilarity. Demography **18**(2), 245–250 (1981). https://doi.org/10.2307/2061096
28. J.F. Jakubs, A distance-based segregation index. Socio-economic planning sciences **15**(3), 129–136 (1981). https://doi.org/10.1016/0038-0121(81)90028-8
29. D.R. James, K.E. Taeuber, Measures of segregation. Sociological methodology **15**, 1–32 (1985). https://doi.org/10.1016/b978-0-12-814199-1.00011-1
30. R.L. Morrill, On the measure of geographic segregation, in *Geography research forum*. Vol. 11. 1991, pp. 25–36. https://doi.org/10.1016/b0-08-043076-7/02591-2
31. D.W.S. Wong, Spatial indices of segregation. Urban studies **30**(3), 559–572 (1993). https://doi.org/10.1080/00420989320080551
32. S.F. Reardon, G. Firebaugh, Measures of multigroup segregation. Sociological methodology **32**(1), 33–67 (2002). https://doi.org/10.1111/1467-9531.00110
33. F.F. Feitosa et al., Global and local spatial indices of urban segregation. International Journal of Geographical Information Science **21**(3), 299–323 (2007). https://doi.org/10.1080/13658810600911903
34. *Spatial econometrics: methods and models*. Vol. 4. Springer Science & Business Media, 2013. https://doi.org/10.2307/143780
35. J. Östh, W.A.V. Clark, B. Malmberg, Measuring the scale of segregation using k-nearest neighbor aggregates. Geographical analysis **47**(1), 34–49 (2015). https://doi.org/10.1111/gean.12053
36. M. Fossett, New methods for measuring and analyzing segregation. Springer Nature (2017). https://doi.org/10.1007/978-3-319-41304-4
37. A. Petrović, M. van Ham, D. Manley, Multiscale measures of population: Within-and between-city variation in exposure to the sociospatial context. Annals of the American Association of Geographers **108**(4), 1057–1074 (2018). https://doi.org/10.1080/24694452.2017.1411245
38. M. Olteanu, J. Randon-Furling, W.A.V. Clark, Segregation through the multiscalar lens. Proceedings of the National Academy of Sciences **116**(25), 12250–12254 (2019). https://doi.org/10.1073/pnas.1900192116

39. G. Le Roux, J. Vallée, H. Commenges, Social segregation around the clock in the Paris region (France). Journal of Transport Geography **59**, 134–145 (2017). https://doi.org/10.1016/j.jtrangeo.2017.02.003
40. C.E. Gehlke, K. Biehl, Certain effects of grouping upon the size of the correlation coefficient in census tract material. Journal of the American Statistical Association **29**(185A), 169–170 (1934). https://doi.org/10.2307/2277827
41. S. Openshaw, The modifiable areal unit problem, in *Quantitative geography: A British view* (1981), pp. 60–69. https://doi.org/10.21428/cb6ab371.5c28c076
42. P.S. Chodrow, Structure and information in spatial segregation. Proceedings of the National Academy of Sciences **114**(44), 11591–11596 (2017). https://doi.org/10.1073/pnas.1708201114
43. B. Barron et al., Extending the Use of Information Theory in Segregation Analyses to Construct Comprehensive Models of Segregation". arXiv preprint arXiv:2212.06980. 2022. https://doi.org/10.1093/oso/9780195139068.003.0018
44. H. Theil, Statistical decomposition analysis: With applications in the social and administrative sciences, in *(No Title)* (1972). https://doi.org/10.1016/0304-4076(75)90038-x
45. S. Kullback, R.A. Leibler, On information and sufficiency. The annals of mathematical statistics **22**(1), 79–86 (1951). https://doi.org/10.1214/aoms/1177729694
46. J. Randon-Furling, M. Olteanu, A. Lucquiaud, From urban segregation to spatial structure detection. Environment and Planning B: Urban Analytics and City Science **47**(4), 645–661 (2020). https://doi.org/10.1177/2399808318797129
47. M. Olteanu, J.-C. Lamirel, When clustering the multiscalar fingerprint of the city reveals its segregation patterns, in *Advances in Self-Organizing Maps, Learning Vector Quantization, Clustering and Data Visualization: Proceedings of the 13th International Workshop, WSOM+ 2019, Barcelona, Spain, June 26-28, 2019 13*. Springer. 2020, pp. 140–149. https://doi.org/10.1007/978-3-030-19642-4_14
48. M.E.J. Newman, Mixing patterns in networks, in *Physical review E* 67.2 (2003), p. 026126.https://doi.org/10.1103/physreve.67.026126
49. M. Olteanu, J. Randon-Furling, Assessing segregation in complex networks through a multi-focal approach, in *CMStatistics*. Vol. 1. 2018, p. 173. https://doi.org/10.1007/3-211-27389-1_123
50. A. Bassolas, V. Nicosia, First-passage times to quantify and compare structural correlations and heterogeneity in complex systems. Communications Physics **4**(1), 1–14 (2021). https://doi.org/10.1038/s42005-021-00580-w
51. S. Sousa, V. Nicosia, Quantifying ethnic segregation in cities through random walks. Nature Communications **13**(1), 1–12 (2022). https://doi.org/10.1038/s41467-022-33344-3
52. L. Gauvin, Modélisation de systèmes socio-économiques à l'aide des outils de physique statistique. PhD thesis. Université Pierre et Marie Curie-Paris VI, 2010
53. J. Von Neumann, The general and logical theory of automata, in *Systems research for behavioral science*. (Routledge, 2017), pp. 97–107. https://doi.org/10.1080/03081078608934933
54. J. Conway et al., The game of life, in *Scientific American* 223.4 (1970), p. 4. https://doi.org/10.1145/1551788.1551852
55. H. Yizhaq, B.A. Portnov, E. Meron, A mathematical model of segregation patterns in residential neighbourhoods. Environment and Planning A **36**(1), 149–172 (2004). https://doi.org/10.1068/a35274
56. D.J. Haw, J. Hogan, A dynamical systems model of unorganized segregation. The Journal of Mathematical Sociology **42**(3), 113–127 (2018). https://doi.org/10.1080/0022250x.2018.1427091
57. L. Gauvin, J. Vannimenus, J.-P. Nadal, Phase diagram of a Schelling segregation model. The European Physical Journal B **70**(2), 293–304 (2009). https://doi.org/10.1140/epjb/e2009-00234-0
58. D. Vinković, A. Kirman, A physical analogue of the Schelling model. Proceedings of the National Academy of Sciences **103**(51), 19261–19265 (2006). https://doi.org/10.1073/pnas.0609371103

59. S. Cheng, Y. Xie, Structural effect of size on interracial friendship. Proceedings of the National Academy of Sciences **110**(18), 7165–7169 (2013). https://doi.org/10.1073/pnas.1303748110
60. S. Currarini, M.O. Jackson, P. Pin, An economic model of friendship: Homophily, minorities, and segregation. Econometrica **77**(4), 1003–1045 (2009). https://doi.org/10.2139/ssrn.1021650
61. A. Douglas Henry, P. Prałat, C.-Q. Zhang, Emergence of segregation in evolving social networks, in *Proceedings of the National Academy of Sciences* 108.21 (2011), pp. 8605–8610. https://doi.org/10.1073/pnas.1014486108
62. X. Dong et al., Segregated interactions in urban and online space, in *EPJ Data Science* 9.1 (2020), p. 20. https://doi.org/10.1140/epjds/s13688-020-00238-7
63. G. Leckie et al., Multilevel modeling of social segregation. Journal of Educational and Behavioral Statistics **37**(1), 3–30 (2012). https://doi.org/10.3102/1076998610394367
64. O. Godechot et al., The great separation: Top earner segregation at work in high-income countries, in *MaxPo Discussion Paper Series* (2020), pp. 20/3. https://doi.org/10.1086/731603
65. M. Olteanu et al., Multidimensional urban segregation: toward a neural network measure. Neural Computing and Applications **32**(24), 18179–18191 (2020). https://doi.org/10.1007/s00521-019-04199-5
66. G. Torrats-Espinosa, Using machine learning to estimate the effect of racial segregation on COVID-19 mortality in the United States, in *Proceedings of the National Academy of Sciences* 118.7 (2021), e2015577118. https://doi.org/10.1073/pnas.2015577118
67. S. Athey et al., Estimating experienced racial segregation in US cities using large-scale GPS data, in *Proceedings of the National Academy of Sciences* 118.46 (2021), e2026160118. https://doi.org/10.1073/pnas.2026160118

Chapter 8
Monocentric or Polycentric City? An Empirical Perspective

Rémi Lemoy

Abstract Do cities have just one or several centers? Studies performing radial or monocentric analyses of cities are usually criticised by researchers stating that cities are actually polycentric, and this has been well known for a long time. Reversely, when cities are studied independently of any center, other researchers will wonder how the variables of interest evolve with the distance to the center, because this distance is known to be a major determinant at the intra-urban scale. Both monocentric and polycentric formalisms have been introduced centuries (respectively, decades) ago for the study of urban areas, and used both on the empirical and the theoretical side in different disciplines (economics, geography, complex systems, physics...). The present work performs a synthesis of both viewpoints on cities, regarding their use in the literature, and explores with data on European urban areas how some cities considered to be the most polycentric in Europe compare to more standard cities when studied through a combination of radial analysis and scaling laws.

8.1 Introduction

Monocentric and polycentric perspectives on cities are frequently opposed in the scientific literature. A radial analysis of cities (monocentric perspective) is quite often automatically criticised by researchers stating that it has been well known for a long time that cities are polycentric—which is certainly true. However, when researchers study cities independently of any center (polycentric perspective), some colleagues will wonder how variables evolve with the distance to the center, because this distance is known to be a major determinant at the intra-urban scale. Both formalisms, polycentric on the one hand, radial or monocentric on the other, have been used for a long time, on the empirical as well as the theoretical side. The aim of this chapter is not to determine which point of view is right (which would be a hopeless endeavour), but rather to explore some empirical facts and understand the relations between these two formalisms and their relation to the empirical reality of cities.

R. Lemoy (✉)
IDEES Laboratory UMR 6266 CNRS, University of Rouen, Mont-Saint-Aignan, France
e-mail: remi.lemoy@univ-rouen.fr

D. Rybski (ed.), *Compendium of Urban Complexity*, Understanding Complex Systems,
https://doi.org/10.1007/978-3-031-82666-5_8

Polycentrism is closely related to urban sprawl, one of the most striking phenomena appearing in the evolution of cities, since the 19th century at least. Many works, for instance [1], have described how households have been able, over time, to live further away from their jobs thanks to the decrease in transport (and commuting) prices through the availability of different means of transport: horse, streetcar, natural combustion engine. As cities sprawl, the distances between nearby cities decrease. Simultaneously, substructures appear, which can be represented with a gravitational mechanism [2]. Cities grow and sprawl through the formation and emergence of secondary centers. New kinds of centrality appear besides main city centers, which can also be linked to fractal approaches (see Chap. 11 or [3]).

We focus in this chapter our empirical analysis on European cities, which we studied in previous work [4, 5] mainly through their radial (center-periphery) profiles of artificial land use—even though we note that most of the literature studies other kinds of data. We study examples of polycentric urban areas, and illustrate this phenomenon using old maps compared to contemporary extensions of cities. French cities are primarily studied since historical data is readily available. Indeed, we use mainly the Cassini map, the first map of the whole kingdom of France, created by several generations of the Cassini family in the 18th century and presented in several figures here.

In the following, we start with a historical approach and a first qualitative observation of some major polycentric areas in Europe (Sects. 8.2 and 8.3). Then we present a quantitative approach using radial analysis and scaling laws and a review of polycentric approaches (Sect. 8.4), before a critical section, open issues and a summary.

8.2 History

On the theoretical side, the oldest quantitative models of cities from urban economics [6] and even the precursor von Thunen model [7] before that have a very monocentric approach to cities, even though this body of literature has shifted to a mainly polycentric view decades ago [1, 8, 9]. However, it is in this disciplinary context that polycentricity was first introduced and studied, both on the empirical and theoretical side. Hence, the empirical side of the economic literature on polycentricity studies rather the location of jobs, between the Central Business District (CBD), employment subcenters and more dispersed locations [10–12]. Indeed, in the Alonso-Muth-Mills (AMM) monocentric model, which is the standard urban economic model [6], the location of the city center is identified with the location of the (main) employment center (CBD). In the subsequent economic literature studying polycentric versions of this AMM model, several employment centers or subcenters are usually considered.

Mentioning pioneers in urban science and planning, we note that in the beginning of the 20th century, Patrick Geddes used the term "conurbation" to describe the process of neighbouring cities tending to mix together as they expand and sprawl [13].

The concept of polycentric city or polycentric urban area really emerges in the scientific literature in the 1970's–80's, when researchers observing suburbanization (urban sprawl) and its link with transportation argue that for an efficient public transport (transit) system, it is better to have secondary centers [14, 15]. A positive view of polycentrism develops at this time. In terms of urban planning, the polycentricity agenda is opposed to urban sprawl. Having several dense centers yields a preservation of open space, as opposed to a continuum of low density urban space. However, it remains clear that the origin of polycentricity lies in urban expansion (and sprawl).

More recently, the complex systems literature close to statistical physics works on the subject of polycentric cities often using rather new data generated by digital footprint [16–18]. City centers and sub-centers are linked with different kinds of activities, not just employment. Explanatory models are also proposed [19].

We note also that different terms have been used in the literature to describe this phenomenon of polycentric urban areas, with different goals and varying success, among which supercity, megaregion, metropolitan region or area, urban region. We rather stick to the term "polycentric" here. We can also note that "polycentrism" and "polycentricity" seem to be equally used with quite similar meanings.

8.3 Empirics

What is a polycentric city? As the name implies, it represents a city with several centers, usually a hierarchy of centers, for instance one main center and several subcenters. However, all cities have subcenters, since economic activities (and human activities in general) are usually clustered in certain areas of cities. Are all cities polycentric, then? To some degree, we can answer in a positive way. This is a legitimate point of view, as it is legitimate to consider that urban areas are organised, in general, around one main central point (monocentric point of view). Any city can be considered as polycentric or monocentric, depending on the way it is studied. A rather close look at local density variations yields a polycentric view, and a rather wide look at the same variables suggests a monocentric view.

Here we wish to go further, and first aim to find good examples of polycentric cities. For comparison, it is probably easier to find good examples of monocentric cities. Cities which are not too close to their neighbours and do not have special physical geography characteristics (coastline, mountainous area, ...) follow quite closely the general rule provided by radial scaling analysis [4, 5]. What would be then a good example of a polycentric city? In the view of researchers and policymakers, there seem indeed to be some cities which are specifically polycentric.

Actually, the examples of polycentric cities mentioned in the literature are usually located in very urbanised regions, which have naturally been expanding or sprawling in the last decades or centuries. Figures 8.1 and 8.2 present the state of urbanization in the 18th century (using data from the Cassini map) for the Lille and Marseille areas respectively, compared with Corine Land Cover 2012 land use data. We see on these maps that cities which were clearly distinct three centuries ago have now spread

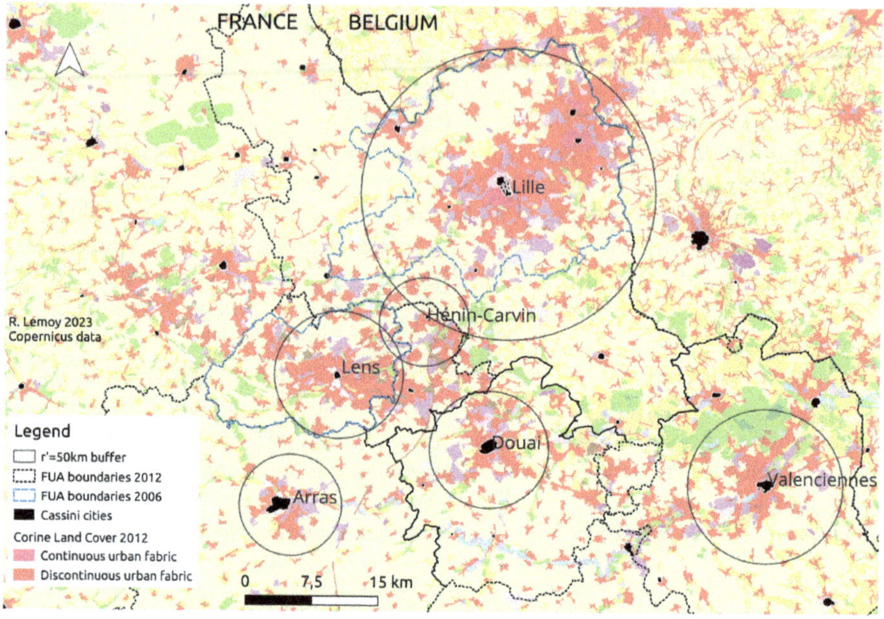

Fig. 8.1 Land use map of the Lille region FUAs (France/Belgium). The standard Corine Land Cover color scheme is used. Buffers represent a rescaled radius $r' = 50\,\text{km}$ around the center. Data from the Cassini map is overlaid, showing cities' extent in the 18th century

so much that the distinctions between their spatial extents are not clear anymore, forming a polycentric urban area—which is also clearly crossing a border in the case of the Lille area. We can also see that contemporary definitions of functional urban areas (FUAs) are contiguous in this case, and subject to changes. We namely present the FUA definitions used for the 2006 and 2012 editions of the Urban Atlas land use dataset. The Lille area is multicentric, since 6 contiguous urban areas are identified by the Urban Atlas 2012 dataset. The Aix-Marseille area is duocentric, the 2 cities being merged in the 2012 version of the data.

In general, cities which are considered good examples of polycentrism in the literature fall into two groups. First, very large cities, like Paris (Fig. 8.7) or London in Europe, with a clear main center, and where subcenters (smaller cities not far away) are also large, and appear more clearly than for smaller cities. In larger cities, one can argue that substructures are also larger, and appear more evidently, thus constituting secondary centers. Second, very dense urban regions, composed of cities of rather similar sizes and close to each other, which have begun to merge through urban sprawl. European examples include the Randstad in the Netherlands (Fig. 8.3—see also [20]—which covers Amsterdam, Rotterdam, Utrecht and the Hague), the Rhine-Ruhr (Fig. 8.4, covering Essen, Dortmund, Wuppertal, Düsseldorf, Mönchengladbach, Köln and Bonn) and Rhine-Main regions (Frankfurt am

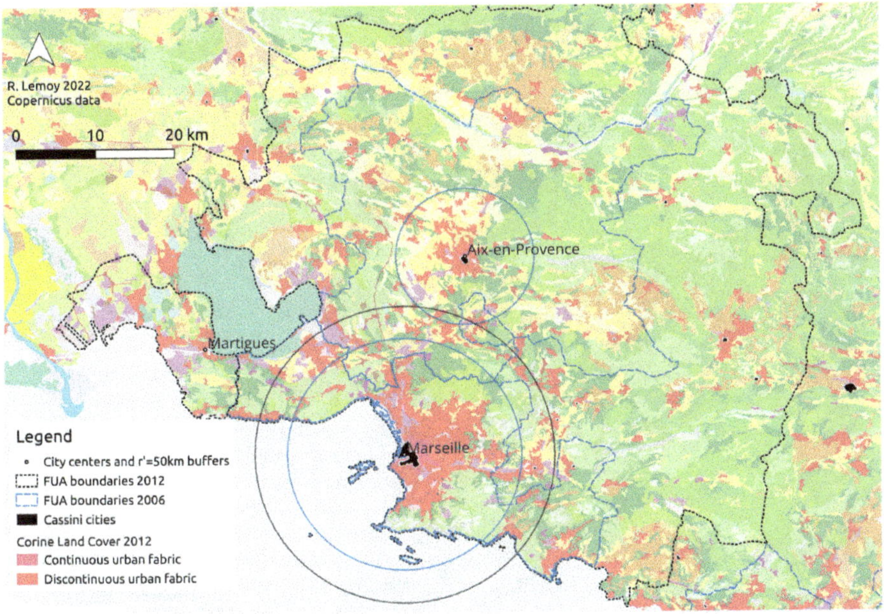

Fig. 8.2 Land use map of Marseille and Aix-en-Provence (France). The standard Corine Land Cover color scheme is used. Buffers represent a rescaled radius $r' = 50$ km around the center. Blue buffers correspond to 2006 data and the black one to 2012. Data from the Cassini map is overlaid, showing cities' extent in the 18th century

Main, Darmstadt, Mainz, Wiesbaden) in Germany, the Lille or Marseille areas in France (Figs. 8.1 and 8.2).

Examples of European polycentric urban regions mentioned in the literature (for instance [21]) also include Greater Dublin and Central Belgium (Brussels, Antwerp, Namur, Ghent, Bruges, Charleroi, Liège), including an area sometimes named the Flemish Diamond. We also add in this group the Central England conurbation (Birmingham, Manchester, Liverpool, Leeds, Sheffield, Leicester, Stoke on Trent, Nottingham, Worcester, Coventry, Lincoln).

All these cities are considered candidates to polycentricity, and marked as such on the map of Fig. 8.6, where the overlap between different urban areas in polycentric regions is apparent. Note that quite many of those polycentric areas are located in the "Blue banana" [22] megalopolis, a highly urbanised area stretching from Liverpool to Milan.

We see from the maps that in Europe at least, polycentric urban areas are, as stated by [23], "urban configurations where a number of historically distinct cities are within contemporary commuting distance". This is especially clear on Figs. 8.1, 8.2, and 8.7, where we can see that contemporary cities are extensions, through urban expansion, of cities already present in the 18th century.

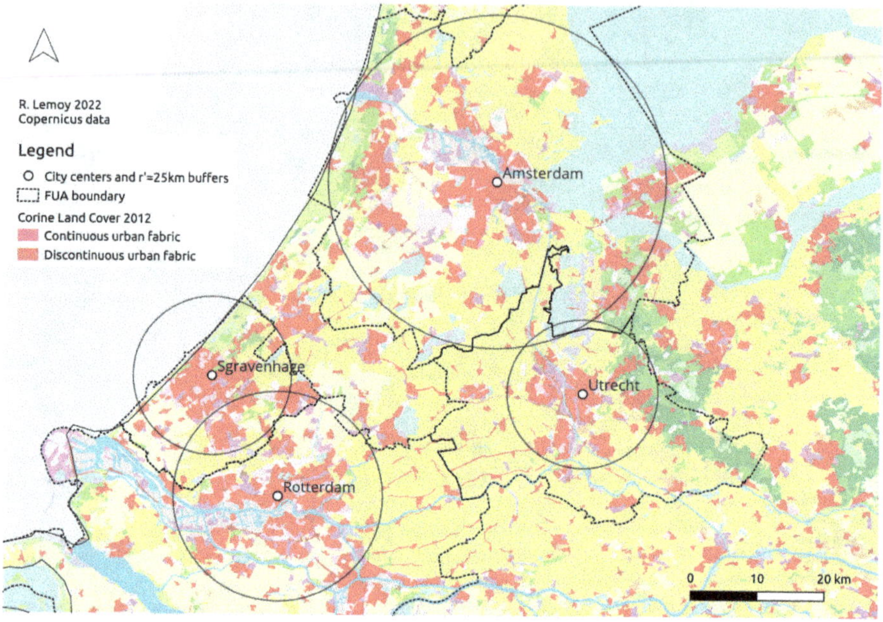

Fig. 8.3 Land use map of the Randstad (the Netherlands). The standard Corine Land Cover color scheme is used. Buffers represent a rescaled radius $r' = 50\,\mathrm{km}$ around the center

8.4 Models

After a first qualitative look at polycentric urban areas, we study them more quantitatively in this section. We first try to formalize more precisely the monocentric and polycentric city frameworks—they remain rather theoretical and unrealistic. For instance, what would be a rigorously monocentric city? In theory, it should have a perfect circular symmetry (around the punctual city center) of the road network, of land uses, water bodies, population density, etc. This is obviously impossible. It would be quite interesting actually (but outside the scope of this chapter) to find which cities are closest to perfect monocentricity. One can however imagine a city which is very convincingly monocentric. Anyway, the monocentric city is rather a theoretical framework which does not easily translate to reality. So we will use it in a pragmatic manner and wonder if many or most cities can be reasonably considered monocentric. Note that monocentric cities can be considered a special case of polycentric cities with only one center. The reverse does not work, obviously.

Monocentric city: A city which clearly dominates its surroundings, without neighbours of comparable size. Secondary centers are small compared to the main center.

Something quite similar can be said of polycentric urban areas (for a different reason). It is rather difficult to find an urban area which is convincingly polycentric.

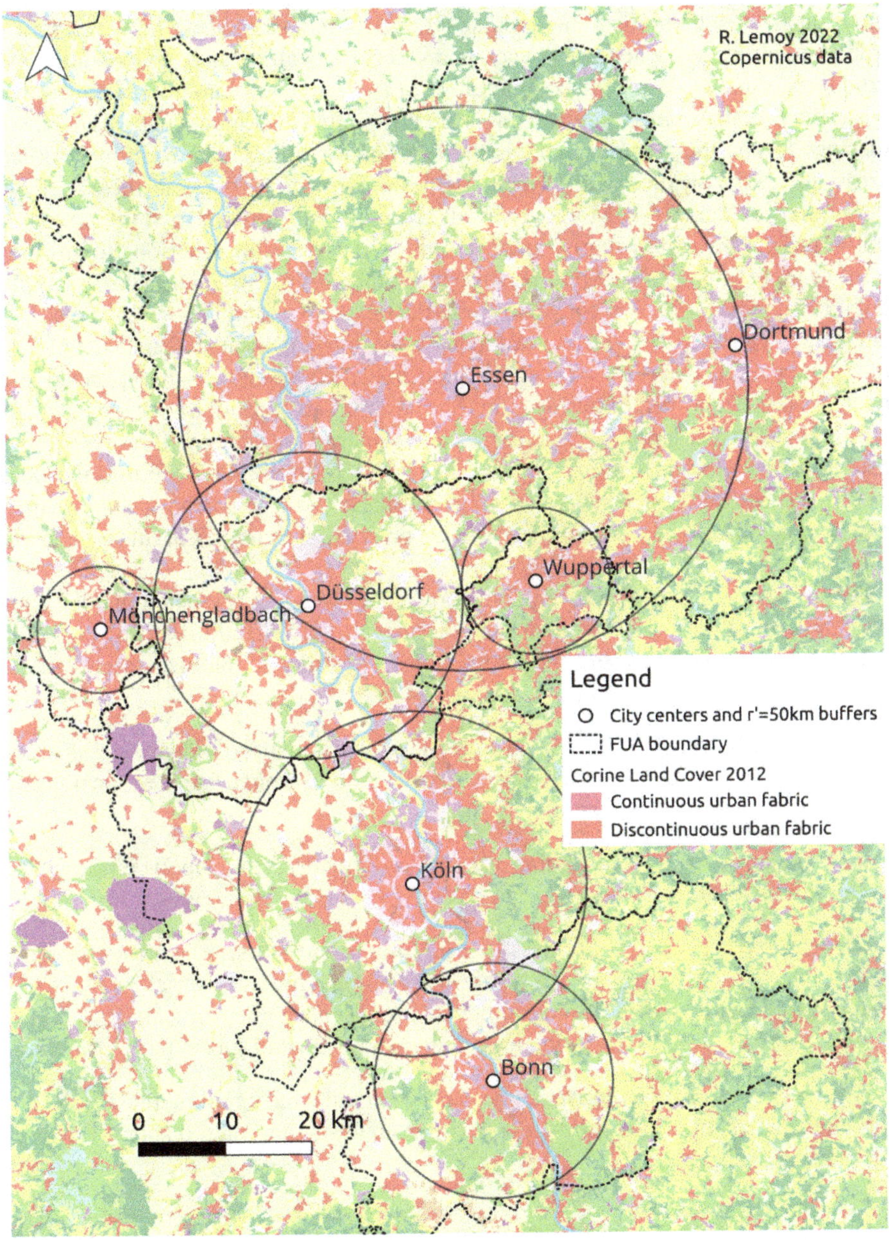

Fig. 8.4 Land use map of the Rhine-Ruhr region (Germany). The standard Corine Land Cover color scheme is used. Buffers represent a rescaled radius $r' = 50$ km around the center

What would this be? A first answer would be simply an urban area which does not have a strictly circular organisation. It should have several centers, of similar or different sizes, which break the circular symmetry. In a strict sense however, this is true of all cities, as argued previously. In a strict sense, all cities are polycentric and no city is monocentric. This is quite close to the general opinion expressed in the scientific literature on the subject: cities are polycentric, not monocentric. However, distance to one main center remains the first spatial differentiation, in most urban areas, and radial profiles of different variables show surprising regularities [4, 5], as developed in Sect. 8.4.1. Moreover, if all cities are polycentric, then the notion of polycentricity is not informative since it is not a discriminating criterion. Some cities must be more monocentric or polycentric than others. What would be a convincingly polycentric city then? We show some classical examples on the maps of Figs. 8.1, 8.2, 8.3, 8.4, and 8.7. But in those examples, polycentricity is rather found at the scale of a region comprising several individual cities, which might be rather monocentric on their own. These neighbouring cities have grown and expanded over long periods of time so as to overlap. The case where these centers have emerged through a rather recent history would be more convincingly polycentric, in the sense that it would not correspond to clearly separated neighbouring cities a few centuries ago which have expanded and recently collided. But we do not find examples of this type in our European dataset.

We find two different cases of polycentric areas though, represented in our examples: the case where a clear dominant center exists (Lille, Marseille, Paris on Figs. 8.1, 8.2, 8.7), and the case where individual cities of rather similar sizes are involved (Randstad or Rhine-Ruhr on Figs. 8.3 and 8.4).

Polycentric urban region	A region where several different individual cities in proximity have started to merge through urban expansion and sprawl.

Consequently, instead of finding very clearly monocentric or polycentric cities, we can define clear monocentric or polycentric perspectives on cities.

Monocentric perspective	Studies cities in a monocentric (radial) manner, focusing on the center-periphery evolution of variables.
Polycentric perspective	Studies cities in a polycentric manner, starting by identifying and counting centers and subcenters.

We now turn to a quantitative analysis of empirical data regarding these polycentric cities, using first a radial or monocentric analysis, and then a short review of polycentric analyses.

8.4.1 Radial Analysis

Many urban phenomena have been studied radially, including of course population density since [24], land use [4, 25], but also building height, air pollution, etc. In previous work, we focus mainly on a radial analysis of urban land use [5], the variable which seems to present most regularity. We use the Urban Atlas 2012 dataset provided

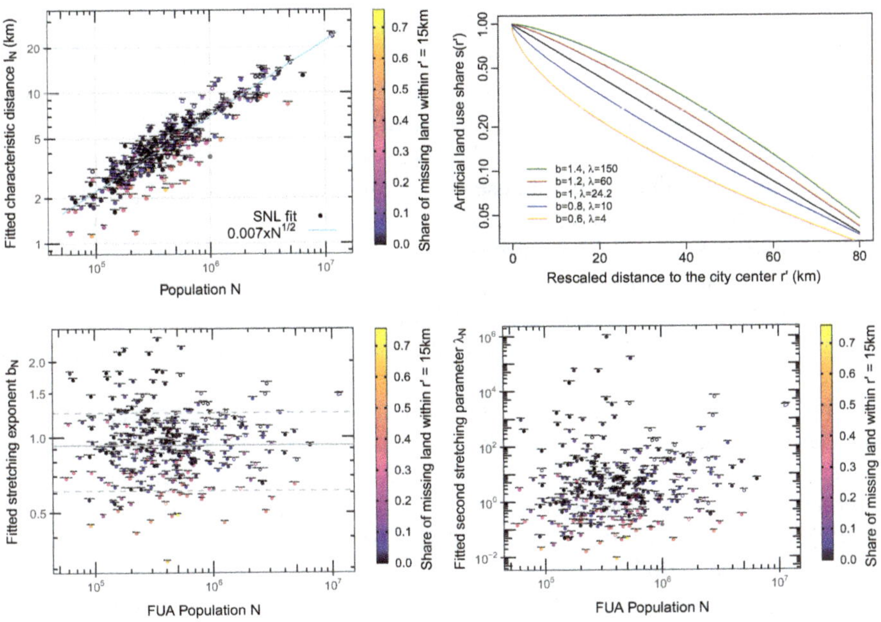

Fig. 8.5 Estimated parameters of the exponential (top left) and stretched exponential (bottom) fits of artificial land use profiles, displayed as functions of the total population N: characteristic distance l_N (top left), stretching exponent b_N (bottom left) and second stretching parameter λ_N (bottom right). Cities which might correspond to polycentric urban regions are marked by a white dot. Top right panel: examples of stretched exponential profiles on a semi-logarithmic graph, with different values of the stretching parameters b and λ_N

by the Copernicus Land Monitoring Service and study only the 300 urban areas which are available both for the 2006 and 2012 versions of the dataset. We show that the radial profile of land use in these urban areas evolves in a very consistent manner from the center (which presents mostly artificial land uses) to the periphery (with mostly natural land uses). More precisely, the share of artificial land $s(r)$ decreases exponentially with the distance to the center r. We use the (main or historical) city hall as the location of the city center, an area where land use is completely artificial, so that $s(0) = 1$. Larger cities, in terms of their population N, have a larger extent in terms of artificial land use, which means that their radial profiles of artificial land use $s_N(r)$ decrease more slowly with distance r to the center than in smaller cities. The radial profile has on average the mathematical form $s_N(r) = \exp(-r/l_N)$, where l_N is the characteristic distance of this exponentially decreasing profile. We fit individual cities' radial profiles using this simple non-linear (SNL) fit (see Fig. 8.5) with a non-linear least squares method.

We also show that this characteristic distance l_N scales like the square root of total city population N. On average, $l_N = l_1\sqrt{N}$, with $l_1 \simeq 7$m the characteristic distance of a (theoretical) city having one inhabitant [5]. This is illustrated on the top panel of Fig. 8.5, and corresponds to a homothetic scaling of radial land use profiles (in the two horizontal physical dimensions), where population N is proportional to artificial surface $S_N \sim \pi l_N^2 \sim N$, since the characteristic radius l_N is proportional to the square root of population, $l_N >\sim \sqrt{N}$. This means that the artificial area per capita is constant and independent of the size of cities. This leads us also to defining a rescaled distance to the center $r' = rk_N$. The rescaling parameter $k_N = \sqrt{N_{\text{Paris}}/N}$ is needed to stretch the radial land use profile $s_N(r)$ of a city of total population N to make it comparable to the profile of Paris, the largest urban area of the dataset with population N_{Paris} (which we use as a reference, without loss of generality). For a city 4 times smaller in terms of population than Paris, like Milano, $k = 2$. Note that this rescaled distance is used for the buffers of Figs. 8.1, 8.2, 8.3, 8.4 and 8.7, which consequently present comparable shares of urban land uses (see also [26] for a quite similar approach).

When we study polycentric cities with this radial approach, we notice, on the top panel of Fig. 8.5, that the polycentric cities mentioned in Sect. 8.3 are not really outliers in this analysis. They follow rather faithfully the general tendency. This is surprising, since one could expect polycentric cities to stand out in a radial (monocentric) analysis. Indeed, polycentricity should break the circular symmetry of cities and yield unusual radial profiles.

To go further, we define an Urban Land Index $\text{ULI} = l_N^2/N$, which gives a measure of the total artificial area (in square meters) per inhabitant. This ULI is independant of city size N and has an average of nearly 50 m^2. It is directly related to the residual of the model presented on the top panel of Fig. 8.5. Its value is actually slightly higher on average for our polycentric candidates cities than for all cities of the database. Its mean value is 49.5 m^2 (resp. 60.1 m^2) for all cities (resp. polycentric candidates), and 54.4 m^2 (resp. 62.5 m^2) when only "continental" cities are considered. Indeed, since coastal cities typically have low values of this Urban Land Index because of the area covered by water, we try to set aside the cities which have more than 20 % missing area (usually covered by water) within their first $r' = 15$ km from the center (that is, 15 km at the scale of Paris).

We see with these values that our polycentric candidates actually stand out from the rest, but not to the point of being outliers, as can be seen on Fig. 8.5. They have higher values of the ULI on average, which is quite understandable since they are located in very urbanised areas, which provides them with more urban land, especially in their periphery, and a higher value of the characteristic distance l_N, compared to other cities of similar population.

We also try to see how much urban areas deviate from the exponentially decreasing radial profile which we observe on average. To this end, we perform a stretched exponential fit of the form

$$s_N(r) \sim \exp(-r^b/\lambda_N)\,,$$

where b is the stretching exponent and λ_N the second fitting parameter. Compared to an exponential function where $b = 1$, values higher than one, $b > 1$, indicate a more concave profile, while values smaller than one, $b < 1$, correspond to a more

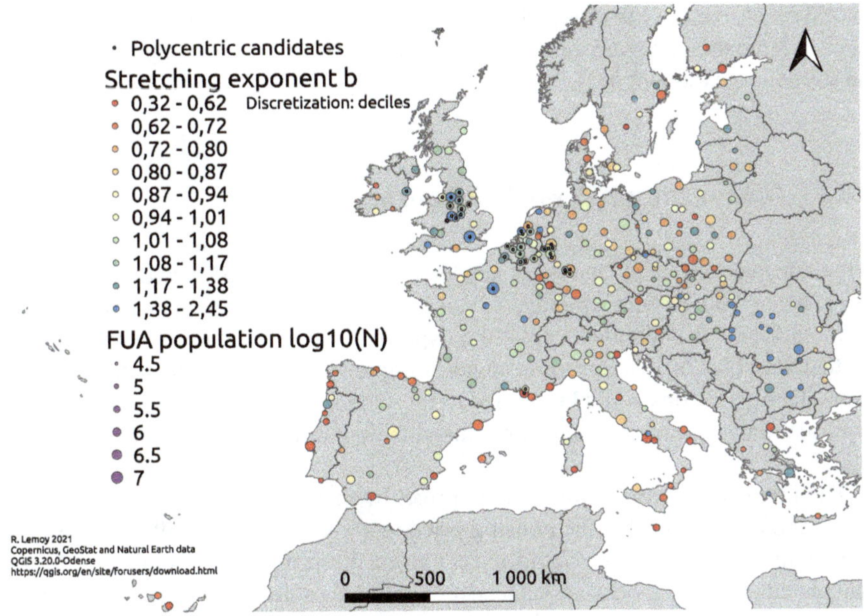

Fig. 8.6 Map of the stretching exponent b of a stretched exponential fit of artificial land use radial profiles for large European urban areas in 2012, showing a rather clear national effect, low values for coastal cities (as expected) and rather large values for cities which are part of polycentric areas. The size of symbols indicates the decimal logarithm of FUA population N. The map is made using Copernicus, GeoStat and Natural Earth data with QGIS 3.20.0-Odense (last accessed 25/06/2024)

convex profile. λ_N is a kind of characteristic distance, but which actually only has the dimension of a distance for $b = 1$. The results are shown in Fig. 8.5, and are also mapped on Fig. 8.6. We observe that the stretching exponent is nicely distributed around $b = 1$, corresponding to an exponential profile, in a lognormal manner (see [5]). Polycentric candidates have clearly higher values of b on average, which means that their radial profile is more convex than the average. This goes with higher values of the second fitting parameter λ_N also.

This can be understood with the top right panel of Fig. 8.5, which displays the shape of stretched exponential radial profiles for different values of b and λ_N, including the average exponential profile ($b = 1$), which appears as a straight line on this semi-logarithmic graph. Indeed, polycentric cities have a higher value of the stretching exponent $b > 1$, which indicates that the share of artificial land is higher in areas close to the city center, compared to cities with no close neighbours. A higher value of the second fitting parameter λ_N is also needed to compensate the higher value of b, as shown by the top right panel of Fig. 8.5. On the contrary, we can note that coastal cities have low values of $b < 1$, and accordingly low values of λ_N. This can be linked to the fact that part of the land is unavailable for artificialization, because it is covered by water. This yields a smaller share of artificial land in central areas of coastal cities, contrary to cities which are part of polycentric areas. We also note that cities in Romania, Bulgaria and the United Kingdom have especially high values of the stretching exponent b.

8.4.2 Intersection of Two Neighboring Cities

What can we expect in theory to appear at the intersection of two neighbouring cities, when studied with a radial analysis? We note in this section that at least three different kinds of phenomena could be expected. These phenomena depend on the studied variables, and we propose some examples of variables which could display them. Such situations could be studied in more detail in future work using pertinent empirical data, in order to gain some insights on the inner workings of polycentric regions.

The first example is that of housing price. In an area between two neighbouring cities, one can expect that the housing price in a given location will be close to the maximum of the price one could expect from the proximity to either of the two considered cities [27], because such an area provides at least the urban opportunities offered by the city with more influence (in terms of size and distance). The same can be expected of urban land uses, and maybe also of population density. Let us note also that [28] performs radial and polycentric analyses of urban land use based on this hypothesis.

One can consider rather the transport cost (in terms of time or distance) to the nearest city center, where job opportunities are concentrated. This variable is of crucial importance in urban economic theory, and studied empirically by [29]. In this case, in the area where two close cities overlap, the transport cost (distance, time, monetary cost) to the nearest center is the minimum of the costs of trips to either of the two neighbouring city centers.

A third example is given by air pollution, for instance the concentration of nitrogen dioxide NO_2, as measured by satellite as an average over a tropospheric column [30]. This air pollution is mainly caused by traffic, through combustion of fuel. As such, it is concentrated in urban areas, but it disperses over regional scales. In this case, in the considered overlapping area of two close cities, one can expect the air pollution to be governed by the sum of the concentrations generated by both cities. A summary is given in Table 8.1.

8.4.3 Polycentric Analysis

We focus here on a radial analysis of polycentric cities, since we use this approach repeatedly in previous research. Indeed, a monocentric city is quite easy to define. The location of the city center is not a very hard problem. For instance, the location of the main or historical city hall can be chosen [4, 31, 32]. The definition of a

Table 8.1 Interactions at the intersection between two neighbouring urban areas

Variable	Intersection	Justification
Housing price, urban land uses, population density	Maximum	Non-additive phenomenon
Transport cost to the center	Minimum	Adding centers adds opportunities and increases accessibility
Air pollution	Sum	Additive phenomenon

polycentric city is much more complex. Once polycentricity is assumed, one needs to define centers and subcenters in empirical settings. This problem is already hard enough to remain quite open after a few decades of research on polycentric cities. There are many competing definitions of centers and subcenters, and methods to locate them.

Thus, a body of scientific literature was built on the analysis of polycentric cities directly. This multidisciplinary literature focuses in particular on the identification and count of centers and subcenters through different techniques. In this strand of literature, radial analysis is usually not considered. The polycentric structure of cities is a preliminary assumption. We review several works which are presumably representative of the different approaches developed in different disciplines. Studies start in economics, where centers are associated to employment centers, following the standard urban economic model. McDonald [33] defines centers using cutoffs on employment density and employment population ratio. This approach with cutoffs is quite wide-spread and used by subsequent literature. For instance, [10] define centers as contiguous "zones" (transportation analysis zones) of sufficiently high employment density and total employment. A drawback of using cutoffs is the introductions of rather arbitrary parameter values, and a part of the literature, in different disciplines at the crossroads of an emerging science of cities, strives to develop non-parametric methods. In an urban area having one main center (CBD), [11], and other works, identify subcenters as significant residuals of the employment density radial profile. This profile has an exponential form, and they look for residuals of a locally/geographically weighted regression of logarithmic employment density (other works use different regression methods). The theoretical model of reference in economics in this case is given by [34], who propose a rather complex extension of the standard urban economic (AMM) model, which predicts, regarding polycentricity, that the number of subcenters grows with population size—a quite understandable stylized fact, which is verified empirically. In geography, [35] study local autocorrelation of logarithmic employment density in Belgium and find that polycentrism is weak. Arribas-Bel and Sanz-Gracia [12] do a related work on US cities at different times, without logarithm. Note that this research in economics, geography and regional science is ongoing, proposing new measures of polycentricity.

More recently, a literature emerged in complex systems with methods sometimes inspired by statistical physics, using new datasets generated by digital footprint. Roth et al. [16] identify centers in London based on real-time public transport data. Louail et al. [17] find hotspots of human activities using mobile phone data in Spain. Their nonparametric method yields a number of subcenters which scales sublinearly with city size, in accordance with a related theoretical model [19]. This rich literature is interesting in order to delineate centers and subcenters, but still leaves room for more empirical work on the question of monocentric versus polycentric cities. Linking these (empirical) polycentric approaches with radial ones in order to obtain new insights on cities and systems of cities would be an interesting research perspective, which is rather unexplored to our knowledge.

8.5 Criticism

Despite discussions in the scientific literature [36, 37], polycentricity has become a normative concept in urban planning, in particular in Europe [38, 39], but also on other continents. Indeed, it is supposed to be associated with increased economic outputs related to agglomeration economies (see Chap. 3). This is also linked with the positive literature about polycentrism from the 1970's–80's [14, 15], where this phenomenon is seen as an efficient way to develop public transportation, with a system of hubs and spokes associated with centers and subcenters. Polycentricity is also seen as a desirable, decentralized model of governance.

However, if in Europe polycentric regions are mostly cities sprawling and merging, as we saw from this study, this policy goal resembles rather a desire to steer the larger, more fundamental and unavoidable process of urban expansion and sprawl. Then, this issue of urban sprawl should probably rather be tackled directly. It is associated with negative outcomes such as losses of arable land, of biodiversity, air pollution, climate change linked to emissions of greenhouse gases, economic vulnerability of periurban households which are dependant on cars and oil-based fuels. Note that this is quite unsettling since polycentricity itself, which we saw here as directly related to urban sprawl, is seen as a positive policy goal. Urban sprawl is usually explained, in particular in the economic literature, by the decreasing price of transport, in particular commuting to work (which is concentrated around city centers).

With this, the question of polycentricity comes back to the much debated question of transport costs. If one considers benefits over the long term (a few decades or centuries), transport and commuting costs are probably too small nowadays, despite recent increases in fuel prices. For instance, it is probably not worth it right now to use a car instead of a bicycle for short urban trips (of less than 3 km for instance), when one takes into account the fact that people in a few generations will have a much more difficult access to those precious fossil fuels (like oil) on which our current economic activity is grounded—in addition to living in a world with a quite different climate. We refer the reader to the discussion in Chap. 4.

We should also mention that the word "polycentric" is used in different contexts related to urban planning, and especially in the context of governance, where it is seen as a positive goal (in line with decentralisation). For instance, a polycentric governance of the European Union (EU)—at this scale, the Blue Banana region mentioned above is the center [40]. This brings us to a discussion on spatial scales, since polycentricity is mentioned at the scale of the EU (continental scale), at the scale of a conurbation like the Rhine-Ruhr region (see Fig. 8.4), with Essen as a possible center (these are both inter-urban scales, the scales of systems of cities) and also at the intra-urban scale (the urban areas of Paris or London for instance).

Are we then speaking of a polycentric city or of a system of cities in proximity? It is not easy to set the limit between both cases, and the urban planning literature is not always clear in this respect. Many studies are carried out at the inter-urban scale, but we have considered polycentricity here mainly at the intra-urban scale. This constitutes the major research issue in the domain, which is simply the robust definition of the system to be studied.

8.6 Open Issues

Cities which are usually described as polycentric, and which we rather describe here as conurbations, are a real scientific challenge mainly when we want to isolate them, that is, in terms of their definition or delineation. Is there a main, central city to whose urban area the whole region could be considered to belong? If we are rather in a case where several urban areas are mixed, which city does this specific neighbourhood belong to?

This question of city (or rather urban area) definition is a major research question and open problem in urban science, and polycentric urban regions are a central challenge in this respect. Municipalities (administrative definitions) are usually not appropriate for scientific studies since they are too arbitrary (really small and numerous in France for instance, larger in Germany). Functional Urban Areas (FUAs) are usually considered a good definition, but many other methods exist, such as morphological definitions.

This issue is illustrated by Fig. 8.7, which presents the results of the definition of the Functional Urban Area (FUA) of Paris using different methods and data. The aim is the same in each case. The FUA is defined based on home-work commuting trips, which indicates the functional extent, in terms of employment area, of cities. However, the results differ significantly, especially in the peripheral parts of the urban area. We can also refer to Fig. 8.2, where two distinct FUAs (Marseille and Aix-en-Provence) in the 2006 edition of the Urban Atlas dataset have been considered as just one in its 2012 edition. Or similarly, different extents of urban areas with the same datasets can be seen on Fig. 8.1.

This issue has of course many implications, as it influences most studies of cities, for instance the distribution of city sizes (Zipf's law for cities, see Chap. 1). Should the secondary cities (centers) around a large city be counted as individual cities, or just be integrated in the large city? With the first option, the city size distribution has a shorter tail and more cities are counted in the bulk of small urban areas. In the

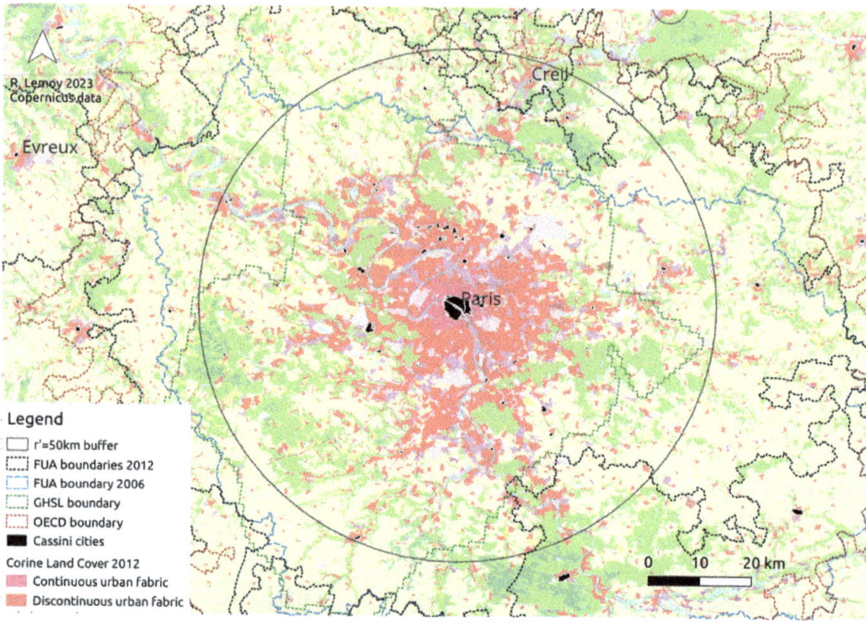

Fig. 8.7 Land use map of the Paris FUA (France). The standard Corine Land Cover color scheme is used. The buffer represents a radius $r' = 50$ km around the center. FUA boundaries are presented: Urban Atlas in its 2006 (blue) and 2012 (black) versions, GHSL (green) and OECD (red). Data from the Cassini map is overlaid, showing cities' extent in the 18th century

second option, the tail tends to be longer with large cities becoming even larger. This has a direct influence on the exponent governing the power-law tail distribution.

In general, the question of the city definition is probably not enough considered in particular in 0-dimensional analyses of cities, where definitions of urban areas from statistical institutes are usually taken as inputs and not really questioned—we should still mention notable exceptions like [41]. For instance, all values of (urban) scaling exponents found in the literature (see Chap. 2) are dependant on this issue of polycentricity and city definition. Studies of cities using percolation techniques also face the problem of polycentric urban regions, which blur the boundaries between individual cities.

We also mention that we mostly considered urban land use as the variable of interest in this chapter (as illustrated by different maps and analyses), but all urban variables are concerned by this issue of polycentricity. And land use happens to be the variable which has the highest spatial scaling exponent (0.5 or 1/2) among the variables studied with radial and scaling analysis so far [4]. Population density has a smaller exponent, around 1/3, and housing price an even smaller one (1/5). As a consequence, the characteristic spatial scales of housing price and population density variations are notably larger than those of land use [27], and polycentricity is an even more serious issue when studying these variables. The influence of a small or medium-size city on housing price and population density is clearly further-reaching than on land use. Land use could actually be the best variable to delineate individual cities, which is rather well illustrated by the different maps presented in this chapter, where individual cities can be quite well distinguished even in clearly polycentric areas.

8.7 Summary

In this chapter, we explore monocentric and polycentric cities in Europe mainly using radial (center-periphery) analysis and spatial scaling laws. We observe that polycentric regions are the result of the urban expansion and sprawl of cities which have been neighbours for centuries and start to merge recently. They can have one clear main center (London, Paris) or a less hierarchical distribution of centers (Randstad, Rhine-Ruhr).

Their radial profiles of urban land use are not really exceptional compared to average cities, although these polycentric cities are on average more urbanised compared to cities of the same size. They also have a more concave radial profile of artificial land use, indicating that more land is urbanised, in the central areas of the city, compared to cities of the same size. This can be opposed to coastal cities, which tend to have more convex profiles because of the areas covered by water.

One key issue which scientists are facing concerning polycentric urban areas is the still open problem of city definition. It is indeed quite difficult to define an urban area in general (where should one set the limit? which criteria should be used?), and more particularly, to separate the individual cities which constitute a polycentric region.

This problem is currently gaining importance as more cities are expanding, sprawling and merging worldwide, constituting more polycentric urban regions. One can wonder how this will go on in the mid or long term. Is the future of urban areas made of ever more urban expansion, which would make this issue of polycentricity even more prominent? Or will cities face a contraction due to new factors (like rising prices of energy, non-renewable resources and transport)?

To conclude, all cities can be considered polycentric, but the radial or monocentric viewpoint is still quite interesting as a simple robust intra-urban model. If only one variable should be kept for a rough or parsimonious intra-urban analysis, very clearly distance to the center is the variable to keep. This discussion amounts largely to a question of viewpoint. When you look at urban areas from far away, you usually identify one main center. But if you look more closely, you see a polycentric structure. The monocentric and polycentric formalisms of cities have been studied for decades already, and they are probably going to coexist for some time more.

Acknowledgements The author acknowledges insightful discussions with Geoffrey Caruso and Gaëtan Laziou, the help of Walid Rabehi and Julien Perret for historical data, and helpful comments from the editor. This study was funded by ANR throught the GreenLand project and Normandy region through RIN project SUCHIES.

References

1. A. Anas, R. Arnott, K.A. Small, Urban spatial structure. J. Econ. Lit. **36**, 1426 (1998). https://www.jstor.org/stable/2564805
2. Y. Li, D. Rybski, J.P. Kropp, Singularity cities. Environ. Plan. B: Urban Anal. City Sci. **48**(1), 43–59 (2021). https://doi.org/10.1177/2399808319843534
3. G. Caruso et al., Morphological similarities between DBM and a microeconomic model of sprawl. J. Geogr. Syst. **13**, 31–48 (2011). https://doi.org/10.1007/s10109-010-0131-7
4. R. Lemoy, G. Caruso, Evidence for the homothetic scaling of urban forms. Environ. Plan. B: Urban Anal. City Sci. **47**(5), 870–888 (2020). https://doi.org/10.1177/2399808318810532
5. R. Lemoy, G. Caruso, Radial analysis and scaling of urban land use. Sci. Rep. **11**(1), 1–8 (2021). https://doi.org/10.1038/s41598-021-01477-y
6. M. Fujita, *Urban Economic Theory* (Cambridge University Press, 1989). https://doi.org/10.1017/cbo9780511625862
7. J. Heinrich Von Thünen, Der isolierte Staat in Beziehung auf Nationalökonomie und Landwirtschaft, in *Gustav Fischer, Stuttgart (reprinted 1966)* (1826). https://doi.org/10.5962/bhl.title.24798
8. P.G. Hartwick, J.M. Hartwick, Efficient resource allocation in a multinucleated city with intermediate goods. Q. J. Econ. **88**(2), 340–352 (1974). https://doi.org/10.2307/1883077
9. M.J. White, Firm suburbanization and urban subcenters. J. Urban Econ. **3**(4), 323–343 (1976). https://doi.org/10.1016/0094-1190(76)90033-4
10. G. Giuliano, K.A. Small, Subcenters in the Los Angeles region. Reg. Sci. Urban Econ. **21**(2), 163–182 (1991). https://doi.org/10.1016/0166-0462(91)90032-i
11. D.P. McMillen, Nonparametric employment subcenter identification. J. Urban Econ. **50**(3), 448–473 (2001). https://doi.org/10.1006/juec.2001.2228
12. D. Arribas-Bel, F. Sanz-Gracia, The validity of the monocentric city model in a polycentric age: US metropolitan areas in 1990, 2000 and 2010. Urban Geogr. **35**(7), 980–997 (2014). https://doi.org/10.2139/ssrn.2474858
13. P. Geddes, *Cities in Evolution* (Williams and Norgate, London, 1915)
14. J.B Schneider, *Transit and the Polycentric City* (The Administration, 1981)
15. P. Gordon, H.L. Wong, The costs of urban sprawl: some new evidence. Environ. Plan. A **17**(5), 661–666 (1985). https://doi.org/10.1068/a170661
16. C. Roth et al., Structure of urban movements: polycentric activity and entangled hierarchical flows. PloS One **6**(1), e15923 (2011). https://doi.org/10.1371/journal.pone.0015923
17. T. Louail et al., From mobile phone data to the spatial structure of cities. Sci. Rep. **4**(1), 1–12 (2014). https://doi.org/10.1038/srep05276
18. M. Lenormand et al., Comparing and modelling land use organization in cities. R. Soc. Open Sci. **2**(12), 150449 (2015). https://doi.org/10.1098/rsos.150449
19. R. Louf, M. Barthelemy, Modeling the polycentric transition of cities. Phys. Rev. Lett. **111**(19), 198702 (2013). https://doi.org/10.1103/physrevlett.111.198702
20. D. Broitman, E. Koomen, The attraction of urban cores: densification in Dutch city centres. Urban Stud. **57**(9), 1920–1939 (2020). https://doi.org/10.1177/0042098019864019
21. P.G. Hall, K. Pain, The polycentric metropolis: learning from mega-city regions in Europe (Routledge, 2006). https://doi.org/10.4324/9781849773911-11
22. R. Brunet, *Les villes européennes: rapport pour la DATAR* (La documentation française, RECLUS, Montpellier, 1989)
23. R.C. Kloosterman, S. Musterd, The polycentric urban region: towards a research agenda. Urban Stud. **38**(4), 623–633 (2001). https://doi.org/10.1080/00420980123921
24. C. Clark, Urban population densities. J. R. Stat. Soc. Ser. A (General) **114**(4), 490–496 (1951). https://doi.org/10.2307/2981088
25. M. Guérois, D. Pumain, Built-up encroachment and the urban field: a comparison of forty European cities. Environ. Plan. A **40**(9), 2186–2203 (2008). https://doi.org/10.1068/a39382
26. W.R. Tobler, Satellite confirmation of settlement size coefficients. Area **1**(3), 30–34 (1969). https://doi.org/10.3133/ofr69285

27. G. Laziou, R. Lemoy, M. Le Texier, Radial analysis and scaling of housing prices in French urban areas. Environ. Plan. B (2024). https://doi.org/10.1177/23998083241281890
28. L. Jiao, Urban land density function: a new method to characterize urban expansion. Landscape Urban Plan. **139**, 26–39 (2015). https://doi.org/10.1016/j.landurbplan.2015.02.017
29. E. Mennicken, Analysis of urban forms in Europe using radial road network simulations. PhD thesis. University of Luxembourg, Esch-sur-Alzette, Luxembourg (2022). https://hdl.handle.net/10993/51924
30. M. Schindler, Spatial modelling of feedback effects between urban structure and traffic-induced air pollution - Insights from quantitative geography and urban economics. PhD thesis. University of Luxembourg, Esch-sur-Alzette, Luxembourg (2016). https://hdl.handle.net/10993/28610
31. S.G. Wilson, *Patterns of Metropolitan and Micropolitan Population Change, 2000 to 2010* (US Department of Commerce, Economics and Statistics Administration, US Census Bureau, 2012)
32. K. Walker, Locating neighbourhood diversity in the American metropolis. Urban Stud. 0042098016643481 (2016). https://doi.org/10.1177/0042098016643481
33. J.F. McDonald, The identification of urban employment subcenters. J. Urban Econ. **21**(2), 242–258 (1987). https://doi.org/10.1016/0094-1190(87)90017-9
34. M. Fujita, H. Ogawa, Multiple equilibria and structural transition of non-monocentric urban configurations. Reg. Sci. Urban Econ. **12**, 161–196 (1982). https://doi.org/10.1016/0166-0462(82)90031-x
35. F. Riguelle, I. Thomas, A. Verhetsel, Measuring urban polycentrism: a European case study and its implications. J. Econ. Geogr. **7**(2), 193–215 (2007). https://doi.org/10.1093/jeg/lbl025
36. N. Bailey, I. Turok, Central Scotland as a polycentric urban region: useful planning concept or chimera? Urban Stud. **38**(4), 697–715 (2001). https://doi.org/10.1080/00420980123684
37. E. Meijers, Measuring polycentricity and its promises. Eur. Plan. Stud. **16**(9), 1313–1323 (2008). https://doi.org/10.1080/09654310802401805
38. European Consultative Forum on the Environment and Sustainable Development. *The European Spatial Development Perspective (ESDP): Comments and Recommendations from the European Consultative Forum on the Environment and Sustainable Development, January 1999*. Office for Official Publications of the European Communities (1999)
39. S. Dühr, Potentials for polycentric development in Europe: The ESPON 1.1.1 project report. Plan. Pract. Res. **20**(2), 235–239 (2005). ISSN: 1360-0583. https://doi.org/10.1080/02697450500414728
40. A. Faludi, Polycentric territorial cohesion policy. Town Plan. Rev. **76**(1), 107–119 (2005). https://doi.org/10.3828/tpr.76.1.9
41. C. Cottineau et al., Diverse cities or the systematic paradox of urban scaling laws. Comput. Environ. Urban Syst. **63**, 80–94 (2017). https://doi.org/10.1016/j.compenvurbsys.2016.04.006

Chapter 9
Urban Climate Through the Lens of Complex System Science

Gabriele Manoli

Abstract Two of the fields that have emerged in the science of cities are urban climatology, historically focused on local to metropolitan-scale meteorological processes, and urban complexity science, focusing on population growth and emergent scaling behaviours across multiple cities. In this chapter, obstacles and opportunities for effectively integrating these two research tracks are discussed. Specifically, it is shown that the integration of fine and coarse-grained approaches from meteorology, hydrology, and statistical physics offers an effective theoretical ground to describe complex cities-biosphere interactions and disentangle the factors leading to the emergence of distinctive urban climates. Such a convergence of disciplines is key to address the complexity of cities and the challenges posed by global urbanization.

9.1 Introduction

Since the birth of science in Western philosophy, tension between two approaches exemplified by Plato and Aristotle [1] continues unabated today. The former deals with abstract and idealized concepts and tends to seek similarities, general patterns, and unifying laws while the latter prefers "down-to-earth" observations, focusing on differences, specific details, and local peculiarities [1, 2]. This dichotomy is often associated with the different perspectives of detail-oriented physics and function-oriented biology [2, e.g.], but it is also apparent in today's urban research agendas. On one hand, the field of urban climate focuses on measuring and modeling mass, momentum, and energy exchanges in urban environments, taking into account the spatio-temporal peculiarities of different building materials, urban geometries, and micro-climatic conditions [3–5, e.g.], often in the form of city-specific studies that are challenging to generalize. On the other, the field of urban complexity, or city analytics [6–10, e.g.], focuses on general patterns and emergent behaviours, neglecting fine-scale peculiarities and often without strong connection with first principles. Such

G. Manoli (✉)
Laboratory of Urban and Environmental Systems, École Polytechnique Fédérale de Lausanne, Lausanne, Switzerland
e-mail: gabriele.manoli@epfl.ch

D. Rybski (ed.), *Compendium of Urban Complexity*, Understanding Complex Systems,
https://doi.org/10.1007/978-3-031-82666-5_9

macroscopic properties of cities are the outcome of complex interactions occurring at the microscopic (local) level that generate seemingly universal power law relations—resembling allometric scaling of living organisms [11, 12]—between coarse-grained indicators Y (e.g. urban infrastructure size, activity) and city "mass" N, i.e. $Y(t) \sim N(t)^{\beta}$, where β is the scaling exponent and N is generally approximated by the city population [6, e.g.].

A first hint on possible links between these two perspectives appeared in the '70s when Oke [13] reported that the intensity of nocturnal canopy urban heat islands (UHIs)—i.e. the air temperature increase observed in cities compared to the rural surrounding—is related to the 1/4th power of N, i.e.

$$\Delta T_{a,max} \sim N^{1/4} \cdot U^{-1/2}, \tag{9.1}$$

where $\Delta T_{a,max}$ is the difference between rural and urban air temperatures (maximum nighttime value under cloudless skies) and U is the regional wind speed. While the dependence on $U^{-1/2}$ points to aerodynamic controls on the exchange of heat [14, 15, e.g.], the $N^{1/4}$ scaling suggests similarity in population size and city structure when analyzed from the perspective of $\Delta T_{a,max}$. In fact, N was used as a proxy for urban infrastructure size and later replaced with more direct measures of urban form, such as the ratio between building height h and street width W or the sky view factor ψ_{sky} in the center of the settlement [3, 16] since they proved more general across multiple sites. However, the UHI based on air temperature, and to a lesser extent to surface skin temperature differences, are the integration of processes that occur from the facet to the regional scales. Hence, from a microclimatological perspective, empirical or otherwise derived relations such as Eq. (9.1) overlook the complexity and heterogeneity of cities, and may seem unwarranted since a large body of literature over the past few decades has advanced understanding and capacity to simulate urban microclimate. This literature has provided a mechanistic description of UHIs including the energy and water balance of both urban and rural surfaces, resolving buildings walls, roofs, roads, trees, grass, and bare soil [4, 17–19, e.g.]. Considering a reference urban volume and neglecting lateral transport processes, the fundamental equations of these approaches are the conservation of energy and mass (water):

$$Q^* + Q_F = Q_H + Q_E + \Delta Q_S \tag{9.2}$$

$$R + I = ET + L + \Delta S, \tag{9.3}$$

where $Q^* = Q_{sw}(1-\alpha) + Q_{lw}^{\downarrow} - Q_{lw}^{\uparrow}$ is the net all-wave radiation, Q_{sw} is the incoming shortwave radiation, α is the mean surface albedo, $Q_{lw}^{\downarrow}$ and $Q_{lw}^{\uparrow}$ are the incoming and outgoing long-wave radiations, Q_F the anthropogenic heat flux, Q_H and Q_E the sensible and latent heat fluxes, respectively, and ΔQ_S is the net storage heat flux. The change in water storage ΔS (for both soil and urban surfaces) depends on rainfall R, irrigation I, evapotranspiration ET, and surface runoff L. Clearly, Eqs. (9.2)–(9.3) are a simplified and lumped representation of complex

3D urban/rural geometries, the description of which would require the solution of heat, mass, and momentum conservation equations by means of computational fluid dynamics (CFD) approaches at scales of fractions of millimeters (e.g. Kolmogorov microscale) and fractions of seconds [20–22, e.g.]. Specifically, the physics of air flow and associated exchanges of heat and humidity within complex three-dimensional urban structures are given by the Navier-Stokes and conservation of heat and moisture equations [20, 21, 23, e.g.].

The aforementioned equations illustrate the diversity of modeling approaches that have been developed to describe the wide range of processes and spatio-temporal scales characterizing urban climate (see Fig. 9.1). Both detailed and simplified models aim at explaining some aspects of real landscapes but they focus on different spatio-temporal scales. On one hand, CFD simulations resolve urban geometries at high resolutions (Fig. 9.1D), on the other urban scaling laws (e.g. Eq. (9.1) and Fig. 9.1B) focus on the general relations that complex systems exhibit at the macroscopic level, where details are smoothed over and emergent patterns arise (Fig. 9.1B). Urban canyon schemes (i.e. Eqs. (9.2)–(9.3)) can be considered mid-level complexity models (Fig. 9.1C) as they are based on physically-based equations applied on spatial scales that are larger than the underlying physical processes. Hence, many approaches often deemed "detailed", such as the urban canyon approximation [3, e.g.] or the use of local climate zones [24, e.g.], are to some degree a coarse-grained representation of far more complex urban structures based on bulk parameterizations of finer scales processes.

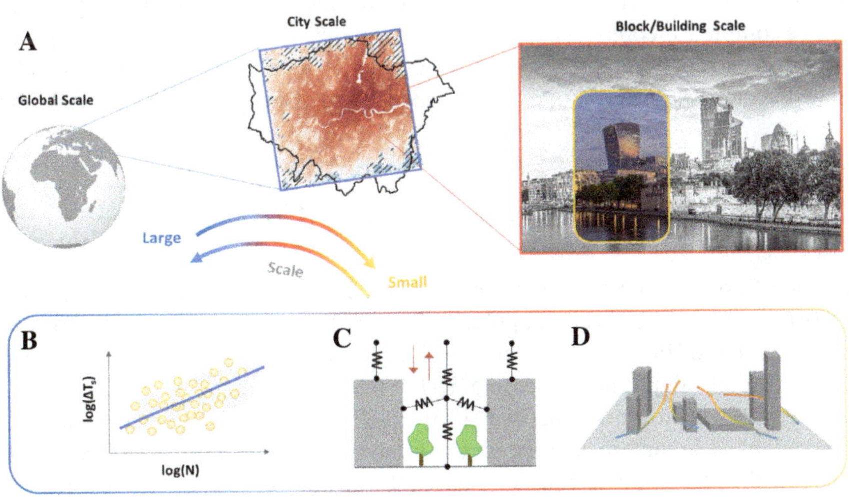

Fig. 9.1 **A** Global, city, and neighborhood/building scales relevant for urban climate studies. Conceptual illustration of different modeling approaches: **B** coarse-grained model, **C** urban canyon schemes, and **D** detailed CFD simulations. *Source* (gray globe and London photo): Wikimedia Commons, CC0 licence

Here, the significance of pursuing different routes and bridging these perspectives is offered as their integration can provide a compass to addressing the complexity of urban systems across scales. The focus remains on the computation of urban-rural differences in canopy and surface temperatures, as their city-to-city variations have been recently explained by combining concepts from statistical physics, urban climate, and complexity science [25–27, e.g.]. It is shown that urban scaling laws, when coupled to energy budget and radiative transfer principles, provide the logical basis for coarse-grained representations of urban-induced warming, thus advancing our understating of cities-biosphere interactions across diverse cities and climatic regions.

9.2 Urban Heat Islands and Complexity Science

Given their implications for energy consumption [28], climate change adaptation plans [29], and public health [30], UHIs have been widely studied over the past decades, considering both air and surface temperature measurements [13, 14, 16, 31–34, e.g.]. Air UHIs—generally referred to as canopy UHIs—are most intense during night-time, while the intensity of surface UHIs, hereafter referred to as SUHIs, generally peak during daytime [4]. The physical drivers of urban warming and its canonical diurnal patterns are reasonably well known, both in terms of air and surface temperatures [4]. However, as detailed in the next sections, ground and satellite observations have revealed global and seasonal trends of UHIs that, until recently, lacked a complete theoretical explanation [14, 16, 35, e.g.]. In this section we present a brief overview of complex system theories applied to urban climate studies and we review the coarse-grained models employed to describe canopy and surface UHIs.

9.2.1 Coarse-Grained Models of Canopy UHIs

The rise of urban complexity research and the discovery of scaling laws linking a variety of urban metrics to city population [6, e.g.] brought the relationship proposed by Oke [13] (i.e. Eq. (9.1)) under new light. However, only a few studies have addressed urban climate processes (e.g. UHIs) from the lens of complexity theories. A recent example is the work by Li et al. [36] who employed urban climate simulations for a variety of urban areas generated with an urban growth model to explore the influence of density and morphology on UHI intensity. For monocentric cities they showed that the daily maximum canopy UHI can be explained by a linear combination of logarithmic city area (A) and logarithmic gross building volume S, that is:

$$\Delta T_{a,max} = a_1 \log A + a_2 \log S + a_3 \,, \tag{9.4}$$

where a_1, a_2, and a_3 are fitting parameters. To extend this result to cities of various shapes, Eq. (9.4) can be generalized by considering the structure of urban sites and their relative distances [36]:

$$\Delta T_{a,max} = c_1 \log A + c_2 \log S + c_3 D + c_4 , \tag{9.5}$$

with

$$D = \frac{1}{N} \sum_{j}^{N} \sum_{i \neq j}^{N} \left(\frac{f_{u,i} w_i}{W_i} \right) d_{i,j}^{-3/2} , \tag{9.6}$$

where c_1, c_2, c_3 and c_4 are model parameters and $d_{i,j}$ is the euclidian distance between urban sites i and j, while f_{ui}, w_i, and W_i are, respectively, the urban fraction, building volume, and street canyon width of urban cell i.

Equations (9.4)–(9.6) are empirical relations linking the canopy UHI to urban morphology metrics but it is also possible to demonstrate that similar scaling arguments arise from basic energy conservation principles. For example, Sobstyl et al. [25] employed an heat radiation model to show that city texture (defined via a building distribution function and the sky view factor) explains city-to-city variations in $\Delta T_{a,max}$. Specifically, they assumed that during nighttime only long-wave radiation contributes to the canopy UHI and urban-rural differences (i.e. $\Delta T_{a,max} = T_u - T_r$) are mainly due to changes in surface area associated with the 3D building surfaces. The ratio between the actual and ground surface area is defined as $\delta = \left[1 + \frac{4Lh}{(L+d)^2}\right]$, where d is the typical distance between buildings, while L and h are their mean linear size and height. Note that these quantities can be quantified using a radial distribution function traditionally used to characterize the atomic structure of crystals and liquids [25]. Under these assumptions, it is possible to define an effective urban temperature $T_u = T_{u,flat}\delta^{1/4}$ defined as the temperature of an equivalent flat surface emitting the same radiation as the 3D urban fabric (whose surfaces have average temperature $T_{u,flat} = \gamma T_r$). Urban rural difference can then be estimated as $\Delta T_a = T_r(\gamma\delta^{1/4} - 1)$, that is:

$$\Delta T_{a,max} = T_r \left[\gamma \left(1 + \frac{4Lh}{(L+d)^2} \right)^{1/4} - 1 \right] . \tag{9.7}$$

Assuming a simplified canyon geometry, it is possible to relate the ratio h/d to the sky-view factor ψ_{sky} [16], which allows to rewrite Eq. (9.7) as:

$$\Delta T_{a,max} = T_r \left[\gamma \left(1 + \frac{2\frac{L}{d} \tan[\arccos(\psi_{sky})]}{(1 + \frac{L}{d})^2} \right)^{1/4} - 1 \right] , \tag{9.8}$$

This equation successfully fit observations but, contrary to the linear relation proposed by [16], suggests a non-linear dependence of $\Delta T_{a,max}$ on ψ_{sky} (Fig. 9.2).

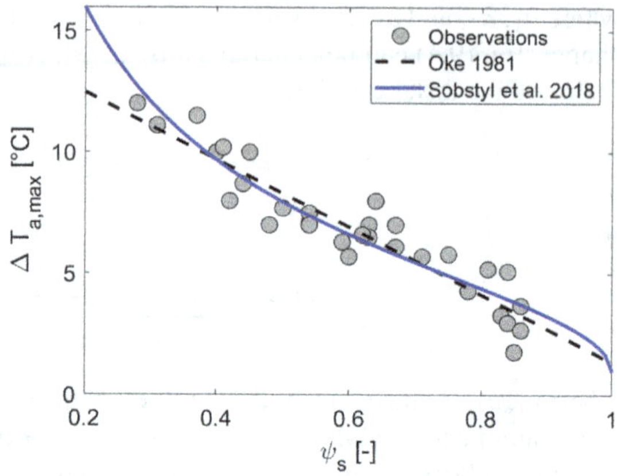

Fig. 9.2 Observed (circles) and simulated (lines) relation between $\Delta T_{a,max}$ and the sky view factor ψ_{sky}. Observations are retrived from [16]. The dashed line represents the linear model by [16] while the solid line represents Eq. (9.8) with $T_r = 40.4°$, $\gamma = 1.024$, and $L/d = 1$ [25]

In general, other relations linking the canopy UHI to the sky view factor exist. For example, Theeuwes et al. [37] proposed a diagnostic equation based on dimensional analysis:

$$\Delta T_{a,max} = (2 - \psi_{sky} - f_{veg}) \left(\frac{Q_{sw} \Delta_r}{U} \right)^{1/4}, \tag{9.9}$$

where f_{veg} is the vegetation fraction, Δ_r is the difference between maximum and minimum temperature in the rural area, and U is wind speed. Note that Eq. (9.9) supports Oke's first intuition of a linear relation between $\Delta T_{a,max}$ and ψ_{sky} but here the scaling exponent of wind speed is different from the $-1/2$ in Eq. (9.1).

9.2.2 *Coarse-Grained Models of Surface UHIs*

Impact of city size and background climate

As first proposed for night-time air temperature (Eq. (9.1)), the proliferation of remote sensing products revealed the existence of similar power law relations between urban population and SUHI intensity at the global scale [33]. Local hydroclimate also contributes to the intensity of SUHIs [14, 15], with higher mean annual precipitation causing higher urban to rural surface temperature differences (ΔT_s). On the hydroclimatic side, both changes in convection efficiency [14] and partition of net radiation in latent/sensible heat fluxes [15, 34, 38] have been proposed as the main drivers of daytime SUHIs. Some studies suggested that SUHI intensity increases linearly with regional precipitation due to changes in rural aerodynamic resistance [14], but remote sensing observations pointed to the existence of a precipitation threshold above which SUHIs become insensitive to precipitation changes [26, 38]. Numerical simulations confirmed possible nonlinear responses of SUHI intensities to precipitation [34] but, unlike previous modelling results, the variability of urban-rural surface temperature differences (ΔT_s) was explained by changes in rural temperature rather than convection efficiency—highlighting a fundamental knowledge gap in understanding the causal links between SUHIs, population, city texture, and background climate.

To shed light on the such global variability of SUHIs across cities and climates, the surface energy balance in Eq. (9.2) can be rewritten considering summertime averages for atmospheric inputs and assuming that changes in energy storage are negligible on time scales exceeding a day (see [26] for details):

$$Q_{sw}(1-\alpha)+\varepsilon_s\sigma(\varepsilon_a T_a^4 - T_s^4) - \rho c_p \frac{T_s - T_a}{r_a} - \lambda ET + Q_F = 0\,, \qquad (9.10)$$

where T_a and T_s [K] are the mean daily air and surface temperatures, respectively, Q_{sw} is the incoming shortwave radiation [W m^{-2}], α is the mean surface albedo [-], σ is the Stefan-Boltzmann constant (σ = 5.67E-08 W m^{-2} K^{-4}), ε_s and ε_a are the land surface and atmosphere emissivities, respectively, ρ [kg m^{-3}] is the mean air density, c_p [J kg^{-1} K^{-1}] is the specific heat of dry air at constant pressure, r_a is the aerodynamic resistance [s m^{-1}], and λ is the latent heat of vaporization [J kg^{-1}]. Note that T_s here is a lumped "skin" temperature comparable with the thermal signal obtainable from remote sensing, and the energy balance is calculated considering a control volume of finite thickness that includes all the sources of anthropogenic heat [14, 26, e.g.]. The intensity of SUHIs can then be derived from Eq. (9.10) considering urbanization as a a first-order perturbation to the rural base state. Specifically, following the derivation by Manoli et al. [26], urban-rural differences in surface temperature can be expressed as:

$$\Delta T_s = \frac{1}{f_s - \frac{\gamma}{a_T} f_a} \Delta S\,; \qquad (9.11)$$

where f_s and f_a [W m^{-2} K^{-1}] are energy redistribution factors associated with surface and air temperature, respectively [39], γ and a_T are parameters accounting for the coupling between T_s and T_a, and ΔS [W m^{-2}] is the differential energy forcing due to urban-induced changes in surface albedo ($\Delta\alpha$), emissivity ($\Delta\varepsilon_s$), evapotranspiration (ΔET), convection efficiency (Δr_a), and anthropogenic heat (ΔQ_F). Specifically,

$$\Delta S = -Q_{sw}\Delta\alpha + \sigma(\varepsilon_a T_a^4 - T_s^4)\Delta\varepsilon_s - \lambda\Delta ET + \rho c_p \frac{T_s - T_a}{r_a^2}\Delta r_a + \Delta Q_F\,. \qquad (9.12)$$

Given the objective of exploring the sensitivity of ΔT_s to background climate and city size we can then employ a set of climate relations linking background meteorological variables (e.g. Q_{sw}, T_a, T_s) and vegetation characteristics (e.g. albedo, vegetation height) to mean annual rainfall ($\bar{R}$). For example, we can assume that $Q_{sw}(\bar{R}) = a_R \cdot exp\left(b_R \cdot \bar{R}\right) + c_R$, where a_R, b_R, and c_R are fitting parameters [26]. Similarly, changes in urban characteristics (e.g. area, mean sky view factor, mean building height, anthropogenic heat) can be linked to city size by employing existing urban scaling laws. For example, following Schläpfer et al. [40], we can estimate mean building height as $h(N) = 1.15 \cdot N^{0.12}$ and use it for calculating the emissivity and roughness of urban surfaces (see [26] for details). Likewise, based on observations from 32 cities across the 5 continents [4], we can calculate the average anthropogenic heat flux as $Q_F(N) = 17.42 \cdot \rho_N(N)^{0.68}$, where ρ_N is population density. With the inclusion of such climate and urban scaling relations, the resulting model in Eq. (9.11) then postulates that the intensity of SUHIs can be coarsely-modelled as a function of population and mean annual rainfall, i.e. $\Delta T_s = \Delta T_s(N, \bar{R})$ where N is an aggregate measure for urban infrastructure size and $\bar{R}$ is a proxy for time-integrated surface-atmosphere exchanges and climatic patterns. The proposed approach combines existing urban scaling laws with the surface energy balance and is deemed "coarse-grained" because "fine-grained" properties of climate, cities, and rural areas are smoothed over in space and time to focus on collective phenomena and global patterns rather than microscopic (i.e. building to block scale) or specific city-climate processes.

In agreement with observations, the intensity of SUHIs modelled by Eq. (9.11) increases linearly at low precipitation regimes but it saturates at precipitation values exceeding around 1200 mm yr^{-1} (see Fig. 9.3). Model inferences suggest that the shape of the relation is largely controlled by changes in evapotranspiration. In wet climates, energy limitations control the upper bound on ET differences between urban and rural environments, thus creating the observed plateau of ΔT_s. In arid regions, water limitations reduce the magnitude of rural ET, thus limiting the contribution of ΔET to ΔT_s and, when urban vegetation is irrigated, the SUHI intensity becomes neutral or even negative. The model also reproduces the scaling of SUHIs with N, showing that the scaling exponent is not constant but varies with the background climatic conditions.

Seasonal hysteresis

Another ubiquitous feature of SUHIs is their seasonality. On seasonal timescales, urban areas exhibit distinctive hysteretic patterns between ΔT_s and background land surface temperature [27]. This seasonal hysteresis has been demonstrated by remote sensing observation from Europe [35] as well as in numerical simulations for the Greater London area [41]. The directionality of hysteresis is always clockwise, but ΔT_s either increases or decreases with T_s depending on the local hydroclimate (i.e. wet versus seasonally-dry, see Fig. 9.4A).

Such hysteretic behaviour can be explained by time lags between urban and rural dynamics [27]. To demonstrate this, Eq. (9.11) can be modified to account for the seasonality of background climate and vegetation. Specifically, changes in background meteorological forcings (e.g. T_a, T_s, R, Q_{sw}) with time t, can be modelled using sine functions $\Gamma(t)$ characterized by mean μ_Γ, amplitude A_Γ, and phase ϕ_Γ (see [27] for details). To compute the seasonality of evapotranspiration and surface albedo in rural areas, we also need to account for the response of vegetation to changes in soil water, as regulated by the water balance presented in Eq. (9.3). Given the stochastic nature of rainfall, Eq. (9.3) can be rewritten as [42, 43]:

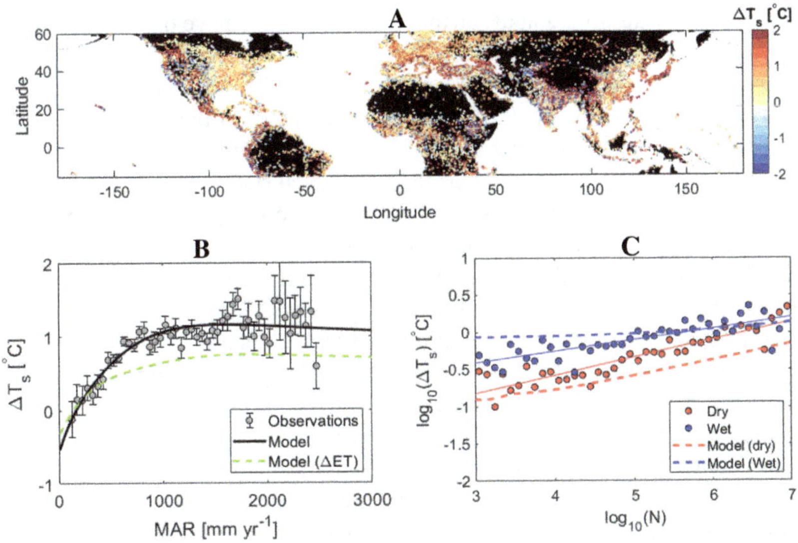

Fig. 9.3 **A** Global map of summertime SUHIs. **B** Observed (symbols) and modelled (solid line) relation between ΔT_s and Mean Annual Rainfall (MAR). The contribution of urban-rural changes in evapotranspiration (ET) to ΔT_s is also shown (green dashed line). **C** Observed (symbols, fitted by solid lines) and simulated (dashed lines) scaling of SUHIs with city population N considering wet and dry climate regions. *Source* modified from [26]

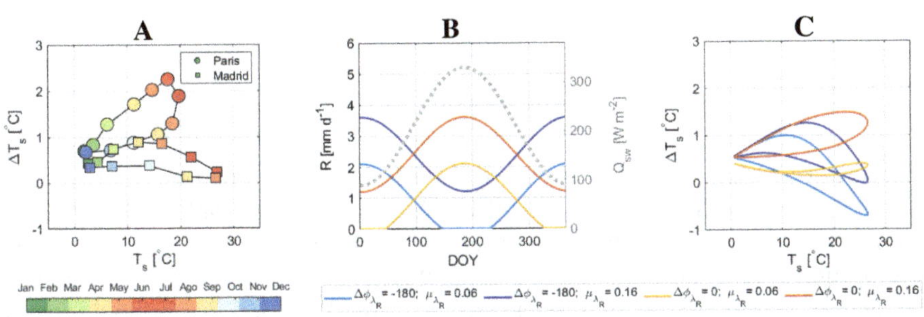

Fig. 9.4 **A** Observed seasonality of SUHIs in Paris and Madrid. **B** Simulated seasonal variations in rainfall R (dashed gray line) and shortwave radiation Q_{sw} (solid lines) and **C** resulting hysteresis of ΔT_s for different rainfall-radiation phase shifts $\Delta\phi_{\lambda_R}$ and mean rainfall frequencies μ_{λ_R}. *Source* modified from [27]

$$\frac{d\langle x(t)\rangle}{dt} = \frac{\lambda_R(t)}{\gamma(t)} - k(t)\langle x(t)\rangle - \frac{\lambda_R(t)}{\gamma(t)} e^{-\gamma(t)(1-\langle x(t)\rangle)}; \quad (9.13)$$

where rainfall is modeled as a marked Poisson process with frequency λ_R, $\langle x(t)\rangle$ is the time-averaged relative soil moisture, $\gamma(t)$ is the soil storage index and $k(t)$ is the normalized potential evapotranpiration (ET_{max}). In this framework, rural evapotranspiration can be calculated as $ET(t) = g_c ET_{max}(t)\langle x(t)\rangle$ (where g_c is the fraction of vegetated area). Assuming that urban and rural vegetation have the same characteristics, changes in ET are then computed considering urban-rural changes in g_c and, if necessary, the addition of irrigation (see [27]). Similarly, seasonal variations in rural albedo can be estimated from $ET(t)$ via the calculation of the leaf area index [27]. This probabilistic approach integrates information on rainfall stochasticity and seasonality to describe the "average" seasonal cycle of surface water and energy fluxes. Note, however, that the selected sine functions represent prototypical examples of intra-annual variability rather than precise conditions observed in the field.

The observed and simulated seasonality of SUHIs is illustrated in Fig. 9.4. On one hand, cities in relatively wet climates (like Paris, Fig. 9.4A) show a concave up hysteresis characterized by a SUHI intensity which is always positive and peaks during summer. On the other hand, cities in seasonally dry regions (as represented by Madrid in Fig. 9.4A) exhibit a concave down curve with peak SUHI in spring and negligible or negative SUHIs during summer/autumn. As illustrated in Fig. 9.4C, the coarse-grained model (Eqs. (9.11)–(9.13)) is able to reproduce such seasonal dynamics of ΔT_s, revealing that the different hysteretic patterns observed result from time-lags between energy (i.e. radiation) and water (i.e. rainfall) availability. Specifically, in wet climates evapotranspiration is in-phase with radiation so that evaporative cooling is maximized in rural regions during summertime (thus causing higher ΔT_s values) while in seasonally dry climates, rainfall is out-of-phase with radiation causing water stress during summertime and a decrease in rural ET that leads to lower SUHI intensities (see [27] for further details).

9.2.3 Horizontal Structure of UHIs

In the previous sections we have focused on UHIs defined as urban-rural differences, thus neglecting the intra-urban variability of the temperature fields. To overcome this limitation, models have been proposed to describe the temperature changes that develop along the urban-rural gradient (from the city center, where the UHI intensity is generally maximum, to the rural land, where the UHI is by definition equal to zero). For example, Streutkler [44, 45] fitted surface temperature observations from the city of Huston, Texas, with a gaussian function of the form:

$$T_s(x, y) = a_0 \cdot e^{-\left(\frac{(x-x_0)^2}{2a_x^2}+\frac{(y-y_0)^2}{2a_y^2}\right)} + a_1 x + a_2 y + T_{s,r}; \tag{9.14}$$

where T_s is surface temperature, $T_{s,r}$ is the mean surface temperature in the rural region (hence, the SUHI intensity can be calculated as $\Delta T_s(x, y) = T_s(x, y) - T_{s,r}$), x and y are the spatial coordinates, and a_1, a_2, a_0, a_x, and a_y are fitting parameters (see Fig. 9.5). A similar approach was followed by Zhou et al. [46] who employed an exponential decay model to examine the changes in SUHI intensity with distance from the city center r in 32 Chinese cities, i.e.

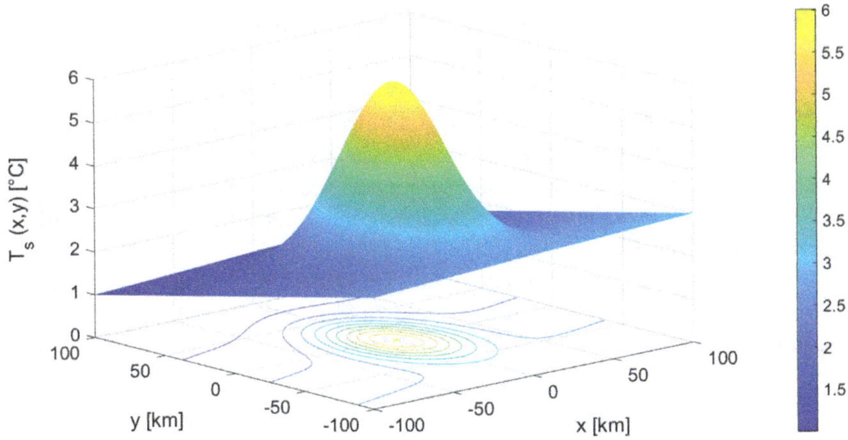

Fig. 9.5 An example of the gaussian surface in Eq. (9.14) with $T_{s,r} = 2\,°C$, $a_x = 20\,km$, $a_y = 30\,km$, $a_0 = 4\,°C$, $a_1 = 0.002\,°C/km$, $a_2 = -0.008\,°C/km$. The city center is located at $(x_0, y_0) = (0, 0)$

$$\Delta T_s(r) = \Delta T_{s,max} e^{-S \cdot r} + T_0 \, ; \tag{9.15}$$

where $\Delta T_{s,max}$ is the maximum UHI intensity, S is the decay rate, and T_0 is the asymptoic value of ΔT_s at $r \to \infty$ (which is close to zero). Interestingly, Zhou et al. [47] demonstrated that the exponential decay model is approximately equivalent to a Gaussian function in euclidean space, so that Eqs. 9.14 and 9.15 can be considered equivalent. Also, the exponential decay model resembles empirical and theoretical results on the changes in urban fraction with distance from the city center [48, 49]:

$$f_u(r) = e^{-\frac{b}{N^{1/2}} r} \, ;$$

where b is a parameter re-scaled by the square root of the total city population N. This similarity points to a linear relation between the SUHI intensity and the urban fraction which have been indeed observed [50] and explained theoretically [51].

Equations (9.14)–(9.15) are based on the fitting of empirical relations to observations and the underlying physical processes are therefore neglected. However, it is possible to recover similar temperature profiles by considering the exchanges of heat and mass over urban and rural areas. For example, Li and Bou-Zeid [52] considered the balance between horizontal advection and vertical turbulent diffusion of heat and water vapor coupled to the surface energy budgets of urban and rural areas. The resulting set of equations was solved analytically, thus providing the following expression for ΔT_a:

$$\Delta T_a(x, z) = (1 - \frac{\beta_u}{\beta_r}) T^*_{rs} f_1(x, z) + g(U_{10})(Q_u - Q_r) f_2(x, z) \, ;$$

where the intensity of the canopy UHI (ΔT_a) is written as a function of the along-wind direction x and the vertical direction z ($0 < x < x_u$ is the urban area, $x < 0$ and $x > x_u$ is the rural area, and $z = 0$ is the land surface), β_u and β_r are the ratios of actual surface specific humidities divided by the saturated surface specific humidities for the urban and rural surfaces, respectively, $Q_u - Q_r$ is the difference in available energy between urban and rural areas (defined as $Q = Q^* - \Delta Q_s$, see Eq. (9.2)), while $f_1(x, z)$ and $f_2(x, z)$ are functions that arise from the solution of the advection-diffusion equations [52].

Following a similar approach, Fan et al. [53] solved the coupled energy budgets of the urban air, urban surface, rural air and rural surface, providing an analytical solution for the horizontal extent of the urban heat dome at night:

$$\frac{D_d}{D} = \sqrt{1 + \frac{(B - A)Q_{H,u}}{(CQ_F - AQ_{H,u})}}; \tag{9.16}$$

where $\frac{D_d}{D}$ is the ratio between the heat dome diameter (D_d) and the urban diameter (D), while $Q_{H,u}$ is the sensible heat flux in the urban area, and A, B, and C are terms accounting for the specific surface area and the heat transfer coefficients of the urban and rural surfaces. The authors [53] also employed a linearized model to calculate the two-dimensional urban heat dome flow at a mean condition of no background wind and showed that $\frac{D_d}{D}$ ranges from 1.5 (from Eq. (9.16)) to 4 (from [54]).

9.3 Criticism and Opportunities

The models illustrated so far provide a coarse-grained description of emergent urban-rural dynamics, but they generally focus on city-scale values (or radial gradients), averaged in space and time (e.g. over daily or monthly timescale). Clearly, this is not sufficient to quantify local microclimatic effects (e.g. on pedestrian thermal comfort) and guide site-specific urban planning solutions as the characteristics of single cities and their spatial heterogeneities can significantly deviate from the emergent behaviours discussed here (by definition, emergence occur when a system exhibits macroscopic properties that are not necessarily observed at the microscopic level [55, e.g.]).

SUHIs are generally retrieved from satellite-derived land surface temperatures (T_s) that suppress rapid atmospheric fluctuations and provide a coarse-grained metric of the urban surfaces as seen by a radiometer [4, 56]. While T_s has direct implications for thermal comfort (due to radiative loading on pedestrians outdoors) and building energy consumption (due to heat conduction through building envelopes), air temperature (T_a) is generally more relevant for thermal comfort, public health, and heat risk studies. Measured (or simulated) air temperature is a point-scale (or fine-grained) metric, which varies appreciably with height within and above surfaces and is influenced by small scale heterogeneities and short-lived fluctuations.

Such space-time dynamics of the temperature field in complex urban settings can be simulated using detailed models (see Sect. 9.1) but, given the computational cost of CFD simulations, existing modelling efforts are limited to building/block scale dynamics, often subjected to simplified geometries and boundary conditions and over the timespan of few hours or days [20, 57, e.g.]. Linking this knowledge on local micro-meteorology with large scale (city to global) studies based on remotely sensed surface temperatures is not an easy task due to the mismatch of scales and the inherent complexity of urban areas. The $T_s - T_a$ relation is complicated by the 3D structure of the urban fabric, especially when dealing with building/block scale processes at hourly or daily resolutions, and with cities located in particular geographical settings (e.g. coastlines, valleys, mountains). As a consequence, the relation between air and surface UHIs (defined as urban-rural differences of average temperatures) is non-trivial at these scales. Clarifying these links between air and surface UHIs and deriving rigorous up-scaling techniques from point to city level is an exciting subject for future research. The challenge is to derive formal averaging and up-scaling procedures to arrive at large (e.g. city) scale quantities starting from the "correct" point-wise physics [58, 59, e.g.].

In general, given this high spatio-temporal variability of the urban climate, a clear distinction should be made between (fine scale) measures for improving local microclimate and thermal comfort [60, 61, e.g.], versus (large scale) mitigation measures for reducing extreme temperatures at the city scale [27, e.g.]. The efficacy of a given strategy (e.g. urban greening) in achieving the two objectives could be very different, and strongly depends on the scale of implementation [62]. Also, climate sensitive urban planning requires a quantification of the overall climatic conditions experienced by residents (i.e. local air temperature, radiative loading, air humidity, and wind speed at the block/building scale from sub-hourly to interannual timescales), as well as their exposure and vulnerability to heat. Hence, the intensity of UHIs is a useful but not sufficient metric to characterize heat stress and guide heat mitigation strategies. General results linking urban form and functions to climate, exposure, and health outcomes are currently missing. In this context, coarse-grained approaches can provide general guidelines [26, 36, e.g.], potentially at the preliminary design stage, but more detailed city-specific studies are necessary to guide local interventions.

In general, it is important to highlight that any attempt to employ scaling theories in the study of urban climate inevitably requires some trade-offs with fine-scale processes, and the applicability of such studies remains limited by the model assumptions and the applicability of the underlying scaling laws that link city properties (e.g. size, population) to urban form and function. This generality comes at cost—it may not inform policy making at the level of individual cities or neighbourhoods. However, they allow for predictions of changes expected in certain city variables and function as a result of a shift in the underlying driver (e.g. how SUHI changes with background climate or how average building height changes with population density) allowing conjectures about broad tendencies to be formulated. Therefore, the focus is purposely distinct from canonical urban climate studies and parameterizations dealing with block/neighbourhood/single-city scale processes [18, 19, 60, 63, e.g.]. Similar coarse-grained global scaling analyses have been critically important in many other fields (such as the Budyko curve in hydrology [64, 65, e.g.]), but they are lagging behind in urban climatology where they can assist in providing a general framework for the increasing number of global modelling and observational studies of cities.

9.4 Summary

Urban complexity research seeks general laws explaining how city growth affects society and the environment [8]. Yet, most of this research has focused on socio-economic indicators and less attention has been paid to the interactions between cities and the biosphere (such as the UHI effect discussed here but also air and water pollution, urban-induced precipitation, carbon and nutrient dynamics, etc). Given the implications for energy consumption, public health, and residents' well-being, urban planning is now called on to reflect on the climatic effects of urban development [66, 67, e.g.]. Given that cities are complex multi-scale systems, this prompts the use of macroscopic (and stochastic) frameworks in addition and combination to current microscopic (and deterministic) approaches. Clearly, this is not the first instance in science when coarse-grained (i.e. spatially averaged models with all parameters treated as effective) and detailed spatially-explicit approaches (i.e. local models where geometry and small-scale processes are explicitly considered) are debated and criticized. Examples from other disciplines where simplicity versus complexity of models [68] was debated include watershed hydrology [69–71, e.g.], ecology [72, e.g.], geomorphology [73, e.g.], boundary-layer dynamics [74, e.g.], and epidemiology [75, e.g.]. Urban climate research can substantially benefit from a convergence of approaches and a constructive discussion on what degree of physical approximation is deemed acceptable to tackle local and global challenges. Specifically, urban climate research can capitalize on the recent advances in big data, city-analytics, and statistical mechanics to explore the "upwards" route from local to global concepts and search for "a rigorous and formal reconciliation of scales" by approaching the problem from two opposite directions [76].

Acknowledgements The author expresses his gratitude to Gaby Katul, Elie Bou-Zeid, Simone Fatichi, Naika Meili, Katty WT Huang, and the editor for the help and insightful discussions.

References

1. P. Ball, Complexity: decoding deep similarities. Nature **545**(7653), 154 (2017). https://doi.org/10.1038/545154a
2. M. Chase, Teleology and final causation in Aristotle and in contemporary science. Dialogue: Can. Philos. Rev. **50**(3) (2011). https://doi.org/10.1017/S0012217311000527
3. G. Mills, Urban climatology: history, status and prospects. Urban Clim. **10**, 479–489 (2014). https://doi.org/10.1016/j.uclim.2014.06.004

4. T.R. Oke et al., *Urban Climates* (Cambridge University Press, 2017). https://doi.org/10.1017/9781139016476
5. I.D. Stewart, Why should urban heat island researchers study history?. Urban Clim. **30**, 100484 (2019). https://doi.org/10.1016/j.uclim.2019.100484
6. L.M.A. Bettencourt et al., Growth, innovation, scaling, and the pace of life in cities. Proc. Natl. Acad. Sci. **104**(17), 7301–7306 (2007). https://doi.org/10.1073/pnas.0610172104
7. M. Batty, The size, scale, and shape of cities. Science **319**(5864), 769–771 (2008). https://doi.org/10.1126/science.1151419
8. L. Bettencourt, G. West, A unified theory of urban living. Nature **467**(7318), 912–913 (2010). https://doi.org/10.1038/467912a
9. D.J. Higham et al., *An Overview of City Analytics* (2017). https://doi.org/10.1098/rsos.161063
10. D. Rybski, E. Arcaute, M. Batty, Urban scaling laws (2019). https://doi.org/10.1177/2399808319886125
11. A. Isalgue, H. Coch, R. Serra, Scaling laws and the modern city. Physica A: Stat. Mech. Appl. **382**(2), 643–649 (2007). https://doi.org/10.1016/j.physa.2007.04.019
12. J.R. Banavar et al., Form, function, and evolution of living organisms. Proc. Natl. Acad. Sci. **111**(9), 3332–3337 (2014). https://doi.org/10.1073/pnas.1401336111
13. T.R. Oke, City size and the urban heat island. Atmos. Environ. (1967) **7**(8), 769–779 (1973). https://doi.org/10.1016/0004-6981(73)90140-6
14. L. Zhao et al., Strong contributions of local background climate to urban heat islands. Nature **511**(7508), 216–219 (2014). https://doi.org/10.1038/nature13462
15. D. Li et al., Urban heat island: aerodynamics or imperviousness?. Sci. Adv. **5**(4), eaau4299 (2019). https://doi.org/10.1126/sciadv.aau4299
16. T.R. Oke, Canyon geometry and the nocturnal urban heat island: comparison of scale model and field observations. J. Climatol. **1**(3), 237–254 (1981). https://doi.org/10.1002/joc.3370010304
17. L. Järvi, C.S.B. Grimmond, A. Christen, The surface urban energy and water balance scheme (SUEWS): evaluation in Los Angeles and Vancouver. J. Hydrol. **411**(3-4), 219–237 (2011). https://doi.org/10.1016/j.jhydrol.2011.10.001
18. C.S.B. Grimmond et al., The international urban energy balance models comparison project: first results from phase 1. J. Appl. Meteorol. Climatol. **49**(6), 1268–1292 (2010). https://doi.org/10.1175/2010JAMC2354.1
19. C.S.B. Grimmond et al., Initial results from Phase 2 of the international urban energy balance model comparison. Int. J. Climatol. **31**(2), 244–272 (2011). https://doi.org/10.1002/joc.2227
20. M.G. Giometto et al., Spatial characteristics of roughness sublayer mean flow and turbulence over a realistic urban surface. Bound.-Layer Meteorol. **160**(3), 425–452 (2016). https://doi.org/10.1007/s10546-016-0157-6
21. Lento Manickathan et al., Parametric study of the influence of environmental factors and tree properties on the transpirative cooling effect of trees. Agric. For. Meteorol. **248**, 259–274 (2018). https://doi.org/10.1016/j.agrformet.2017.10.014
22. Q. Li et al., Revisiting the relation between momentum and scalar roughness lengths of urban surfaces. Q. J. R. Meteorol. Soc. (2020). https://doi.org/10.1002/qj.3839
23. M. Llaguno-Munitxa, E. Bou-Zeid, Shaping buildings to promote street ventilation: a large-eddy simulation study. Urban Clim. **26**, 76–94 (2018). https://doi.org/10.1016/j.uclim.2018.08.006
24. M. Demuzere, B. Bechtel, G. Mills, Global transferability of local climate zone models. Urban Clim. **27**, 46–63 (2019). https://doi.org/10.1016/j.uclim.2018.11.001
25. J.M. Sobstyl et al., Role of city texture in urban heat islands at nighttime. Phys. Rev. Lett. **120**(10), 108701 (2018). https://doi.org/10.1103/PhysRevLett.120.108701
26. G. Manoli et al., Magnitude of urban heat islands largely explained by climate and population. Nature **573**(7772), 55–60 (2019). https://doi.org/10.1038/s41586-019-1512-9
27. G. Manoli et al., Seasonal hysteresis of surface urban heat islands. Proc. Natl. Acad. Sci. **117**(13), 7082–7089 (2020). https://doi.org/10.1073/pnas.1917554117
28. M. Santamouris et al., On the impact of urban heat island and global warming on the power demand and electricity consumption of buildings–a review. Energy Build. **98**, 119–124 (2015). https://doi.org/10.1016/j.enbuild.2014.09.052

29. F. Estrada, W.J. Wouter Botzen, R.S.J. Tol, A global economic assessment of city policies to reduce climate change impacts. Nat. Clim. Change **7**(6), 403–406 (2017). https://doi.org/10.1038/nclimate3301
30. C. Mora et al., Global risk of deadly heat. Nat. Clim. Change **7**(7), 501–506 (2017). https://doi.org/10.1038/nclimate3322
31. C. Sue, B. Grimmond, T.R. Oke, Aerodynamic properties of urban areas derived from analysis of surface form. J. Appl. Meteorol. Climatol. **38**(9), 1262–1292 (1999). http://dx.doi.org/10.1175/1520-0450(1999)038%3C1262:APOUAD%3E2.0.CO;2
32. M.L. Imhoff et al., Remote sensing of the urban heat island effect across biomes in the continental USA. Remote Sens. Environ. **114**(3), 504–513 (2010). https://doi.org/10.1016/j.rse.2009.10.008
33. N. Clinton, P. Gong, MODIS detected surface urban heat islands and sinks: global locations and controls. Remote Sens. Environ. **134**, 294–304 (2013). https://doi.org/10.1016/j.rse.2013.03.008
34. G. Yaofeng, D. Li, A modeling study of the sensitivity of urban heat islands to precipitation at climate scales. Urban Clim. **24**, 982–993 (2018). https://doi.org/10.1016/j.uclim.2017.12.001
35. B. Zhou, D. Rybski, J.P. Kropp, On the statistics of urban heat island intensity. Geophys. Res. Lett. **40**(20), 5486–5491 (2013). https://doi.org/10.1002/2013GL057320
36. Y. Li et al., On the influence of density and morphology on the Urban Heat Island intensity. Nat. Commun. **11**(1), 1–9 (2020). https://doi.org/10.1038/s41467-020-16461-9
37. N.E. Theeuwes et al., A diagnostic equation for the daily maximum urban heat island effect for cities in northwestern Europe. Int. J. Climatol. **37**(1), 443–454 (2017). https://doi.org/10.1002/joc.4717
38. D. Zhou et al., Climate–vegetation control on the diurnal and seasonal variations of surface urban heat islands in China. Environ. Res. Lett. **11**(7), 074009 (2016). https://doi.org/10.1088/1748-9326/11/7/074009
39. Z. Zeng et al., Climate mitigation from vegetation biophysical feedbacks during the past three decades. Nat. Clim. Change **7**(6), 432–436 (2017). https://doi.org/10.1038/nclimate3299
40. M. Schläpfer, J. Lee, L. Bettencourt, Urban skylines: building heights and shapes as measures of city size. arXiv preprint arXiv:1512.00946 (2015). https://doi.org/10.48550/arXiv.1512.00946
41. B. Zhou et al., Assessing seasonality in the surface urban heat island of London. J. Appl. Meteorol. Climatol. **55**(3), 493–505 (2016). https://doi.org/10.1175/JAMC-D-15-0041.1
42. A. Porporato, E. Daly, I. Rodriguez-Iturbe, Soil water balance and ecosystem response to climate change. Am. Nat. **164**(5), 625–632 (2004). https://doi.org/10.1086/424970
43. X. Feng, A. Porporato, I. Rodriguez-Iturbe, Stochastic soil water balance under seasonal climates. Proc. R. Soc. A: Math. Phys. Eng. Sci. **471**(2174), 20140623 (2015). https://doi.org/10.1098/rspa.2014.0623
44. D.R. Streutker, A remote sensing study of the urban heat island of Houston, Texas. Int. J. Remote Sens. **23**(13), 2595–2608 (2002). https://doi.org/10.1080/01431160110115023
45. D.R. Streutker, Satellite-measured growth of the urban heat island of Houston, Texas. Remote Sens. Environ. **85**(3), 282–289 (2003). https://doi.org/10.1016/S0034-4257(03)00007-5
46. D. Zhou et al., The footprint of urban heat island effect in China. Sci. Rep. **5**(1), 11160 (2015). https://doi.org/10.1038/srep11160
47. B. Zhou, D. Rybski, J.P. Kropp, On the relation between UHI intensity and city proximity. arXiv preprint arXiv:1710.01726 (2017). https://doi.org/10.48550/arXiv.1710.01726
48. R. Lemoy, G. Caruso, Evidence for the homothetic scaling of urban forms. Environ. Plan. B: Urban Anal. City Sci. **47**(5), 870–888 (2020). https://doi.org/10.1177/2399808318810532
49. Y. Li, D. Rybski, J.P. Kropp, Singularity cities. Environ. Plan. B: Urban Anal. City Sci. **48**(1), 43–59 (2021). https://doi.org/10.1177/2399808319843534
50. R. Bassett et al., Urbanisation's contribution to climate warming in Great Britain. Environ. Res. Lett. **15**(11), 114014 (2020). https://doi.org/10.1088/1748-9326/abbb51
51. C.T.F. Chiu et al., An analytical approximation of urban heat and dry islands and their impact on convection triggering. Urban Clim. **46**, 101346 (2022). https://doi.org/10.1016/j.uclim.2022.101346

52. D. Li, E. Bou-Zeid, Synergistic interactions between urban heat islands and heat waves: the impact in cities is larger than the sum of its parts. J. Appl. Meteorol. Climatol. **52**(9), 2051–2064 (2013). https://doi.org/10.1175/JAMC-D-13-02.1
53. Y. Fan et al., Horizontal extent of the urban heat dome flow. Sci. Rep. **7**(1), 11681 (2017). https://doi.org/10.1038/s41598-017-09917-4
54. A. Lemonsu, V. Masson, Simulation of a summer urban breeze over Paris. Bound.-Layer Meteorol. **104**, 463–490 (2002). https://doi.org/10.1023/A:1016509614936
55. S. Kivelson, S.A. Kivelson, Defining emergence in physics. npj Quantum Mater. **1**(1), 1–2 (2016). https://doi.org/10.1038/npjquantmats.2016.24
56. C. Adderley, A. Christen, J.A. Voogt, The effect of radiometer placement and view on inferred directional and hemispheric radiometric temperatures of an urban canopy. Atmos. Meas. Tech. **8**(7), 2699–2714 (2015). https://doi.org/10.5194/amt-8-2699-2015
57. J. Allegrini, J. Carmeliet, Simulations of local heat islands in Zürich with coupled CFD and building energy models. Urban Clim. **24**, 340–359 (2018). https://doi.org/10.1016/j.uclim.2017.02.003
58. A.G Hunt, S. Manzoni, Percolation theory, the effective-medium approximation and upscaling, in *Networks on Networks: The Physics of Geobiology and Geochemistry* (Morgan & Claypool Publishers, 2015). https://doi.org/10.1088/978-1-6817-4159-8ch2
59. M.F. Schmid et al., Volume averaging for urban canopies. Bound.-Layer Meteorol. **173**(3), 349–372 (2019). https://doi.org/10.1007/s10546-019-00470-3
60. N. Meili et al., An urban ecohydrological model to quantify the effect of vegetation on urban climate and hydrology (UT&C v1. 0). Geosci. Model Dev. (2020). https://doi.org/10.5194/gmd-13-335-2020
61. I. Pigliautile, A.L. Pisello, E. Bou-Zeid, Humans in the city: representing outdoor thermal comfort in urban canopy models. Renew. Sustain. Energy Rev. **133**, 110103 (2020). ISSN: 1364-0321. https://doi.org/10.1016/j.rser.2020.110103
62. J. Yang, E. Bou-Zeid, Scale dependence of the benefits and efficiency of green and cool roofs. Landscape Urban Plan. **185**, 127–140 (2019). https://doi.org/10.1016/j.landurbplan.2019.02.004
63. D. Li, E. Bou-Zeid, Quality and sensitivity of high-resolution numerical simulation of urban heat islands. Environ. Res. Lett. **9**(5), 055001 (2014). https://doi.org/10.1088/1748-9326/9/5/055001
64. P. Gentine et al., Interdependence of climate, soil, and vegetation as constrained by the Budyko curve. Geophys. Res. Lett. **39**(19) (2012). https://doi.org/10.1029/2012GL053492
65. W.R. Berghuijs, R.A. Woods, M. Hrachowitz, A precipitation shift from snow towards rain leads to a decrease in streamflow. Nat. Clim. Change **4**(7), 583–586 (2014). https://doi.org/10.1038/nclimate2246
66. Y. Rydin et al., Shaping cities for health: complexity and the planning of urban environments in the 21st century. The Lancet **379**(9831), 2079–2108 (2012). https://doi.org/10.1016/S0140-6736(12)60435-8
67. K.C. Seto et al., Sustainability in an urbanizing planet. Proc. Natl. Acad. Sci. **114**(34), 8935–8938 (2017). https://doi.org/10.1073/pnas.1606037114
68. C. Paola, M. Leeder, Simplicity versus complexity. Nature **469**(7328), 38–39 (2011). https://doi.org/10.1038/469038a
69. J. Michaud, S. Sorooshian, Comparison of simple versus complex distributed runoff models on a midsized semiarid watershed. Water Resour. Res. **30**(3), 593–605 (1994). https://doi.org/10.1029/93WR03218
70. H. Bormann et al., Spatially explicit versus lumped models in catchment hydrology–experiences from two case studies, in *Uncertainties in Environmental Modelling and Consequences for Policy Making* (Springer, 2009), pp. 3–26. https://doi.org/10.1007/978-90-481-2636-1_1
71. S. Fatichi et al., An overview of current applications, challenges, and future trends in distributed process-based models in hydrology. J. Hydrol. **537**, 45–60 (2016). https://doi.org/10.1016/j.jhydrol.2016.03.026

72. M.R. Evans et al., Do simple models lead to generality in ecology?. Trends Ecol. Evol. **28**(10), 578–583 (2013). https://doi.org/10.1016/j.tree.2013.05.022
73. W.E. Dietrich et al., Geomorphic transport laws for predicting landscape form and dynamics. Geophys. Monogr.-Am. Geophys. Union **135**, 103–132 (2003). https://doi.org/10.1029/135GM09
74. G. Bellon, B. Stevens, Using the sensitivity of large-eddy simulations to evaluate atmospheric boundary layer models. J. Atmos. Sci. **69**(5), 1582–1601 (2012). https://doi.org/10.1175/JAS-D-11-0160.1
75. J.L. Aron, Simple versus complex epidemiological models. Appl. Math. Ecol. Springer 176–192 (1989). https://doi.org/10.1007/978-3-642-61317-3_7
76. D. Or, The tyranny of small scales–on representing soil processes in global land surface models. Water Resour. Res. (2019). https://doi.org/10.1029/2019WR024846

Chapter 10
Designing Complexity? The Role of Self-Organization in Urban Planning and Design

Anat Goldman and Efrat Blumenfeld-Lieberthal

Abstract This chapter explores the concept of self-organization in urban planning and design, highlighting its role in shaping the unique characteristics of cities. It examines how various socio-economic, cultural, and political factors contribute to the development of distinct architectural styles, emphasizing the morphological patterns and self-organization principles. The chapter addresses the emergence of scaling laws and fractal geometry in urban forms, using historical and contemporary examples to illustrate these concepts. The discussion also delves into the cognitive aspects of urban design, examining how the physical layout of cities influences cognitive maps and perceptions of urban environments, and how these perceptions, in turn, influence urban design. Through the prism of self-organization, it demonstrates the dynamic interplay between individual and collective actions and the shaping of the urban landscape. This analysis offers insights into the complex, self-organizing systems that define urban spaces, emphasizing the interdependencies among architectural design, urban planning, and human cognition in shaping cityscapes.

10.1 Introduction

This chapter introduces conceptually self-organization phenomena, i.e. the emergence of scaling laws in buildings and in open urban spaces. Cities are complex entities by their nature, and their landscape is affected by many socio-economic, cultural, political, and other influences and forces. They all contribute to processes of self-organization that result in the creation of unique architectural styles that characterize cities. These styles are often based on morphological patterns that govern the cityscape. This applies to the architectural design of buildings (i.e. the proportions of the built volumes, their positioning on the plots, the design of the façades, etc.) as well as to the design and the distribution of the open spaces in the city.

A. Goldman · E. Blumenfeld-Lieberthal (✉)
The Azrieli School of Architecture, Tel Aviv University, Tel Aviv, Israel
e-mail: efratbl@tauex.tau.ac.il

D. Rybski (ed.), *Compendium of Urban Complexity*, Understanding Complex Systems,
https://doi.org/10.1007/978-3-031-82666-5_10

We start by introducing the term self-organization and explain in what way it is connected to architecture and urban design. Then, we demonstrate how self-organization is found in the morphology of individual buildings. Next, we present a cognitive model that explains how self-organization between the environment and human cognition creates a distinct city form. Finally, we provide real-world examples of this model in actual cities.

In 1961, Jane Jacobs published her canonical book "The death and life of great American cities" [1] in which she criticized modern urban planning while presenting its effect on humans and urban life. In the last chapter of her book, she asks what kind of problem a city poses, and answers that cities are, in fact, problems of organized complexity where the numerous variables that create the city are not cluttered, they are "interrelated into an organic whole" [1, p. 565]. In "A city is not a tree" Christopher Alexander (1965) [2] compares cities to a semi-lattice, a topological structure of a set of elements and their connections. Indeed, as Bar-Yam (1998) points out in his book "Dynamics of Complex Systems" a dictionary definition of the term "Complex" is: "consisting of interconnected or interwoven parts" [3, p. 1]. This implies that in order to understand a complex system we must understand not only its parts but also their interrelations and interdependencies. Thus, complex systems can be understood as systems that contain numerous components and interactions between them, where the behavior of the system as a whole cannot be deduced from the behavior of its components [3]. Complex systems can exhibit collective behavior that occurs due to the numerous interactions between the system's components. This collective behavior can be referred to as 'order' and the phenomenon where order appears due to the individual behavior of the system's components is called 'emergence'. This phenomenon is also related to one of the main attributes of complex systems, namely their self-organization. In the following, we explain the term 'self-organization' and demonstrate how it manifests itself in unique buildings as well as in cityscape, created by the assembling of numerous buildings.

10.2 Introduction to Self-Organization Phenomena

Self-organization refers to the spontaneous emergence of complex structures and patterns in a system without the need for external control or guidance. It is a process in which the components of a system spontaneously arrange themselves into a pattern or structure, often resulting in the emergence of new properties and functions at the global level. Self-organization is a common feature of many natural and artificial systems, and it can play a key role in the development and evolution of complex systems. Examples of self-organization can be found in a wide range of fields, including biology, physics, chemistry, computer science, and sociology [4–8].

The term "self-organization" first appeared in 1947 [9], though related ideas are considerably older. Krugman (1996) [10] explained the multi-disciplinarity of the term. "What links the study of embryos and hurricanes, of magnetic materials and collections of neurons, is that they are all self-organizing systems: systems that, even when they start from an almost homogeneous or almost random state, spontaneously form large-scale patterns" [10, p. 3]. He observed the way economies organize themselves in space and through time. Self-Organized criticality (SOC) describes a phenomenon where a system naturally develops into a critical state, where minor events can lead to a significant, often disproportionate response [11]. SOC is characterized by several features including scale-free distribution of features, i.e. patterns within the system are similar at different scales. Their size distribution follows a power-law indicating a scenario where small sizes are highly frequent while exceptionally large sizes are uncommon.

In 'Self-Organization and the City', Portugali [12] describes self-organization as "The phenomena by which a system self-organizes its internal structure independent of external causes, is a fundamental property of open and complex systems. Such systems also exhibit phenomena of non-linearity, instability, fractal structures and chaos—phenomena which are intimately related to the general sensation of life and urbanism at the end of the 20th century" [12, p. 49].

Self-organization in urban design is often expressed in terms of self-similarity. This refers to the repetition of resembling design elements or patterns at different scales within a city or a neighborhood. In urban design, self-similarity creates a cohesive and coherent urban form, as the repetition of unique elements or patterns can help to create a sense of unity and order within the built environment. Notably, this phenomenon occurs despite the wide range of individual design choices made by various architects, urban planners, and residents. Each of these contributors, in their own unique and decentralized way, collectively shapes the city's design.

The above mentioned repetition can relate to building types and designs, street patterns, or public spaces. One representation of urban self-similarity is the fractal characteristics of certain urban attributes, where geometric shapes within the cityscape re-appear at different scales. In urban design, fractal patterns create complex and intricate patterns that sometimes resemble nature in terms of their aesthetics.

Self-similarity in urban design may create a sense of coherence and unity within a city or a neighborhood, helping to create a sense of place. This can be especially important in large, complex urban environments, where the local repetition of certain elements can help to orient and guide people as they move through the city. However, self-similarity may also lead to a lack of diversity in the built environment, as the repetition of certain elements or patterns can result in a monotonous and uniform appearance. This can be especially problematic in terms of the imageability of the city and wayfinding within it. To address this, urban planners usually use landmarks that create a sense of direction within the city.

One of the key principles of self-organization in urban planning is the idea that cities and neighborhoods are complex systems that emerge and evolve through the interactions and decisions of individuals and groups. This means that the structure and form of a city or neighborhood is only partly determined by a central authority or master plan, but to a large extent emerges spontaneously as a result of the decisions and actions of the people who live and work there. Thus, the collective behavior of the complex system can be found in the morphological and social characteristics of cities and the buildings within them. Here, we start by addressing the physical aspects of self-organization in buildings and cities that are expressed either by scaling laws or by the emergence of physical patterns that characterize a specific city.

Scaling laws are mathematical relationships that describe how a certain property or characteristic of a system changes as the size or scale of the system is varied. They are often expressed as power-laws, in which the property of interest is proportional to some power of the size or scale of the system. Such scaling laws appear also in distributions. Unlike many other distributions, a power-law distribution describes an uneven distribution, in which even extremely large values of the considered quantity can occur with non-vanishing probability. Power-law distributions of different attributes of different systems describe self-similarities at different scales of the systems. For example, Zipf's law shows that the size of cities in different countries distributes according to a power-law with an exponent close to 1. This means that the largest city is twice as big (in terms of its population) as the second largest city, three times the size of the third largest city, and so on (for more information about city size distributions see Chap. 1).

Power-laws, however, are not only limited to describing statistical attributes (e.g. population, income) but also physical geometrical forms such as fractals. A fractal, as defined by Benoît Mandelbrot (1982) [13], is an object whose spatial form is not smooth in the sense of Euclidean geometry, thus termed "irregular", and whose irregularity repeats itself geometrically across scales.

The fractal geometry theory was conceived by the mathematician Benoît Mandelbrot in the 1970's. His famous study "The Fractal Geometry of Nature" [13] is considered the basis to this theory which complements Euclidean geometry. Mandelbrot coined the word fractal based on the Latin adjective "fractus." He chose this word because the corresponding Latin verb "frangere" which means "to break" or "to create irregular fragments." A fractal is a geometrical shape or pattern made of parts, which are similar to the overall pattern. It exhibits self-similarity, which means that each small portion, when magnified, can reproduce a larger portion (for elaboration see Chap. 11).

Batty and Longley (1994) [14] demonstrated how fractal geometry might be applied to cities in many ways. Cities demonstrate fractal structure as their functions are self-similar across a variety of scales. When growing, cities do not occupy the area compactly. The gaps between built areas are the result of organic development as well as physical and planning restrictions.

While there is a consensus that cities are complex systems by their nature [12, 15], buildings, may not seem like natural complex systems, as they are composed of simple components (shapes, volumes, materials, and so on) that form a closed system. However, these shapes and volumes often do exhibit the characteristics of complex systems because their plans integrate numerous components including local regulations, building technologies and materials, different construction, design, and the personal preferences of the entrepreneurs and planners (architects and engineers that took part in the planning process).

And indeed, it was found that buildings also exhibit fractal characteristics of their plans, elevations, and volumes. It is important to note, that while mathematical fractals are infinite, similarly to fractals in nature, architectural fractals are not, as buildings are finite entities.

10.3 Fractal Geometry in Planning and Design

While self-similarity is an important concept, it is not always a feature that is created by the collective behavior of people who live and use the city. In the next paragraphs, we address the use of fractal geometry intentionally or unintentionally by architects.

We can define fractal architecture as characterized by the repetition of geometric shapes at different scales, resulting in complex patterns. Such patterns are in analogy to those found in nature (e.g. trees, leaves, spirals). Many architects who incorporated such patterns into their architecture are said to be inspired by nature (see below). In addition to the aesthetic aspects of fractal architecture, there are also practical benefits that can be gained by incorporating fractal principles into buildings. The repetition of simple shapes at different scales usually creates strong and stable structures that are efficient in terms of material. This can be referred to as Biomimicry architecture or as Biomimicry models, followed by the application of their forms, processes, systems, and strategies in addressing human challenges.

Examples of fractal geometry in architecture can be found in the early monuments such as the *Kandariya Mahadeva Temple* in India (around 1000 AD) which is one of the Khajuraho temples that present self-similarity at different scales. Capo (2004) [16] utilizes fractal analysis techniques to demonstrate that the classical orders possess a fractal-like property, characterized by the persistent presence of detailed patterns on increasingly finer scales [17].

Shapes that resemble well-known fractals are also found when examining medieval cities. Their surrounding walls often resemble the Koch Curve [14]. Churches that were built during this time also present geometrical self-similarity in their façades, where the column, arches, and flying buttresses re-appear at different scales.

Moving forward to the Renaissance, we can find fractal characteristics in plans such as Michelangelo's and Bramante's designs for the *Saint Peter's Cathedral*, which also resemble the Koch curve [18]. Salingaros (2006) [19] posits that ornamentation can play a crucial role in imparting self-similar characteristics to architecture by introducing detailed elements at even the smallest scales of the architectural form. And indeed, ornamentation is a key factor in the fractal-like characteristics of building styles like the Baroque and Rococo (1600–1800). During these periods, the façades of many churches in Europe presented fractal attributes where the main structure of the pillars that hold the pediments re-appeared at different scales (see Fig. 10.1). This structure can be found in the Jesuit church *Il Gesù* (Giacomo Della Porta, 1584), the *Church of S. Susanna* in Rome (1597–1603), the *Church of St. Peter and Paul* in Krakow, Poland (1605–1619 to 1630) and many more.

Since fractal geometry is referred to as the geometry of nature, it is not surprising that buildings, designed by architects who were deeply affected by natural shapes

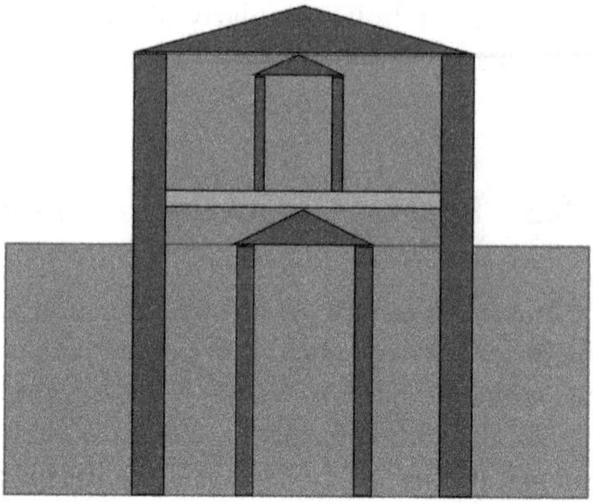

Fig. 10.1 A general form of a façade of a Baroque church

and phenomena, are considered as having fractal characteristics. Frank Lloyd Write (1867–1959), one of the most famous architects, believed that natural relations comprise of "… the part is to the whole as the whole is to the part and which all devoted to a purpose consistently" [20]. And indeed, his work is often referred to as containing fractal characteristics. Eaton (1998) [21] posits that the fractal nature of the Wright's *Palmer House* plan is apparent from the recurrence of an equilateral triangle as a geometric module at seven different scales, ranging from the complete structure of the building to specific elements within it. The major critique regarding Eaton's argument states that if the Palmer House plan is considered fractal, then any floor plan could be considered fractal as well, given that most houses contain spaces of various sizes which exhibit self-similarity as exemplified by the Palmer House plan [22]. Harris (2007) [23] agreed that the mere presence or repetition of similar shapes in a plan is not considered sufficient proof of fractal architecture, even in the work of an organic architect. He demonstrated the fractality of the Palmer House by developing a fractal shape that, by repeating it at different scales, after three iterations generates a similar pattern to the Palmer House plan (see [23, Fig.3]).

Approximately, at the same time, on the other side of the Atlantic Ocean in Barcelona, Spain, Antoni Gaudí started the *Segrada Familia* (1882). He was also known for being inspired by nature and nature's shapes. His masterpiece encompasses many different fractals. The arches are of different sizes, and the columns resemble tree trunks that divide into smaller and smaller branches. Staircases and ornaments such as spiral and flower-shaped fractals cover the interior of the church (see Fig. 10.2).

In modern architecture, we find fractals as well. Despite the fact that modern technologies allow for new forms, we still find examples of using fractals in different types of architectural design. Both the *Tribunal judiciaire de Melun*, France (1992), designed by Jourda and Perraudin Architects, and the *Stuttgart Airport Terminal*, Germany (1991), designed by Von Gerkan, Marg+Partner, represent examples using a tree-like design for the constructional columns. A similar example can be found in the *Tote Restaurant* in Mumbai (2013), designed by the Serie Architects, UK.

Fig. 10.2 Examples for fractals in the Segrada Familia

Other examples, that are less explicit in the way they employ fractal geometry are the façade of the *Grand Egyptian Museum*, in Cairo (2014), designed by Heneghan-Peng Architects, where the façade is based directly on the pattern of the Sierpinski Pyramid; and the façade of the *Lideta Mercato* in Addis Ababa, Ethiopia (2017), designed by Xavier Vilalta which resembles the Sierpinski Cube. For further elaboration on fractals in architecture please see [24] and [17].

10.4 Self-Similarity and Spatial Perception

As mentioned earlier, self-organization refers to the emergence of patterns in a system with no need for external or central control or guidance. After the first part of this chapter deals with the geometric expressions of self-organization in architecture, the second part will focus on the cognitive expressions of self-organization in urban design. While individual buildings usually present self-similarity in their morphology, cityscape is more complex, and its self-similarity is sometimes hidden from the eye, i.e. it is conceived only at the cognitive level, because the scale of an urban area is too large for a human eye to capture at once.

The urban street network serves as an example. When visiting a new city, we usually study the main streets and the network they form to help us navigate. Although these streets constitute only a small fraction of the total streets in a city, they are sufficient to help us create a mental map that helps us find our way. Jiang (2007) [25] claimed that less than 1% of the streets in a city, which are the most connected ones, are sufficient to create the city's skeleton. In later works [26, 27], he claimed that the image of the city (which can be referred to as a mental map) can be generated from the perspective of the different elements that form the city. He suggested that the image of the city is the outcome of the scaling laws that govern size distributions of urban artifacts or locations.

But what are these mental maps that Jiang refers to? And how do they emerge?

Tolman (1948) [28] introduced the concept of "cognitive map," which refers to the internal representation in the mind of a living organism (human or animal) of their external environment. These maps aid the organisms in navigating and exploring their surroundings. Lynch's "The Image of the City" (1960) [29] showed how people perceive urban environments. He proposed that the image of a city is composed of five elements: edges, nodes, paths, districts, and landmarks. He also claimed that the image of the city is based on three attributes: identity, structure, and meaning. Identity differentiates the city from other entities, structure refers to the spatial or pattern relationship with the observer/other objects, and meaning can be either practical or emotional to the observer.

Portugali [12, 30] defines the relationship between people and the environment as an inter-representational network (IRN), composed of external and internal, cognitive and emotional representations. These representations are linked in a feedback loop, so that external representations become internal through sensory absorption and internal representations become external through verbal, visual, and gestural expression. To explain the events within the cognitive system itself, Portugali draws on the theory of synergetics represented by the letter "S" in a new model which he calls SIRN [31]. According to this theory, awareness, cognition, cognitive mapping, and interaction between internal and external representations are all complex systems that develop in line with synergetic principles in self-organization.

Cognitive map research shows that each person has a unique representation of the same space. This representation is related, among other things, to early familiarity with the place, modes of arrival, early social perceptions, and so on. The cognitive process involves interactions between internal representations that continue to form and change over time, according to the level of familiarity with the environment, and to produce a new external representation [12]. Thus, the internal and external representations are described as ad-hoc identities created from the dynamics of the system and the context.

The cognitive process that leads to our spatial perception is dynamic and leads to the expression of multiple representations according to the task at hand. The process is described by Haken and Portugali (1996) [32] in a variable-based general model that portrays the effect of information acquisition by an agent, its processing, and the dissemination of the processed information.

The model is described as consisting of three sub-parts that relate to the number of agents involved in the process. The first refers to an individual, the second to a meeting of a number of agents, and the third to the interaction between a large number of agents connected to a collective information repository (also referred to as culture or society).

Regarding the current chapter, a third model, the Interpersonal-Information Reservoir Model, is of particular significance. This model describes a cognitive-collective process involving a large number of agents. The importance of this process is due to an understanding that every agent's action in the urban environment contributes to redesigning the information that the city provides to other agents, i.e. commuting to work, moving to a new house, shopping, constructing new buildings, building infrastructure, etc. In this sense, the city can be viewed as a common reservoir produced collectively by the actions of many individuals, including collective architecture. This reservoir has a physical expression in a variety of forms, i.e. text, objects, buildings, and even whole cities.

In his book "Invisible Cities" (1978) [33], Italo Calvino outlines the description of cities in the conversation between Marco Polo and Kublai Khan. Polo portrays the city Zoe as follows: "The man who is traveling and does not yet know the city awaiting him along his route wonders what the palace will be like, the barracks, the mill, the theater, the bazaar. In every city of the empire every building is different and set in a different order: but as soon as the stranger arrives at the unknown city and his eye penetrates the pine-cone of pagodas and garrets and haymows, following the scrawl of canals, gardens, rubbish heaps, he immediately distinguishes which are the princes' palaces, the high priests' temples, the tavern, the prison, the slum. This—some say—confirms the hypothesis that each man bears in his mind a city made only of differences, a city without figures and without form, and the individual cities fill it up." [33, p. 34].

Calvino's perception of the image of the city resembles Portugali's model as he too, describes an inner representation of a general notion of the city we carry within us.

Blumenfeld Lieberthal et al. (2018) [34] tried to extract the common image of a city that we all bear in our minds. They emphasized the city's morphological structure, with a focus on its statistical characteristics and how they are perceived. The authors ran a survey where they asked the participants to rank different configurations of artificial urban morphologies in order to identify the morphological attributes that form the inner representation of cities.

The artificial environments they used were created using a computational method where they controlled and changed different morphological characteristics. These environments were based on two basic street-networks: a non-orthogonal network that corresponds to European cities that were developed bottom-up (i.e. without an overall urban plan), and an orthogonal grid, which corresponds to North American cities that were developed top-down (see [2] for elaboration). Additionally, they used two different size distributions that obeyed either power-law distributions (with different exponents) or Poisson distributions (with different values of λ). Environments created with power-law distributions exhibited stark differences between largest and smallest sizes (e.g. street length, building height), when the exponent was large, and moderate difference when the exponent was small. Power-law distributions, when referring to the morphology of the urban entities, are related to fractal morphology. By doing so, the authors examined weather environments that resemble organic morphologies are perceived as resembling urban environments (which are actually man-made artifacts).

The Poisson distribution created a rather uniform set of sizes, where the majority of entities are near the average size, and there is a sharp decline in sizes at both the smallest and largest ends, leading to a limited range of sizes. In the present context, a high value of λ leads to a more uniform distribution of sizes, characterized by a narrow size range, whereas a low value of λ expands this range and results in higher values for entity sizes.

The authors found that orthogonal street grids were the most significant predictor of an environment's similarity to an urban setting. However, power-law distributions of entity sizes (of street length and width as well as building height) can also serve as predictors. The combination of these predictors of urban similarity suggests that the mental image of a city is not defined by a single feature, but rather by a combination of top-down planning (represented by the orthogonal grid) and bottom-up processes (represented by the power-law distributions). These findings extend Calvino's (1978) [33] and Portugali's (1996) [35] 'inner representation of the city', as it seems that not only the differences between the urban entities form our mental image of the city, but also the morphological and statistical relations between them.

Imageability, a concept introduced by Kevin Lynch in his influential book "The Image of the City" (1960) [29], refers to the quality of a physical environment that makes it clear, memorable, coherent, and recognizable to people. Lynch argued that the imageability of the urban environment is essential in helping individuals forming a mental map of the city, enabling them to navigate and understand the urban space successfully. Shushan et al. (2016) [36], used virtual reality to test which morphological configuration creates better imageability of urban areas. They too showed that heterogeneous environments (e.g., environments with fractal characteristics) offer higher levels of information and greater imageability compared to homogeneous environments, which are characterized by normal distributions.

To summarize, the human cognitive system can be conceptualized as a complex and open system with characteristics of emergence and self-organization. First, complex systems operating in self-organization are present at various levels, ranging from individual human cognition to the relationships between a group of people in a small community or in a society, and the urban environment, which is also an open and complex self-organizing system. Second, it emphasizes the dynamic process in which urban landscapes are actively created and changed over time, due to the complexity of the process, involving the generation of individual products, connections between individuals, and ultimately the collective action of society in a constructive process.

10.5 Self-Organization and Cityscapes

Self-organization is also present in cities given the way we perceive them as many cities have a unique style characterizing them. When we see the cityscape of cities like New York, Paris, or Jerusalem we can usually identify these cities even if we have not visited them. The question—of what makes a place look like itself—occupied researchers from various disciplines. Doersch et al. (2012) [37] approached this question from a computer science point of view and developed a method to automatically identify the most distinctive elements that form the image of a specific area (in their example—the city of Paris). They claimed that the "look and feel" of a city largely lies on a set of stylistic elements that consist of the visible details that we are exposed to in our daily urban life, rather than on the few famous landmarks (that correspond to Lynch's elements). While we may assume the Hausmann Boulevards, the Eifel Tower, or the Louvre might be the best representative of the city, the authors found that it is rather the balconies, windows with their railings, and the traditional blue/green/white street signs which were the most informative features identifying the city as Paris. This can be explained as a self-organized phenomenon related to the SRIN model presented earlier.

Portugali (2011) [30] approached this issue from another perspective and explained the exemplary phenomenon of the butterfly effect of balconies in Tel Aviv as self-organization. The city of Tel Aviv, established in the early 1920s, is characterized by many Bauhaus buildings with balconies. Once the first resident decided to expand their apartment by closing the balcony and converting it into a half-room, others followed him/her, and the cityscape of Tel Aviv (and later other cities in Israel) changed. To counteract this trend, the municipality of Tel Aviv imposed taxes on all balconies, regardless of whether they were open or closed (unlike previously when only closed balconies were taxed). This caused developers to construct buildings with closed balconies, as they enlarged the potential size of the apartments. However, with the advent of postmodern architecture, balconies regained popularity as decorative features. To meet the demands of this changing architectural style, city planners granted permits for the construction of open balconies, which could not be closed, leading to the prevalence of "jumping balconies" in the urban landscape of Israel (Fig. 10.3). This style gained popularity to the stage that a new hotel in Tel Aviv was built with jumping balconies (even though it is obvious that none of their guests will ever close the balcony in order to enlarge their rooms).

Goldman (2022) [38] followed Haken and Portugali's SIRN model (1996) [32], in order to identify common characteristics of shared residential buildings in Tel Aviv. These types of buildings constitute the majority of buildings in central Tel Aviv and therefore represent reliably the common and mundane architecture of everyday life. Based on the assumption that architects, as well as residents, are affected by the environment they plan within, a local appearance of the city can be detected.

Goldman (2022) [38] found that the typical building has several repeating characteristics. First, it has a box shape of three to four stories; second, it is built on pillar-supported level; and third, balconies are a routine feature appearing in every floor on the main façade. Thus, creating a vertical sequence. She grouped the residential buildings into several classes and sub-classes, based on the topology of their façades. The different typologies can be examined by grouping them into classes with similar characteristics. For example, Group A includes type A and a, Group B includes types B and b, etc. (see Fig. 10.4). Goldman [38] found that the balcony represents the recognizable symbol of the Tel Aviv residential building (as 100 percent of the buildings she studies had balconies). Although the balcony characterizes the building in a Mediterranean environment, it is not a local invention [39]. It has become a key component in Tel Aviv since its establishment, and balconies facing the main street characterize each of the different types of residential buildings. In addition to the climate role of the balcony in ventilating the apartment, it also serves as a key design element in creating the building's primary façade. The reason for this is that the buildings, being densely positioned in the city center, create a hierarchy in the architectural treatment of the façades. Usually, only one façade of the building

Fig. 10.3 Examples for jumping balconies [38]

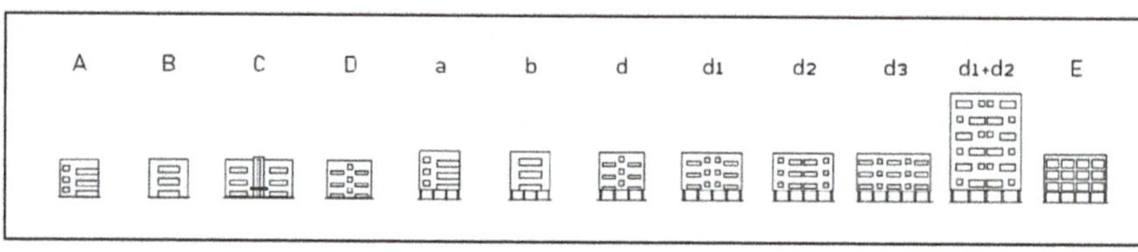

Fig. 10.4 The typologies of buildings characterizing the center of Tel Aviv [38]

faces the street (with the exception of corner buildings which have two street-facing sides). The sides of the buildings only offer a view on the front façades and most are obscured by the plot's depth, with the rear façade not visible from the street. Hence, the main design focus of architects is on the street façade, and balconies play a key role in creating visual depth and serving as a focal point. Despite the city's efforts to direct the design of balconies and prevent their enclosure, the appearance of balconies in the 1930s was diverse. In practice, balconies were built in various configurations without considering legal restrictions but rather based on the accumulative everyday actions of residents that preferred converting the balcony's open space into additional in-house space.

In addition to the balconies being a key element in shaping the cityscape of Tel Aviv, the Pilotis were another key element that appeared in most of the façades of

Tel Aviv from 1939 onwards. Figure 10.4 shows that 75% of the different typologies that Goldman (2022) [38] found in the Tel Aviv residential buildings have pilotis that elevate the buildings from the ground floor.

The use of columns (pilotis) in construction arose mainly due to the technological revolution in the building industry that took place at the end of the 19th century in Europe and the US, which involved the adoption of reinforced concrete as the primary building material. In Tel Aviv, the adoption of reinforced concrete started in the early 1930s, with 1933 being a significant year. Building on columns was a ground-breaking and aesthetically pleasing concept (following Le Corbusier's pilotis in *Villa Savoye*) that allowed for the separation of the building mass from the ground and the increase in height of heavy structures. This resulted in the creation of an "air pocket" between the ground level and the street and within the building mass itself, which now starts 2.5 meters above the street level on the first floor. Additionally, using reinforced concrete in columns, beams, and girders offered advantages for subsequent floors of the building, such as reduced wall thickness, leading to larger spaces, and wider and larger openings (especially windows) compared to before.

However, at first, the idea of building with columns was not favored by the municipality and it took several years until it was legally established in early 1939 after the number of requests to integrate columns into the ground floor of residential buildings had increased. The decision, adopted by the municipality to embed columns at the ground level facing the street was a compromise that blended modernist concepts, as represented by architects, with the city's goal of reducing the necessity of enforcing building code violations. This approach was adopted in recognition of the continued occurrence of these violations, and the belief that it was better to address them in a straightforward manner. Resembling the balcony affair, here too the municipality had to adopt new regulations to address the order that emerged bottom-up (i.e. self-organized), from planners and people who lived in the city. Despite the evolution of building materials, urban regulations, and architectural styles, elevated buildings on pilotis are also built in Tel Aviv today, and still represent the cityscape.

10.6 Summary

To summarize, cityscape that forms the image of cities in our minds is created by self-organizing processes that combine bottom-up forces (e.g., architects, planners, and residents of cities) with top-down ones (e.g. urban regulations). The results of these forces and the processes they lead to are the unique styles of different cities, that are expressed at different scales, i.e. the building design, the size distribution of streets length and width, etc. In this chapter, we reviewed self-similarity and self-organization in both buildings and cityscape and showed that while the representations of these phenomena are usually morphological, they affect our spatial perception of our surroundings and create a circular causality where the built environment affects the way we act within it, and the way we act within it affects, in return, the built environment. This complexity is typical for urban systems at different scales, and their relations with humans.

Acknowledgements The authors thank Prof. Juval Portugali for insightful conversations and co-tutoring Anat's PhD research.

References

1. J. Jacobs, *The Death and Life of Great American Cities* (Vintage, New York, 1961). ISBN: 9783839413272-099
2. C. Alexander, A city is not a tree. Ekistics **139**, 344–348 (1965)
3. Y. Bar-Yam, S.R. McKay, W. Christian, Dynamics of complex systems (studies in nonlinearity). Comput. Phys. **12**(4), 335–336 (1998). https://doi.org/10.1063/1.4822633
4. E. Karsenti, Self-organization in cell biology: a brief history. Nat. Rev. Mol. Cell Biol. **9**(3), 255–262 (2008). https://doi.org/10.1038/nrm2357
5. M. Perc, Self-organization of progress across the century of physics. Sci. Rep. **3**(1), 1720 (2013). https://doi.org/10.1038/srep01720
6. J.-M. Lehn, Towards complex matter: supramolecular chemistry and self-organization. Eur. Rev. **17**(2), 263–280 (2009). https://doi.org/10.1017/S1062798709000805
7. M. Mamei et al. Case studies for self-organization in computer science. J. Syst. Architect. **52**(8–9), 443–460 (2006). https://doi.org/10.1016/j.sysarc.2006.02.002
8. L. Leydesdorff, *A Sociological Theory of Communication: The Self-Organization of the Knowledge-Based Society* (Universal-Publishers, 2001)
9. W.R. Ashby, Principles of the self-organizing dynamic system. J. Gen. Psychol. **37**(2), 125–128 (1947). https://doi.org/10.1080/00221309.1947.9918144
10. P. Krugman, *The Self Organizing Economy* (Wiley, 1996)
11. P. Bak, C. Tang, K. Wiesenfeld, Self-organized criticality. Phys. Rev. A **38**(1), 364 (1988). https://doi.org/10.1103/PhysRevA.38.364
12. J. Portugali, *Self-Organization and the City* (Science & Business Media. Springer, Berlin, Heidelberg, 2000)
13. B.B. Mandelbrot, *The Fractal Geometry of Nature*, vol. 1 (WH Freeman, New York, 1982)
14. M. Batty, P.A. Longley, *Fractal Cities: A Geometry of Form and Function* (Academic Press, San Diego, CA and London, UK, 1994). http://www.fractalcities.org/
15. M. Batty, *Cities and Complexity: Understanding Cities with Cellular Automata, Agent-Based Models, and Fractals* (MIT Press, 2007). ISBN: 9780262524797
16. D. Capo, The fractal nature of the architectural orders. Nexus Netw. J. **6**, 30–40 (2004). https://doi.org/10.1007/s00004-004-0004-9
17. Y. Joye, A review of the presence and use of fractal geometry in architectural design. Env. Plan. B **38**(5), 814–828 (2011). https://doi.org/10.1068/b36032
18. G.L. Hersey, *The Monumental Impulse: Architecture's Biological Roots* (MIT Press, Cambridge, MA, 2001)
19. N.A. Salingaros, *A Theory of Architecture* (Umbau-Verlag Harald Püschel, Solingen, Germany, 2006)
20. P.J. Meehan, *Truth Against the World: Frank Lloyd Wright Speaks for an Organic Architecture* (Wiley, New York, 1987)
21. L. Eaton, Frank Lloyd Wright and Fractals, in *Nexus: Architecture and Mathematics*, ed. by K. Williams (Edizioni Dell'Erba, Fuchecchio, Italy, 1998), pp. 23–38
22. Y. Joye, A tentative argument for the inclusion of nature-based forms in architecture. Dissertation (Ghent University, Ghent, Belgium, 2007)

23. J. Harris, Integrated function systems and organic architecture from Wright to Mondrian. Nexus Netw. J. **9**(1), 93–102 (2007). https://doi.org/10.1007/s00004-006-0031-9
24. I.M. Rian, M. Sassone, Tree-inspired dendriforms and fractal-like branching structures in architecture: a brief historical overview. Front. Archit. Res. **3**(3), 298–323 (2014). https://doi.org/10.1016/j.foar.2014.03.006
25. B. Jiang, A topological pattern of urban street networks: universality and peculiarity. Phys. A **384**(2), 647–655 (2007). https://doi.org/10.1016/j.physa.2007.05.064
26. B. Jiang, The image of the city out of the underlying scaling of city artifacts or locations. Ann. Am. Assoc. Geogr. **103**(6), 1552–1566 (2013). https://doi.org/10.1080/00045608.2013.779503
27. B. Jiang, Head/tail breaks: a new classification scheme for data with a heavy-tailed distribution. Prof. Geogr. **65**(3), 482–494 (2013). https://doi.org/10.1080/00330124.2012.700499
28. E.C. Tolman, Cognitive maps in rats and men. Psychol. Rev. **55**(4), 189 (1948). https://doi.org/10.1037/h0061626
29. K. Lynch, *The Image of the City* (MIT Press, Cambridge, MA, 1960)
30. J. Portugali, *Complexity, Cognition and the City*. Understanding Complex Systems (Springer, Berlin, Heidelberg, 2011)
31. H. Haken, Synergetics, in *Self-Organizing Systems*, ed. by F.E. Yates et al. (Springer, Boston, MA, 1987). Chap. 21, pp. 417–434. https://doi.org/10.1007/978-1-4613-0883-6_22
32. H. Haken, J. Portugali, Synergetics, inter-representation networks and cognitive maps, in *The Construction of Cognitive Maps*, ed. by J. Portugali (Springer, Dordrecht, 1996), pp. 11–43. https://doi.org/10.1007/978-0-585-33485-1_3
33. I. Calvino, *Invisible Cities* (Houghton Mifflin Harcourt, 1978)
34. E. Blumenfeld-Lieberthal, N. Serok, E.L. Milner, What makes us think it's a city?. Built Environ. **44**(2), 177–191 (2018). https://doi.org/10.2148/benv.44.2.177
35. J. Portugali, Inter-representation networks and cognitive maps, in *The Construction of Cognitive Maps*, ed. by J. Portugali (Springer, Dordrecht, 1996), pp. 11–43. https://doi.org/10.1007/978-0-585-33485-1_2
36. Y. Shushan, J. Portugali, E. Blumenfeld-Lieberthal, Using virtual reality environments to unveil the imageability of the city in homogenous and heterogeneous environments. Comp., Environ. Urban Syst. **58**, 29–38 (2016). https://doi.org/10.1016/j.compenvurbsys.2016.02.008
37. C. Doersch et al., What makes Paris look like Paris? ACM T. Graphic. **31**(4) (2012). hal-01053876. https://doi.org/10.1145/2830541
38. A. Goldman, The Patterns of Morphological Changes of the Residential Environment in Central Tel-Aviv. in Hebrew. Dissertation (Tel Aviv University, Tel Aviv, Israel, 2022)
39. R. Fox, The Arab House in Eretz Israel: a renewed examination. Katadre **89** (1998). in Hebrew, pp. 83–126

Chapter 11
Fractality of Cities

Hadrien Salat

Abstract This chapter defines first heuristically and formally the mathematical notion of fractality and explains why introducing it became necessary to study highly irregular geometric sets, in particular those resulting from self-organization processes such as cities. It then reviews the main applications of the notion to the study of urban morphology, as well as the theories attempting to explain its origin in the specific context of cities. Next, it reviews the literature discussing the validity of presenting a single fractal exponent to describe an entire city, together with a brief recall of the discussion around the same issue for the related notion of urban scaling. This naturally leads to the definition of the mathematical notion of multifractality which solves both issues by allowing to combine, for a given urban measure, a multitude of local scaling exponents with a multitude of local fractal exponents into a single spectrum of values.

11.1 Introduction

Cities attempt to harness space to maximize the efficiency of human activity. Their positive development therefore requires a way to accurately measure space and to finely understand the underlying spatial relations between their components. To that end, planners and geographers have traditionally relied on classical geometry. However, it became more and more apparent throughout the 20th century that natural and man-made phenomena, including geography [1] and later cities [2], could create spatial patterns that cannot be described accurately by classical geometry. In the 1980s, Mandelbrot [3] codified a new framework to measure space, called *fractal geometry*. It was initially based on the Hausdorff dimension, which was invented to study highly irregular sets in mathematics. A set is usually called *fractal* when its dimension is not an integer. Fractal geometry soon became a staple of studying urban morphology [4].

H. Salat (✉)
Alan Turing Institute, British Library, London, England, UK
e-mail: hadrien.salat@ucl.ac.uk

D. Rybski (ed.), *Compendium of Urban Complexity*, Understanding Complex Systems,
https://doi.org/10.1007/978-3-031-82666-5_11

A few theories have emerged to explain the origin of fractality within urban environments [5, 6]. They are usually tied to the observation that phenomena that result from self-organization processes, i.e. from a sum of local interactions and decisions, such as cities, often exhibit self-similarity. Self-similarity is the property of being similar to one-self when viewed at different scales. It is a property that creates canonical fractal patterns [7]. In fact, fractality has been so often associated to underlying self-similarity and in so many domains, that both notions are often mistakenly conflated. In reality, many natural phenomena fit the fractality criterion without exhibiting self-similarity [8]. We therefore need to define the notion with sufficient precision before we can effectively apply fractal geometry to study urban morphology. This will be done first heuristically, then formally, in Sect. 11.2.

In Sect. 11.3, we present the fractal framework applied to studying urban morphology by reviewing the existing literature from a historical perspective. We will find that fractality can help to create realistic models of cities, that it can be linked to the efficiency of human activities, and that it is still a relevant and currently studied tool. We then present the different theories explaining the origins of fractality in urban systems in the fourth section. In Sect. 11.5, we discuss some of the essential arguments around the usefulness, pertinence and universality of the notion for urban science. We will show that a more complex and flexible notion, *multifractality*, is necessary to characterize more accurately and more usefully patterns of human activity inside cities. This new notion will be briefly defined in the last section, then framed within the context of urban science.

The full mathematical formalism of both fractality and multifractality is complex and many methodological variants have been developed to adapt the fundamental notions to different practical contexts. A complete mathematical development of the notions is therefore beyond the scope of this chapter. We instead intend to guide the reader towards an intuitive, yet accurate, understanding of the notions; towards materials that will allow direct use in practice; and finally towards an understanding of the central arguments and open research questions that shape the discussion around fractality within urban science.

11.2 Fractals

Since fractals have only been formalized recently, they do not have a single clear-cut and consensual definition. The meaning of fractality and ways to measure it are often re-defined for specific contexts. Historically, Mandelbrot [3] was the first to introduce an extensive framework that applies existing mathematical definitions of dimension for highly irregular spaces to the study of spatial patterns that occur in the natural world but that are not well categorized by classical Euclidean geometry. We first propose an intuitive first principles approach. The aim is to understand precisely what fractal geometry aims to characterize and why Mandelbrot chose his formalism. We then give a brief introduction of the fundamental mathematical principles that support the notion.

11.2.1 Heuristic Approach

Starting with the most simple geometric objects, lines, disks and balls are easy to describe. The Euclidean space is associated to a coordinate system that identifies all points it contains. Mathematical objects that are simple continuous deformations of these objects are also easy to describe, since they can inherit a coordinate system from the parent Euclidean coordinate system, as long as the continuous mapping between the objects is known explicitly. It is then natural to define the notion of *dimension* of an object as the minimum number of coordinates that is necessary to create a full coordinate system for it. This number is often referred to as the *degree of freedom* of the coordinate system. A line has dimension one, a disk two and a ball three. We could then extend the definition of dimension to sets of discontinuous objects by using a discrete index that acts as a supplementary dimension. For example, a set containing an infinite number of parallel lines could be considered of dimension two, since identifying any point inside the set requires one signed integer, identifying on which line the point lies, and one real number, identifying where the point lies on the line. However, is two a good representation of the dimension of such a set? If the lines are far apart from one another, then they do not fill the plane they lie on and should not have the same dimension as the plane itself. In other words, the signed integer identifying the lines does not have the same continuity quality that the two dimensions of the plane have. To enforce this idea of covering space in contrast to counting the degrees of freedom, a collection of disconnected objects could be better described by the dimension of its biggest object. In that case, the set of parallel lines should be of dimension one. What if the lines are now so densely packed that they *seem* to cover the entire plane?

The *Minkowski* and *Hausdorff* dimensions were created to answer the need for a definition of dimension that measures how much space is filled. Sometimes, the dimension is not well represented by an integer. For example the dimension of the set of densely packed parallel lines should be greater than one, but smaller than two. In that case, these dimensions do not attempt to round the dimension to the closest integer, but try to measure as precisely as possible where the dimension should fall in-between two integers. Hence, the term *fractal* was coined to characterize a set whose dimension is better represented by a fraction than by an integer. However, it is desirable that these new dimensions match the other usual definitions of dimension when they are well defined. Beyond the etymology, fractals can be defined as any set for which a Hausdorff dimension can be calculated, including sets of integer dimension. The definitions of the Minkowski and Hausdorff dimensions are technical and cannot be summarized easily. A full formal definition is therefore delayed until the next section.

Mandelbrot's main contribution was to identify that many natural phenomena are spatially distributed according to patterns that closely resemble geometric sets that have a non-integer Hausdorff dimension, see Fig. 11.1. The Hausdorff dimension is the most general in mathematics, however it is usually too strict to be calculated outside of the mathematical world. A number of simpler dimensions, in particular

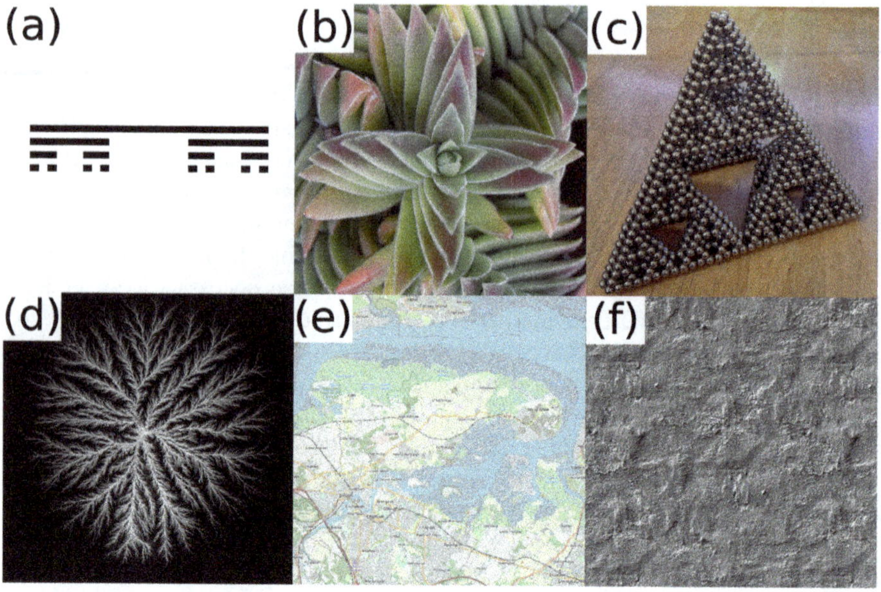

Fig. 11.1 Examples of fractal sets. **a** A mathematical self-similar fractal: the middle-third Cantor set; **b** A natural self-similar fractal: a Fractal Plant (CC-BY 2.0); **c** A man-made self-similar fractal: a Sierpinsky pyramid (CC0); **d** A quasi-self-similar fractal: a Lichtenberg figure left by lightning captured in epoxy (CC-BY 4.0); **e** An irregular natural fractal: the coast of Britain, Northern Kent (OpenStreetMap (ODbL)); **f** An irregular fractal: paint texture (Inkscape)

the Minkowski, or *box-counting*, dimension, have been used instead to characterize real-life patterns. These definitions are usually constructed to all match for a particular type of simple fractals, the *self-similar* fractals. Self-similarity, or scale invariance, is the property that a whole object is similar to parts of itself. Consider the middle-third Cantor set (Fig. 11.1a). It is made of a segment from which the middle third has been removed. The middle third of each remaining sub-segment is then also removed, and this process is repeated iteratively an infinite number of times. The hollowed out segment is similar to both the left and right hollowed child sub-segments, albeit at a scale three times bigger. Self-similarity renders the computation of all usual fractal dimensions straightforward when it exits. Mandelbrot found that this property occurs abundantly in natural phenomena, as illustrated in Fig. 11.1b, d. It occurs so often, in fact, that the term fractal is often confused with self-similarity. However, there exist sets that have non-integer Hausdorff or Minkowski dimensions and are not self-similar, such as the coast of Britain (Fig. 11.1e) or paint textures (Fig. 11.1f). These are also fractal sets since they fill space more than their geometric dimension suggests. Numeric values for all these examples are given in the next section, after the computation methods are presented.

Self-similarity in real-life fractals is often due to the strong scaling properties of the underlying processes that created them. A homogeneous function f is a function that has the property $f(a \cdot x) = a^{\alpha} \cdot f(x)$, for some α. Such functions transform objects x into scaled copies of themselves. It is therefore not surprising that processes that can be quantified by homogeneous functions, such as power laws, materialize into self-similar spatial patterns. This is the case, for example, of the urban environment. Chapter 2 details the scaling laws underlying many aspects of urban life. However, we also present here a particular tool, *multifractals*, that combines both of these intricate aspects into a singular output called the *multifractal spectrum*. Beyond these cases, as with natural fractals, all urban fractals are not necessarily self-similar. It can be beneficial, for example, to measure the degree of irregularity of city borders, lying at the intersection between the impervious built environment and the natural environment, with a fractal dimension. In the context of cities, we propose to call fractal any pattern for which the less restrictive box-counting dimension can be calculated. We expect this dimension to usually be between zero and two, since cities are constrained by the surface on which they are built, although extending the notion to 3D cities could be a next step in future analysis.

11.2.2 More Formal Approach

In mathematics, infinity can be reached at different rates, allowing for comparisons at the limit. For example, it is well known that $\lim_{n\to+\infty} n^2/(n+1) = +\infty$. This is due to the squared quantity growing at a much faster rate than the non-squared quantity. Similarly, for mathematical spaces, we can count how many elements of a certain diminishing size are needed to cover the entire space. If that number n increases much faster than the size ε of the elements diminishes, then the total size of the covering, that is the sum of the sizes of all elements or $n\varepsilon$, will converge to $+\infty$. Otherwise, if $n \to +\infty$ much slower than $\varepsilon \to 0$, then the total size of the covering will converge to 0. If both quantities change at a similar rate, then the product converges to a certain positive value. We want this value to become the dimension of the space. Indeed, the size of the elements that can be used to cover the space is proportional to their diameter raised to the power of their dimension. For example, the size of a segment of a line of length l is l^1, the size of a square whose edges are of length l is l^2, etc. As with the $n^2/(n+1)$ example, the exponent gives the strength of the convergence of the size of the covering. Fractals are sets for which the convergence of $n\varepsilon$ to a positive value is obtained for coverings made of abstract elements of non-integer dimension. The different definitions that we will now overview are based on this concept.

The simplest and most used definition of a fractal dimension is the Minkowski dimension. It calculates directly the relation between the minimum number of boxes necessary to cover a space S and their size. Denote $N(\varepsilon)$ the minimum number of boxes of side length ε required to cover the space S. Then, the box-counting dimension is defined as

$$d_{\mathcal{M}}(S) := \lim_{\varepsilon\to 0} \frac{\log N(\varepsilon)}{\log(1/\varepsilon)},$$

when the limit exists. The limit exists in particular when, approximately, $N(\varepsilon) \propto \varepsilon^{-d_{\mathcal{M}}(S)}$, that is when the covering is equivalent to a covering by square boxes of dimension $d_{\mathcal{M}}(S)$. In practice, since it is not possible to implement a covering by boxes of non-integer dimension, the equation can be considered as an indirect way of estimating the dimension by studying the variations under ε instead. It is often assumed that the boxes are square-shaped elements forming a grid on top of the space, although this is not a requirement, and any partition of the space by elements of same area could work. These assumptions and approximations are strong, prompting the need for a dimension that is always defined.

To understand the more general Hausdorff dimension, start by considering again the simplest spaces for which a dimension is defined. The simplest example are the well known vector spaces. A vector space is a set with two binary operations, addition and scalar multiplication, that satisfy a number of regularity axioms. A *basis* of a vector space is made of sub-elements that are linearly independent and can identify all other sub-elements through linear combinations. It therefore forms a system of coordinates. Every vector space has at least one basis and all bases of a vector space have the same cardinality, defining its dimension. This notion can be extended to vector spaces of infinite dimension. The *Lebesgue covering* dimension extends the notion to topological spaces.[1] Topological spaces are sets of points for which a notion of *neighborhood* can be defined. These spaces are usually vector spaces, manifolds, i.e. smooth deformations of vector spaces, and metric spaces, i.e. spaces where neighborhoods are defined by proximity according to a mathematical distance. The Lebesgue covering dimension is the smallest integer n such that for every open cover of the set, there is a sub-cover in which every point of the set lies in the intersection of no more than $n + 1$ covering elements. A simple example is given by the real line $\mathbb{R}$. Consider $x \in \mathbb{R}$. To include x in a cover by open intervals of $\mathbb{R}$, then one of these intervals needs to be $I_x =]x - \varepsilon_1, x + \varepsilon_2[$, with $\varepsilon_1, \varepsilon_2 > 0$. To include the rest of the line in a sub-cover, the interval to the right of I_x cannot be $]x + \varepsilon_2, \cdot[$, otherwise $x + \varepsilon_2$ would not be covered. However, it does not need to go beyond x, since x is already included in I_x. Therefore, any interval $]\alpha, \cdot[$, with $x < \alpha < x + \varepsilon_2$, can be considered. The same applies symmetrically to the left of I_x. So, the covering must have non-empty intersections ($n + 1 \geq 2$), however no points must be included in more than two open intervals ($n + 1 = 2$ is admissible), hence the Lebesgue covering dimension of the line is $n = 1$.

The Hausdorff dimension is a refinement of this idea for metric spaces that allows n to be non-integer. In contrast to the box-counting dimension, it does not require all elements of the covering to have the same size. Let X be a metric space. If $S \subset X$ and $d \in [0, \infty[$, define

$$\mathcal{H}^d(S) := \lim_{\varepsilon \to 0} \left(\inf_{U_i} \left\{ \sum_{i=1}^{\infty} \operatorname{diam}(U_i)^d, \bigcup_{i=1}^{\infty} U_i \supseteq S \text{ and } \operatorname{diam}(U_i) < \varepsilon \right\} \right),$$

[1] Note that there exist concurrent approaches to define the dimension of topological spaces such as *inductive* dimensions.

where U_i are open balls, the Hausdorff measure of X. The part within the brackets is the minimum "total size" of a covering by open balls of diameter at most ε. The exponent d accounts for the "size" of an equivalent ball of non-integer dimension d. This measure therefore calculates the rate at which the minimum number of d-dimensional elements required to cover the entire space grows compare to the size of the covering elements when this size is diminished to an infinitesimally small quantity. For example, if we were to cover a square S_{sq} with segments of lines (of dimension $d = 1$), we would need a "doubly" infinite number of such segments to achieve the result, since we need to cover the square both vertically and horizontally. Hence, the number of segments outgrows the rate at which their size diminishes, and $\mathcal{H}^{d=1}(S_{sq})$ is infinite. If, however, we were to try to cover the square S with cubes ($d = 3$), then the size of the square is infinitely smaller than that of the cube, which has an entire dimension in excess, so $\mathcal{H}^{d=3}(S_{sq})$ is 0. There is a singular value of d for which $\mathcal{H}^d(S)$ is neither $+\infty$ or 0, and this number defines the Hausdorff dimension $d_{\mathcal{H}}$ of the space. Explicitly,

$$d_{\mathcal{H}} := \inf\{d \geq 0 : \mathcal{H}^d(X) = 0\}.$$

This value $d_{\mathcal{H}}$ represents the abstract dimension that a finite number of elements must have to cover the entire space without excessive overlapping, making it the most robust formalization of the idea introduced at the beginning of the section.

When a set is self-similar, it is made of a number of copies of itself at a different scale. Then, the value returned by both the Hausdorff dimension and the box-counting dimension is the logarithm of the number of scaled sub-copies divided by the logarithm of the return to scale factor. For example, the middle-third cantor set is made of two sub-copies of itself at scale 1/3 (Fig. 11.1a). Its dimension is therefore $\log(2)/\log(3) \approx 0.63$. For the Sierpinsky pyramid (Fig. 11.1c), there are 4 sub-copies at scale 1/2, hence the dimension is $\log(4)/\log(2) = 2$, an integer. Despite being apparently self-similar, the plant in Fig. 11.1b is tricky to evaluate, since the scaling is not sufficiently constant with each iteration. The Minkowski and Hausdorff dimensions are not strictly equivalent in the general case however. For example, a dense set of points on a line would have dimension 0 according to the Hausdorff definition, but dimension 1 according to the box-counting method. The Hausdorff dimension is always lower or equal to the box-counting dimension. It is also always defined, even when the box-counting dimension cannot be calculated. Other similar definitions, usually equivalent for strictly self-similar spaces, include the *packing dimension*, the *lacunarity dimension* and the *correlation dimension*. These are not described here as they are less often encountered in the literature.

A much more detailed presentation of the notions mentioned above can be found in [9]. In order to run simple fractal analyses of images, the plugin `Fraclac` for `ImageJ` offers a very simple platform. For example, it returns a box-counting dimension of 1.69 for the Lichtenberg image (Fig. 11.1d) and a box-counting dimension of 1.82 for the paint texture (Fig. 11.1f). The coast of England (Fig. 11.1e) cannot be analyzed without heavy pre-processing. Its dimension is usually reported as

1.25 [1]. The software `Fractalyse` provides more advanced options. The package `fractaldim` can be used to run analyses in the `R` language. The box-counting dimension is simple to code and packages usually exist for common programming language. In most cases, the user may find useful to tailor their approach to the specific set under study. It can be, in particular, convenient to modify the shape of the "boxes". If, for example, one wants to study the distribution of a phenomenon in relation to groups of people rather than strictly in relation to the physical space separating people, then using administrative boundaries rather than boxes could be a preferable alternative.

11.3 Fractals for Urban Morphology

Urban morphology is the physical form of the city and its evolution through time. Its fractality has been well documented since the mid-1980s. Batty and Longley [4] first introduced fractality as a way to recreate realistic patterns of cities. The authors had identified that urban models were focused on activities and were lacking a proper representation of the physical form of the city. They proposed to cascade self-similarity processes to generate patterns of land-use that are visually similar to real cities and consistent with their basic metrics. Batty, Longley, and Fotheringham [10] and Batty, and Longley [11] then proposed to add diffusion-limited aggregation growth processes [12], which are known to naturally generate realistic fractal patterns [8]. In addition to developing fractal growth processes, Frankhauser [13] uses extensively the fractal dimension as a metric to control the realism of urban models. The seminal paper by Batty [6] synthesizes the introduction of fractality as a method to develop the planning of future city growth.

Fractality has also been used descriptively, in particular to study the complexity of geographical and urban boundaries [14–18], even before the notion was fully formalized [2]. An example of a fractal geographical boundary is given in Fig. 11.1e. Breaks within the fractal dimension of the built area can be used to define and characterize the true boundaries of a city, which are becoming more and more masked by urban sprawl and historic administrative boundaries becoming less relevant. Evidence that urban boundaries are non self-similar fractals, with a dimension varying both through scales and time, has been found. The two-dimensional fractal aspects of human settlements are approached by Frankhauser [19], in particular as a tool to compare cities [20]. Collecting fractal dimensions of the built area can inform about the historical evolution of a city. For example, Benguigui et al. [21] identify that Tel Aviv went through a period of homogenization, in particular from 1985, with fractal exponents for different areas converging towards the same value. De Keersmaecker, Frankhauser, and Thomas [22] use fractality to describe how Belgian cities have sprawled, a phenomena often negatively associated with social segregation, traffic congestion and energy inefficiency.

A city can also be seen as a network, both physically, for example when the edges are defined by streets or transportation links and nodes by intersections, and abstractly, for example when the edges are defined by flows of people and nodes by venues for activities. Such a network creates a geometric object that can be analyzed fractally [23–25]. Recent examples include Derrible and Kennedy [26], who study the metro transportation system of 33 cities and find that these are scale-free, in other words are self similar fractals, with dimensions comprised between 2.10 and 5.52 [see also 27, for additional empirical evidence]; and Tannier et al. [28], who study accessibility through the road network of Besançon across fractal and non-fractal simulated urban developments and find that introducing fractal development rules has a positive impact on accessibility.

These fractal descriptions are still being investigated, for example by Molinero [29], who compiles case studies for the road network, the occupation of space, the fractal dimension of the city, the average heights, the maximum heights and the interactions or GDP of a city. The author uses these examples to explain how the fractal geometric shape of cities can induce the scaling nature of human activities, an idea also supported by West [30] and West [31]. We will explore the converse in the next section and see if the fractal geometric patterns in cities could instead be the result of the scaling nature of human activities. Batty and Longley [32] offer a comprehensive review of early works on fractal cities, in particular boundaries and growth processes. References containing a detailed synthesis of more recent applications of fractality to urban morphology include [33, 34].

11.4 The Origins of Fractality in Urban Systems

Salingaros and West [5] and [20] propose an explanation of the fractality of cities as a result of combining random bottom-up individual and collective processes with top-down planning, creating different morphological scales. Schweitzer and Steinbrink [35] believe that fractal patterns emerge through self-organization from contradictory decisions at individual level. For example, a conflict resulting from trying to simultaneously minimize the distance to the city center and to the countryside would create irregularities. This theory could explain why a city is a fractal in the sense of a highly irregular space (a space for which there exists a non-integer box-counting dimension), however it cannot explain the existence of self-similarities.

Pumain [36] observes that scaling properties in cities emerge from the spatial organization of urban systems. Pumain however contests the possibility that they could be the result of random growth processes or space optimization principles alone. Instead, the role of upgrading the transportation systems should be considered. Indeed, each new mode of transportation brings a sudden increase in speed and efficiency, redefining the scale of human activities within a city and within systems formed by several cities. This can be related to Banavar, Maritan, and Rinaldo [37], who show that the efficiency of transportation networks itself exhibits scaling relationships with their sizes that originate from the form of these networks, an observation spurred by an analogy with networks in biological systems, such as the cardiovascular or respiratory networks [7]. Batty [6] assembles these different arguments and proposes that the origin of fractality is a result of growth from bottom-up processes resulting from an intense competition for space, in which scaling emerges from the constraints imposed on the underlying transportation network. In particular, space constraints induce a higher increase in the number of links compared to the number of nodes, which influences the form and efficiency of the transportation network itself. These ideas have continued to garner support until now [38–43].

We see that fractal patterns are thought to emerge from the scaling properties of the phenomena that underlie them, in particular the transportation network. It might therefore be necessary to study independently how scaling has emerged for each phenomenon. For example, Gomez-Lievano, Patterson-Lomba, and Hausmann [42] add new arguments to explain the scaling that is observed in the rank-size distribution of cities. A first one is that the more complex a phenomenon is, the larger number of complementary factors must be simultaneously present for it to happen. A second one is that the Darwinian selection process transforms the distribution into a logarithmic distribution exacerbating extreme values. Applied to cities, this would mean that the selection processes caused by the attractivity of cities induce a logarithmic distribution of the number of factors depending on population size, which is the very definition of a power law. We refer the reader to Chap. 2 for more insights into the origin of urban scaling laws, see also [44].

11.5 Fractals and Scaling Exponents Contested Universality

In addition to the morphological self-similarity identified in the previous sections, numerous power laws were found independently in city characteristics, see Chap. 2. We have briefly explained in Sect. 11.2 the links between fractality and power laws. These have been empirically confirmed and theorized for cities by Ribeiro et al. [45], and later by Molinero and Thurner [43] and Molinero [29]. The ubiquity of scaling laws suggests that scaling is a vital tool to understand and analyze urban environments. However, the broader significance of these independent results and their ability to continue predicting future behaviors remains unclear. It has been pointed out that the scaling exponents depend heavily on local parameters around the studied areas. For example, [46] have shown that the morphological self-similarity may be different within different neighborhoods of a city, while [47] have highlighted that varying the definition of city boundaries can induce large variations of the calculated scaling exponents for many urban indicators. Nevertheless, that same study by [46] also evidences that the exponents calculated at the scale of an entire city are in good agreement for many cities across borders within Europe, suggesting that they may be driven by some common planning imperatives that overtake the national contexts. Understanding the universality of the identified fractal and scaling exponents is crucial, as it determines their practical reusability.

We refer to Chap. 2 for a full discussion of the shortcomings of scaling analysis in urban systems. However, we recall a few arguments that are necessary to progress the discussion. For [48], "the processes relating urbanization to economic development and knowledge creation are very general, being shared by all cities belonging to the same urban system and sustained across different nations and times." [see also 49, 50]. These observations are the starting point for the theory developed by [42] mentioned above, who found them consistent with their own findings using U.S. data on education, employment, innovation, disease and crime. However, the exponents were not found consistent when calculated at a finer scale [27, 47, 51, 52]. In particular, The problem of the arbitrarity of administrative boundaries and their induced inconsistencies in urban analysis is well known as the Modifiable Areal Unit Problem (MAUP). It impacts empirical scaling laws. To tackle this problem, [47] propose different clustering and percolation methods to identify the "natural" boundaries of cities, and to recover, among other results, consistent scaling exponents. This approach was later complemented by [53] and [54]. In addition, a general lack of rigor in scaling exponent computations was criticized [55]. The authors vigorously alert on possible "unforeseen, disastrous consequences" that could result from wrong scaling estimates.

These various inconsistencies may be due to the fact that cities may be more complex than a simple exponent can capture. The morphology of cities has been pointed out as requiring more than one fractal exponent for a full characterization [46, 56, 57]. For example, [56] explains this by the fact that contemporary patterns result from several independent past processes, each tied to their own exponent. The fact that [46] find different exponents for the morphology of different parts of one same city suggests that cities may have been shaped by several intertwined local self-similar phenomena, each possessing their own scaling exponent. Cities may be more than *mono*-fractal and *mono*-scaling, they may be *multi*-fractal. Multifractals were created soon after fractality was codified by [3]. They are aimed at studying these more complex situations. As we will see in the next section, significant work has already been completed to apply them to cities. It has been shown in particular by [58] that they could be used to study the MAUP rather than be influenced by it. We will see in the next section how the multifractal framework can solve some of the problems mentioned above.

11.6 Combining Scaling and Fractals with Multifractals

Multifractals extend fractals to mathematical measures [59]. Measures can be understood as layers on top of the physical space that quantify the whole space and each of its sub-parts, in a way that ensures that the combined value of two disjoint sub-parts is the sum of the values associated to each sub-part. More tangibly, house prices can be seen as a measure, since each house is sold at a certain price, so that buying two houses will cost the sum of their individual prices; similarly, energy consumption measures the kWh of electricity that each house consumes, so that the consumption of two houses will be the sum of their individual consumption. Deciding whether an urban indicator is a suitable measure requires an understanding of the context. If an estate agent is able to negotiate a discount on house bundles or if there are heating energy savings due to some houses being terraced rather than detached, then the strict summation hypothesis is contradicted. However, from the point of view of a city planner who needs to understand how much public housing is needed and how building new housing will impact the social layout of a particular area, these subtleties can be neglected. Some indicators than are nearly measures can be adapted. For example, the average income per inhabitant inside an area is not a measure, since the average incomes in different places cannot be summed. However, the total income earned by the entire population of each area is.

Assuming a measure has been properly defined for a particular context, the measure is considered *multifractal* if the measure of a ball around any point is a power law of the radius of the ball. The power law exponent can be different for different areas. The fractal dimension of the sets formed by all areas that share a common power law exponent can be calculated. The curve representing these fractal dimensions versus their corresponding exponents is called the *multifractal spectrum*. It has a definite shape and contains rich statistical information on the spatial heterogeneity generated by the measure. The different fractal dimensions describe the morphological variety, while the power law exponents describe the scaling variety of the urban measure. These different notions are illustrated in Fig. 11.2. Multifractality is surprisingly common in complex systems resulting from self-organization with random processes, such as cities. In addtion, the multifractal framework can be applied to studying the heterogeneity of non-spatial urban sets. For example, time series can be seen as a measure layered on top of time rather than space. The framework therefore finds applications in a large variety of subjects, such as the complexity of urban morphology, the stability and efficiency of financial markets, and the spatial patterning of inequality in cities.

11.6.1 Technical Elements

A multifractal is a mathematical measure that obeys a double scaling/fractal condition. Succinctly, a *mathematical measure* is a set function μ over a space X that is non-negative, equal to 0 for the empty set, and that satisfies countable additivity. In equations,

$$\forall A \subset X, \mu(A) \geq 0\,,$$
$$\mu(\emptyset) = 0\,,$$
$$\mu(\cup A_k) = \sum \mu(A_k)\,,$$

where $\{A_k\}_{k=1}^{\infty}$ is a collection of pairwise disjoint sets. In practice, the space X needs to satisfy a number of properties to be measurable, the detail of which goes beyond the scope of this chapter. Understanding the meaning of the properties of a measure

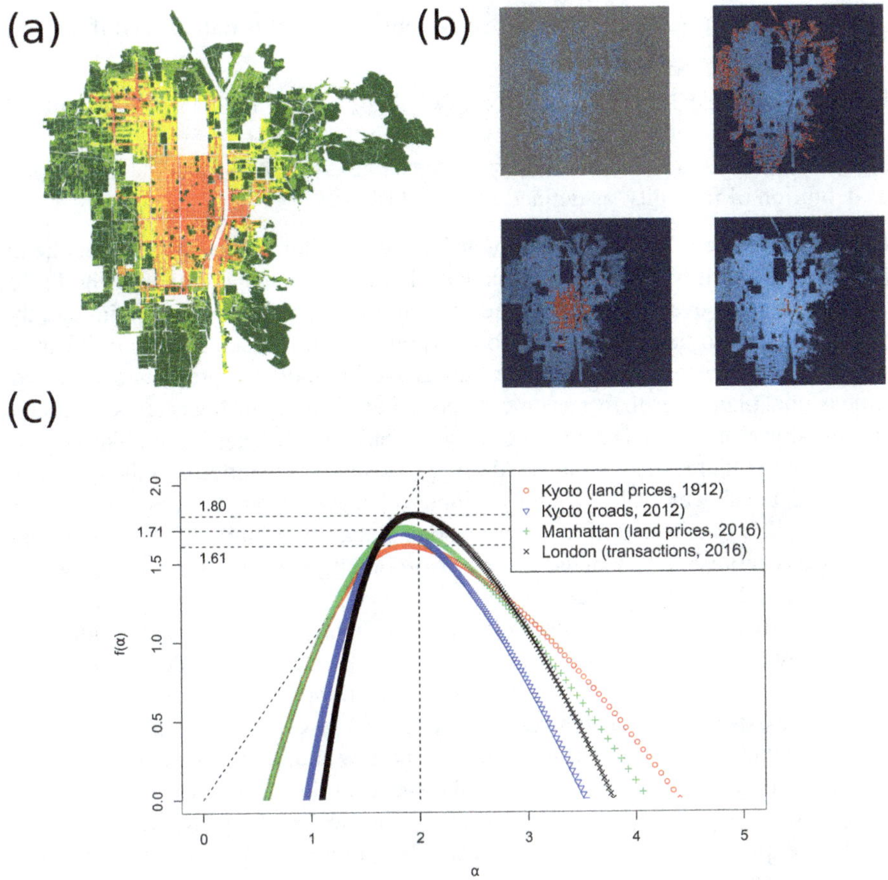

Fig. 11.2 Multifractal land prices. **a** The land prices in 1912 Kyoto are a measure (green are the lowest prices and red the highest); **b** Kyoto land prices are in fact multifractal. Different values of the local exponents are represented in red inside each sub figure; **c** The corresponding multifractal spectrum for Kyoto land prices in 1912 is compared to the spectrum for similar datasets for Kyoto in 2012, and London and Manhattan in 2016. The x-axis corresponds to the local power law exponents and the y-axis corresponds to the fractal dimensions. *Source* all image-parts are taken from Salat et al. [60] (CC-BY 4.0)

should suffice. Let us see a simple example. Any land plot has a non-negative financial value, buying no land is free, and if several disjoint plots of land are bought together, then the price of the entire lot will be the sum of the individual price of each plot of land. Hence, land prices are a measure (see Fig. 11.2a). The GDP per inhabitant and population density within an administrative boundary are not measures, since the mean GDP and mean population density of two administrative units need to be averaged and not summed. The total GDP instead of the GDP per inhabitant within the boundary is a measure. Note that the dimension of the space X is usually two in the context of urban studies, although it can be any value in theory.

Assuming a measure has been properly defined, then it is multifractal if it obeys the following two conditions.

1. The measure scales locally, i.e. for any ball $B(x, r)$ of radius r with a center x and contained within X, $\mu(B(x, r)) \sim r^{\alpha_x}$, for some $\alpha_x \in \mathbb{R}$;
2. the set of all centers x for which α_x is the same is a fractal for some appropriate definition of fractality, as defined in Sect. 11.2.

The curve representing the fractal dimensions versus their corresponding exponents α_x defines the multifractal spectrum and will always have a shape similar to Fig. 11.2c. If α_x is the same everywhere, then the spectrum collapses to a single point and the measure is said to be *monofractal.* To interpret the multifractal spectrum, it can be useful to consider what a local power law around a point x represents over a two-dimensional plane. It is in fact a conic shape centered on x. The lower α_x is compared to 2, the slower horizontal sections converge to the apex (the shape looks like a bowl). At 2, the cone flattens into a plane. Above 2, the shape is inverted and the higher α_x is above 2, the slower the horizontal sections widen away from the apex. Illustrations and a detailed interpretation of the α exponents can be found in [60]. The fractal dimensions represent how densely the corresponding shapes fill the space (with a maximum of 2 in our planar example).

The left-hand side of the spectrum therefore represents high value points surrounded by progressively lower value points and the right-hand side represents the converse. Consider house prices as an example. If the spectrum has a squared shape, then there exists many isolated heterogeneous points (exceptionally high or low value houses for their neighborhood) in the space and few situations where heterogeneity is gradual around a particular house. If the spectrum is asymmetrical, for example skewed to the right, then the most expensive houses are typically surrounded by similarly expensive houses, whereas cheaper houses are often found isolated among more expensive houses. These principles applied to Fig. 11.2 are explained in [60]. The multifractal framework therefore allows to ally both form and function to study urban morphology.

To compute the spectrum in practice, consider the mass of the measure μ_i contained within an area indexed by i. The first condition for multifractality can be written $\mu_i \sim r^{\alpha_i}$. To express the second condition, we define a function $f : \alpha \mapsto f(\alpha)$ that maps the scaling exponents to their fractal dimensions. The number of areas $N(\alpha_i)$ whose scaling is close to α_i can then be written $N(\alpha_i) \sim \rho(\alpha_i)d\alpha_i r^{-f(\alpha_i)}$ for some density function ρ. Now, consider for $q > 0$, the quantity

$$
\begin{aligned}
Z(q) := \sum_i \mu_i^q &\sim \int_\alpha N(\alpha) r^{\alpha q} \\
&\sim \int \rho(\alpha) r^{\alpha q - f(\alpha)} d\alpha \, .
\end{aligned}
$$

This technique is called the method of moments [59, 61]. It is a widely use technique in mathematics that works akin to a magnifying glass, since the exponentiation by

q allows to isolate specific parts of a phenomenon. In our case, only the value of α that minimizes

$$\tau(q) := \alpha q - f(\alpha)$$

makes a significant contribution to $Z(q)$. Then,

$$\begin{aligned} Z(q) &\approx r^{\tau(q)} \int_{\alpha} \rho(\alpha) d\alpha \\ &= r^{\tau(q)} . \end{aligned}$$

The quantity $Z(q)$, which can be directly observed, therefore provides an estimation of α and $f(\alpha)$, since

$$\alpha(q) = \frac{d\tau(q)}{dq} ,$$

and

$$f(\alpha(q)) = \alpha(q)q - \tau(q) ,$$

so that the full spectrum can be calculated from the computation of $\tau(q)$ for $-\infty < q < +\infty$.

The related *generalized dimension*, defined by

$$D_q := \begin{cases} \lim_{r\to 0} \left[\frac{1}{q-1} \frac{\log Z(q)}{\log r} \right] = \frac{\tau(q)}{q-1} & \text{if } q \neq 1 \\ \lim_{r\to 0} \frac{-\sum_i \mu_i \log \mu_i}{-\log(r)} & \text{if } q = 1 \end{cases} ,$$

is often encountered in the literature. D_0 is simply the usual box-counting dimension of the space where the measure is defined. D_1 is referred to as the *information dimension*. It relates to Shannon's entropy and captures how even the data density is. Higher values of D_1 indicate a more uniform density. D_2 is similar to the correlation dimension, which is the probability of pairs of independent events occurring in the same area. It measures how scattered the data is. Higher values of D_2 indicate higher compactness.

An exact multifractal is unlikely to occur naturally in urban systems. For this reason, it is often necessary to use approximate methods. For example, the gliding-box technique allows supersampling [62], while the multiplier method allows extracting more realistic partial local scaling exponents [63]. Images have been used as a measure [25, 64]. They can be used either as the trivial measure (0 for a blank pixel, 1 for a dark pixel), or as grayscale images. The fundamental unit of the structure is then given by a pixel, which depends on the image resolution. It is not clear how the resolution interferes with calculations. Additionally, a large number of boxes that only contain few points can lead to a large overestimation of the mass exponents [65]. One method that can resolve this problem is the sandbox method introduced by Tél, Fülöp, and Vicsek [66]. Other methods of data acquisition, such as censuses, pose a similar problem since administrative boundaries are also artificial and can potentially interfere with the calculations. These form of data are to be taken as approximation of an underlying phenomenon that scales continuously. Computing multifractal spectra is a technical and sprawling matter and the interested reader is referred to the literature for more details. A formal mathematical introduction of multifractal measures can be found in [9, Chap. 17]. A review of the most common practical methods can be found in [67]. Extra care must be taken when performing multifractal analysis, since any application outside of the mathematical world is likely to pose challenges.

11.6.2 Context in Urban Science

One of the first attempts to identify multiple fractal dimensions to characterize the morphology of an urban area comes from [57]. At this early stage, their study is limited to a general classification into five exponents. Later papers, by [68] and [25], explicitly use multifractality as a tool to study the complexity of street networks. Their approach is inspired by the successful application of the methodology to study the features of river and eye blood vessel networks. They report positive multifractality for the density of intersections in the street networks of Cordoba in Spain and of London in the UK. Their methodology relies on analyzing the networks as black and white images, more precisely as trivial measures equal to one if a pixel contains an intersection and zero otherwise. We have mentioned that when the multifractal spectrum collapses to a unique point, the measure is monofractal. This is what [25] report for the evolution of London's street network through time, between 1786 and 2010. They explain this phenomenon by the homogenization effect of constrained boundaries created by the green belt around London. Benguigui et al. [21] report similar results for the built pattern of Tel Aviv after 1985. The same convergence towards monofractality is also observable, although not reported as a result, in the work of [68], where a wider spectrum is obtained for the historic Roman-Arab-Christian parts of Cordoba compared to the parts redeveloped in the 1950s. This phenomenon is again observed by [69], who study the multifractality of the urban impervious surface in each pixel of landsat images of Shanghai between 1997 and 2010, and by [60], who study house and land prices over the last century in London, New York and Kyoto.

There have been additional reports of multifractality for the dynamics of city-size and settlement distributions [70–72], for city form and population growth processes [73, 74], and for agricultural land use [75]. These studies are primarily descriptive. By contrast, Chen [76] proposes a method that exploits multifractality to distinguish rural areas from urban areas. Similarly, [58] introduce an approach that uses multifractality to measure the Modifiable Areal Unit Problem, which is the fact that analyses in geography are dependent on artificially created partitions of space that may lead to inconsistent results. They hypothesize that the quantity of information contained inside each areal unit would be monofractal if administrative boundaries were well adapted to the studied measure. Multifractality could then be understood as a measure of the default to monofractality resulting from an inadequate partition of space.

Fractals have been widely used in connection to the study of economic efficiency and the detection of risks of a financial crisis. Most of the work related to that topic is focused on time-series, a tool that does not take into account the spatial nature of a city, but can take into account the temporal nature of its financial markets [e.g. 77–86]. However, [87, 88] and [60] have managed to apply the multifractal methodology to the spatial patterning of land prices. Hu et al. [87, 88] test two different methods in the city of Wuhan in China. A first attempt to predict the future evolution of land prices was made. The scarcity of the data, however, only allowed them to train their model over a limited range of three close years (2001, 2004 and 2007). The second method explains the land price distribution from the geographical background. Finally, [60] found multifractality for the land prices of Kyoto in Japan, both in 1912 and 2012, as well as for land lot prices of New York in 2015, and house prices of London in the 2000s. They compare the real land prices to extrema patterns simulated from the Kyoto land price distributions. They find that the patterns from the early 20th century (Kyoto) result in a wide multifractal spectrum, close to the maximum case when compared to the simulated data; while the patterns from the 21st century (Kyoto, New York and London) result in a narrow spectrum, close to the minimum case when compared to the simulated data, confirming this idea of loss of multifractality in recent times.

A different, more recent and fruitful, application of multifractals to urban science consists in characterizing spatial inequality in cities. [89, 90] and [91] have created a software, *Fractalopolis*,[2] to help to reduce urban sprawl, which is thought to have a negative impact on accessibility and segregation, by creating optimized strategies of urban development. The software iterates a multifractal pattern that takes into account a number of user defined accessibility, environmental and social rules. The goal is to find the most efficient way to limit sprawl while maintaining the quality of life of the citizens by ensuring that the chosen rules remain in place. Finally, an extensive study of the multifractality of various urban measures that are levers of inequality has been undertaken [92]. These measures are real estate prices, energy consumption, income and job accessibility. A better description of the heterogeneous patterns that result from these measures is obtained compared to classical inequality and segregation indicators, such as the Gini coefficient, the Neighborhood Sorting Index, the Centile Gap Index, and the Ordinal Variation Ratio Index. In particular, it was shown that the multifractal spectrum offers a simultaneous description of the exposure and the clustering dimensions of segregation, as defined by [93], for continuous variables, which is usually not achieved by classic indicators.

In conclusion, multifractals appear to solve the issues identified in Sect. 11.5, in relation to the lack of precision and accuracy of monofractals, which have a single fractal dimension associated to a single scaling exponent. The field is still new, but has already found some applications that could have a direct impact on urban development planning and policy making.

[2] https://sourcesup.renater.fr/www/fractalopolis/ (last accessed: 19.6.2024).

11.7 Summary

Fractals can be defined in different ways depending on the context. At the most general level, they are irregular spaces for which a dimension, usually non-integer, can be defined. They are often associated with self-similarity, although self-similarity is not a mandatory property for a space to be considered fractal. This property conflates the most common definitions of fractality. For practical use, the box-counting dimension, which extracts the rate at which space is filled by the fractal at different scales, is usually retained. In the urban morphology context, fractals are used to characterize city boundaries, city growth patterns, and street and transportation networks. They allow comparison between cities, the creation of realistic city models, can be used to measure efficiency, and they can explain why human activity exhibits scaling properties. They are thought to be the result of urban growth from random bottom-up processes and competition for space. Self-similarity in particular is thought to emerge from the different speeds imposed by the transportation network.

A critical review of fractal exponents and scaling laws in urban environments has revealed that a single exponent is usually not sufficient to characterize a city and that exponents may largely vary between cities, although a clear convergence towards more uniformity is observed in the recent development of cities. In order to achieve a more accurate description of cities and to link fractal form and scaling laws, the multifractal framework has been proposed. Multifractals extend the notion of fractals by studying the fractal dimension of sets where a specified urban measure behaves locally as a power law of similar exponent. Multifractals have been shown to arise naturally in self-organized systems such as cities. Applications to the study of street patterns and inequality measures have been found.

References

1. B. Mandelbrot, How long is the coast of Britain? statistical self-similarity and fractional dimension. Science **156**(3775), 636–638 (1967). https://doi.org/10.1126/science.156.3775.636
2. M.F. Goodchild, Fractals and the accuracy of geographical measures. J. Int. Assoc. Math. Geol. **12**(2), 85–98 (1980). https://doi.org/10.1007/BF01035241
3. B.B. Mandelbrot, *The Fractal Geometry of Nature* (W.H. Freeman, San Francisco, CA, 1982). ISBN: 978-0-7167-1186-5
4. M. Batty, P. Longley, The fractal simulation of urban structure. Environ. Plann. A **18**(9), 1143–1179 (1986). https://doi.org/10.1068/a181143
5. N. Salingaros, B. West, A universal rule for the distribution of sizes. Environ. Plann. B **26**(6), 909–923 (1999). https://doi.org/10.1068/b260909
6. M. Batty, The size, scale, and shape of cities. Science **319**(5864), 769–771 (2008). ISSN: 0036-8075, 1095-9203. https://doi.org/10.1126/science.1151419
7. G.B. West, A general model for the origin of allometric scaling laws in biology. Science **276**(5309), 122–126 (1997). ISSN: 00368075, 10959203. https://doi.org/10.1126/science.276.5309.122
8. H.E. Stanley, N. Ostrowsky (Eds.), *On Growth and Form: Fractal and Non-fractal Patterns in Physics* (Springer, Dordrecht, The Netherlands, 1986). ISBN: 978-0-89838-850-3. https://doi.org/10.1007/978-94-009-5165-5

9. K.J. Falconer, *Fractal Geometry: Mathematical Foundations and Applications*, 2nd edn. (Wiley, Chichester, UK, 2003). ISBN: 978-0-470-84861-6 / 978-0-470-84862-3
10. M. Batty, P. Longley, S. Fotheringham, Urban growth and form: scaling, fractal geometry, and diffusion-limited aggregation. Environ. Plann. A: Econ. Space **21**(11), 1447–1472 (1989). https://doi.org/10.1068/a211447
11. A.S. Fotheringham, M. Batty, P.A. Longley, Diffusion-limited aggregation and the fractal nature of urban growth. Papers Reg. Sci. Assoc. **67**(1), 55–69 (1989). https://doi.org/10.1007/BF01934667
12. T.A. Witten, L.M. Sander, Diffusion-limited aggregation, a kinetic critical phenomenon. Phys. Rev. Lett. **47**(19), 1400–1403 (1981). https://doi.org/10.1103/PhysRevLett.47.1400
13. P. Frankhauser, Fractal geometry of urban patterns and their morphogenesis. Discrete Dyn. Nat. Soc. **2**(434163) (1998). https://doi.org/10.1155/S1026022698000107
14. M. Batty, P.A. Longley, Urban shapes as fractals. Area **19**(3), 215–221 (1987). https://www.jstor.org/stable/20002475
15. M. Batty, P.A. Longley, Fractal-based description of urban form. Environ. Plann. B: Plann. Des. **14**(2), 123–134 (1987). https://doi.org/10.1068/b140123
16. P. Longley, M. Batty, On the fractal measurement of geographical boundaries. Geogr. Anal. **21**(1), 47–67 (1989). https://doi.org/10.1111/j.1538-4632.1989.tb00876.x
17. C. Tannier et al., A fractal approach to identifying urban boundaries. Geogr. Anal. **43**(2), 211–227 (2011). https://doi.org/10.1111/j.1538-4632.2011.00814.x
18. C. Tannier, I. Thomas, Defining and characterizing urban boundaries: a fractal analysis of theoretical cities and Belgian cities. Comput. Environ. Urban Syst. **41**, 234–248 (2013). https://doi.org/10.1016/j.compenvurbsys.2013.07.003
19. P. Frankhauser, The fractal approach. A new tool for the spatial analysis of urban agglomerations. Popul. Eng. Sel. **10**(1), 205–240 (1998). http://www.jstor.org/stable/2998685
20. P. Frankhauser, Comparing the morphology of urban patterns in Europe. A fractal approach, in *European Cities: Insights on Outskirts, Structures*, ed. by P. Borsdorf, P. Zembri (ESF COST Office, Brussels, Belgium, 2004), pp. 79–105. ISBN: 9782110856630
21. L. Benguigui et al., When and where is a city fractal?. Environ. Plann. B: Plann. Des. **27**(4), 507–519 (2000). https://doi.org/10.1068/b2617
22. M.-L. De Keersmaecker, P. Frankhauser, I. Thomas, Using fractal dimensions for characterizing intra-urban diversity: the example of brussels. Geogr. Anal. **35**(4), 310–328 (2003). https://doi.org/10.1111/j.1538-4632.2003.tb01117.x
23. L. Benguigui, The fractal dimension of some railway networks. J. Phys. I Fr. **2**(4), 385–38 (1992). https://doi.org/10.1051/jp1:1992151
24. L. Benguigui, A fractal analysis of the public transportation system of Paris. Environ. Plann. A: Econ. Space **27**(7), 1147–1161 (1995). https://doi.org/10.1068/a271147
25. R. Murcio et al., Multifractal to monofractal evolution of the London street network. Phys. Rev. E **92**(6) (2015). ISSN: 1539-3755, 1550-2376. https://doi.org/10.1103/PhysRevE.92.062130
26. S. Derrible, C. Kennedy, The complexity and robustness of metro networks. Physica A: Stat. Mech. Appl. **389**(17), 3678–3691 (2010). ISSN: 03784371. https://doi.org/10.1016/j.physa.2010.04.008
27. R. Louf, C. Roth, M. Barthelemy, Scaling in transportation networks. PLoS ONE **9**(7), e102007 (2014). https://doi.org/10.1371/journal.pone.0102007
28. C. Tannier et al., Spatial accessibility to amenities in fractal and nonfractal urban patterns. Environ. Plann. B: Plann. Des. **39**(5), 801–819 (2012). https://doi.org/10.1068/b37132
29. C. Molinero, A fractal theory of urban growth. Front. Phys. **10** (2022). https://doi.org/10.3389/fphy.2022.861678
30. G.B. West, Size, scale and the boat race; conceptions, connections and misconceptions, in *Hierarchy in Natural and Social Sciences*, ed. by D. Pumain, vol. 3 (Springer, Berlin, Heidelberg, 2006), pp. 71–80. ISBN: 978-1-4020-4126-6. https://doi.org/10.1007/1-4020-4127-6_4
31. G.B. West, *Scale: The Universal Laws of Growth, Innovation, Sustainability, and the Pace of Life in Organisms, Cities, Economies, and Companies* (Penguin Press, New York, NY, 2017). ISBN: 978-1-59420-558-3

32. M. Batty, P. Longley, *Fractal Cities: A Geometry of Form and Function* (Academic Press, London, UK, 1994). ISBN: 978-0-12-455570-9
33. M. Batty, *Cities and Complexity: Understanding Cities with Cellular Automata, Agent-based Models, and Fractals* (MIT Press, Cambridge, MA, 2005). ISBN: 978-0-262-02583-6
34. M. Batty, *The New Science of Cities* (MIT Press, Cambridge, MA, 2013). ISBN: 978-0-262-01952-1
35. F. Schweitzer, J. Steinbrink, Analysis and computer simulation of urban cluster distributions, in *Fundamental Principles of Urban Growth*, ed. by K. Humpert, K. Brenner, S. Becker (Müller + Busmann, Wuppertal, Germany, 2002), pp. 142–157. ISBN: 9783928766517
36. D. Pumain, Scaling Laws and Urban Systems. Santa Fe Institute Working Papers. Paper No. 2004-02-002, https://www.santafe.edu/research/results/working-papers/scaling-laws-and-urban-systemsURL (2004)
37. J.R. Banavar, A. Maritan, A. Rinaldo, Size and form in efficient transportation networks. Nature **399**, 130–132 (1999). ISSN: 0028-0836. https://doi.org/10.1038/20144
38. S. Arbesman, J.M. Kleinberg, S.H. Strogatz, Superlinear scaling for innovation in cities. Phys. Rev. E, Stat., Nonlinear, Soft Matter Phys. **79**(1), 016115 (2009). ISSN: 1539-3755. https://doi.org/10.1103/PhysRevE.79.016115
39. W. Pan et al., Urban characteristics attributable to density-driven tie formation. Nat. Commun. **4**, 1961 (2013). ISSN: 2041-1723. https://doi.org/10.1038/ncomms2961
40. L.M.A. Bettencourt, The origins of scaling in cities. Science **340**(6139), 1438–1441 (2013). ISSN: 0036-8075, 1095-9203. https://doi.org/10.1126/science.1235823
41. K. Yakubo, Y. Saijo, D. Korosak, Superlinear and sublinear urban scaling in geographical networks modeling cities. Phys. Rev. E, Stat., Nonlinear, Soft Matter Phys. **90**(2), 022803 (2014). ISSN: 1550-2376. https://doi.org/10.1103/PhysRevE.90.022803
42. A. Gomez-Lievano, O. Patterson-Lomba, R. Hausmann, Explaining the prevalence, scaling and variance of urban phenomena. Nat. Hum. Behav. **1**(1), 0012 (2016). ISSN: 2397-3374. https://doi.org/10.1038/s41562-016-0012
43. C. Molinero, S. Thurner, How the geometry of cities determines urban scaling laws. J. R. Soc. Interface **18**(176) (2021). https://doi.org/10.1098/rsif.2020.0705
44. F.L. Ribeiro, D. Rybski, Mathematical models to explain the origin of urban scaling laws. Phys. Rep. **1012**, 1–39 (2023). https://doi.org/10.1016/j.physrep.2023.02.002
45. F.L. Ribeiro et al., A model of urban scaling laws based on distance dependent interactions. R. Soc. Open Sci. **4**(160926), 160926 (2017). https://doi.org/10.1098/rsos.160926
46. I. Thomas, P. Frankhauser, D. Badariotti, Comparing the fractality of European urban neighbourhoods: do national contexts matter? J. Geogr. Syst. **37**, 189–208 (2012). https://doi.org/10.1007/s10109-010-0142-4
47. E. Arcaute et al., Constructing cities, deconstructing scaling laws. J. R. Soc. Interface **12**(102), 20140745 (2015). https://doi.org/10.1098/rsif.2014.0745
48. L.M.A. Bettencourt et al., Growth, innovation, scaling, and the pace of life in cities. Proc. Natl. Acad. Sci. **104**(17), 7301–7306 (2007). ISSN: 0027-8424, 1091-6490. https://doi.org/10.1073/pnas.0610172104
49. H. Youn et al., Scaling and universality in urban economic diversification. J. R. Soc. Interface **13**(114), 20150937 (2016). ISSN: 1742-5689, 1742-5662. https://doi.org/10.1098/rsif.2015.0937
50. L.M.A. Bettencourt, J. Lobo, Urban scaling in Europe. J. R. Soc. Interface **13**(116), 20160005 (2016). ISSN: 1742-5689, 1742-5662. https://doi.org/10.1098/rsif.2016.0005
51. J.C. Leitão et al., Is this scaling nonlinear?. R. Soc. Open Sci. **3**(7), 150649 (2016). https://doi.org/10.1098/rsos.150649
52. C. Cottineau et al., Diverse cities or the systematic paradox of urban scaling laws. Comput. Environ. Urban Syst. **63**, 80–94 (2017). https://doi.org/10.1016/j.compenvurbsys.2016.04.006
53. A.P. Masucci et al., On the problem of boundaries and scaling for urban street networks. J. R. Soc. Interface **12**(111), 20150763 (2015). https://doi.org/10.1098/rsif.2015.0763
54. E. Arcaute et al., Cities and regions in Britain through hierarchical percolation. R. Soc. Open Sci. **3**(4), 150691 (2016). https://doi.org/10.1098/rsos.150691

55. R. Louf, M. Barthelemy, Scaling: lost in the smog. Environ. Plann. B: Plann. Des. **41**(5), 767–769 (2014). https://doi.org/10.1068/b4105c
56. B.T. Milne, Lessons from applying fractals models to landscape patterns, in *Quantitative Methods in Landscape Ecology*, ed. by M.G. Turner, R.H. Gardner (Springer, New York, 1990), pp. 200–235. https://doi.org/10.1007/978-3-030-96611-9_10
57. S. Encarnação et al., Fractal cartography of urban areas. Sci. Rep. **2**, 527 (2012). https://doi.org/10.1038/srep00527
58. F. Semecurbe, C. Tannier, S. Roux, Spatial distribution of human population in France: exploring the modifiable areal unit problem using multifractal analysis. Geogr. Anal. **48**(3), 292–313 (2016). https://doi.org/10.1111/gean.12099
59. M. Ghil, R. Benzi, G. Parisi (eds.), *Turbulence and Predictability in Geophysical Fluid Dynamics and Climate Dynamics* (North-Holland Publ. Co., Amsterdam/New York, 1985), p. 449. ISBN: 9780444869364
60. H. Salat et al., Uncovering inequality through multifractality of land prices: 1912 and contemporary Kyoto. PLoS One (2018). https://doi.org/10.1371/journal.pone.0196737
61. H. Atmanspacher, H. Scheingraber, G. Wiedenmann, Determination of $f(\alpha)$ for a limited random point set. Phys. Rev. A, General Phys. **40**(7), 3954–3963 (1989). ISSN: 0556-2791. https://doi.org/10.1103/PhysRevA.40.3954
62. Q. Cheng, The gliding box method for multifractal modeling. Comput. Geosci. **25**(9), 1073–1079 (1999). ISSN: 0098-3004. https://doi.org/10.1016/S0098-3004(99)00068-0
63. A.B. Chhabra et al., Direct determination of the $f(\alpha)$ singularity spectrum and its application to fully developed turbulence. Phys. Rev. A **40**(9), 5284–5294 (1989). ISSN: 0556-2791. https://doi.org/10.1103/PhysRevA.40.5284
64. S.G. De Bartolo, R. Gaudio, S. Gabriele, Multifractal analysis of river networks: sandbox approach. Water Resour. Res. **40**(2) (2004). https://doi.org/10.1029/2003WR002760
65. R. Riedi, An improved multifractal formalism and self-similar measures. J. Math. Anal. Appl. **189**(2), 462–490 (1995). https://doi.org/10.1006/jmaa.1995.1030
66. T. Tél, Á. Fülöp, T. Vicsek, Determination of fractal dimensions for geometrical multifractals. Physica A: Stat. Mech. Appl. **159**(2), 155–166 (1989). ISSN: 0378-4371. https://doi.org/10.1016/0378-4371(89)90563-3
67. H. Salat, R. Murcio, E. Arcaute, Multifractal methodology. Physica A: Stat. Mech. Appl. **473**, 467–487 (2017). ISSN: 03784371. https://doi.org/10.1016/j.physa.2017.01.041
68. A.B. Ariza-Villaverde, F.J. Jiménez-Hornero, E. Gutiérrez De Ravé, Multifractal analysis of axial maps applied to the study of urban morphology. Comput. Environ. Urban Syst. **38**, 1–10 (2013). ISSN: 01989715. https://doi.org/10.1016/j.compenvurbsys.2012.11.001
69. Q. Nie, J. Xu, Z. Liu, Fractal and multifractal characteristic of spatial pattern of urban impervious surfaces. Earth Sci. Inf. **8**(2), 381–392 (2015). ISSN: 1865-0473, 1865-0481. https://doi.org/10.1007/s12145-014-0159-1
70. G. Haag, The rank-size distribution of settlements as a dynamic multifractal phenomenon. Chaos, Solitons Fractals **4**(4), 519–534 (1994). ISSN: 09600779. https://doi.org/10.1016/0960-0779(94)90063-9
71. Y. Chen, Y. Zhou, Multi-fractal measures of city-size distributions based on the three-parameter Zipf model. Chaos, Solitons Fractals **22**(4), 793–805 (2004). ISSN: 09600779. https://doi.org/10.1016/j.chaos.2004.02.059
72. Y. Chen, Multifractals of central place systems: models, dimension spectrums, and empirical analysis. Physica A: Stat. Mech. Appl. **402**, 266–282 (2014). ISSN: 03784371. https://doi.org/10.1016/j.physa.2014.01.061
73. Y. Chen, A wave-spectrum analysis of urban population density: entropy, fractal, and spatial localization. Discrete Dyn. Nat. Soc. **2008**, 1–22 (2008). ISSN: 1026 0226, 1607-887X. https://doi.org/10.1155/2008/728420
74. Y. Chen, J. Wang, Multifractal characterization of urban form and growth: the case of Beijing. Environ. Plann. B: Plann. Des. **40**(5), 884–904 (2013). ISSN: 0265-8135, 1472-3417. https://doi.org/10.1068/b36155

75. D. Wang et al., Multifractal analysis of land use pattern in space and time: a case study in the Loess Plateau of China. Ecol. Compl. **7**(4), 487–493 (2010). ISSN: 1476945X. https://doi.org/10.1016/j.ecocom.2009.12.004
76. Y. Chen, Defining urban and rural regions by multifractal spectrums of urbanization. Fractals **24**(01), 1650004 (2016). ISSN: 0218-348X, 1793-6543. https://doi.org/10.1142/S0218348X16500043
77. B.B. Mandelbrot, A multifractal walk down wall street. Sci. Am. **280**(2), 70–73 (1999). https://doi.org/10.1038/scientificamerican0299-70
78. J.-F. Muzy, J. Delour, E. Bacry, Modelling fluctuations of financial time series: from cascade process to stochastic volatility model. Eur. Phys. J. B **17**(3), 537–548 (2000). ISSN: 1434-6028. https://doi.org/10.1007/s100510070131
79. E. Bacry, J. Delour, J.-F. Muzy, Modelling financial time series using multifractal random walks. Physica A: Stat. Mech. Appl. **299**(1-2), 84–92 (2001). ISSN: 03784371. https://doi.org/10.1016/S0378-4371(01)00284-9
80. T. Di Matteo, T. Aste, M.M. Dacorogna, Scaling behaviors in differently developed markets. Physica A: Stat. Mech. Appl. **324**(1-2), 183–188 (2003). https://doi.org/10.1016/S0378-4371(02)01996-9
81. T. Di Matteo, Multi-scaling in finance. Quant. Finance **7**(1), 21–36 (2007). https://doi.org/10.1080/14697680600969727
82. J. Barunik et al., Understanding the source of multifractality in financial markets. Physica A: Stat. Mech. Appl. **391**(17), 4234–4251 (2012). issn: 03784371. https://doi.org/10.1016/j.physa.2012.03.037
83. R. Morales, T. Di Matteo, T. Aste, Non-stationary multifractality in stock returns. Physica A: Stat. Mech. Appl. **392**(24), 6470–6483 (2013). ISSN: 03784371. https://doi.org/10.1016/j.physa.2013.08.037
84. R.J. Buonocore, T. Aste, T. Di Matteo, Measuring multiscaling in financial time-series. Chaos, Solitons Fractals **88**, 38–47 (2016). ISSN: 09600779. https://doi.org/10.1016/j.chaos.2015.11.022
85. R.J. Buonocore, T. Aste, T. Di Matteo, Asymptotic scaling properties and estimation of the generalized Hurst exponents in financial data. Phys. Rev. E **95**(4) (2017). https://doi.org/10.1103/PhysRevE.95.042311
86. Z.-Q. Jiang et al., Multifractal analysis of financial markets: a review. Rep. Prog. Phys. **82**(12), 125901 (2019). https://doi.org/10.1088/1361-6633/ab42fb
87. S. Hu et al., Multifractal characterization of urban residential land price in space and time. Appl. Geogr. **34**, 161–170 (2012). ISSN: 01436228. https://doi.org/10.1016/j.apgeog.2011.10.016
88. S. Hu et al., Modeling land price distribution using multifractal IDW interpolation and fractal filtering method. Landscape Urban Plann. **110**, 25–35 (2013). ISSN: 01692046. https://doi.org/10.1016/j.landurbplan.2012.09.008
89. C. Yamu, P. Frankhauser, Spatial accessibility to amenities, natural areas and urban green spaces: using a multiscale, multifractal simulation model for managing urban sprawl. Environ. Plann. B: Plann. Des. **42**(6), 1054–1078 (2013). ISSN: 0265-8135, 1472-3417. https://doi.org/10.1068/b130171p
90. C. Yamu, A. van Nes, An integrated modeling approach combining multifractal urban planning with a space syntax perspective. Urban Sci. **1**(4), 37 (2017). ISSN: 2413-8851. https://doi.org/10.3390/urbansci1040037
91. P. Frankhauser et al., An integrated multifractal modelling to urban and regional planning. Comput. Environ. Urban Syst. **67**, 132–146 (2018). ISSN: 01989715. https://doi.org/10.1016/j.compenvurbsys.2017.09.011
92. H. Salat, Multifractality applied to the study of spatial inequality in urban systems. PhD thesis. University College London, UK, UCL (2019)
93. D.S. Massey, N.A. Denton, The dimensions of residential segregation. Soc. Forces **67**(2), 281–315 (1988). https://doi.org/10.1093/sf/67.2.281

Chapter 12
Entropy and the City: Origins, Trajectories and Explorations of the Concept in Urban Science

Vinicius M. Netto, Otavio M. Peres, and Caio Cacholas

Abstract Entropy is arguably one of the most powerful concepts to understand the world, from the behavior of molecules to the expansion of the universe, from how life emerges to how hybrid complex systems like cities come into being and continue existing. Yet, despite its widespread application, it is also one of the most misunderstood concepts across the sciences. This chapter seeks to demystify entropy and its main interpretations, along with some of its explorations in the context of cities. It first establishes the foundations of the concept by describing its trajectory since its inception in thermodynamics and statistical mechanics in the 19th century, its different incarnations from Boltzmann's pioneering formulation and Shannon's information theory to its absorption in biology and the social sciences, until it reaches a nascent urban science in the 1960s. The chapter then identifies some of the main domains in which entropy has been explored to understand cities as complex systems—from entropy-maximizing models of spatial interaction and applications as a measure of urban form, diversity, and complexity to a tool for understanding conditions of self-organization and urban sustainability.

V. M. Netto (✉)
Research Centre for Territory, Transports and Environment (CITTA), Faculty of Engineering (FEUP), University of Porto, Porto, Portugal
e-mail: vmnetto@fe.up.pt

O. M. Peres
Faculty of Architecture and Urbanism (FAUrb), Federal University of Pelotas (UFPel), Pelotas, Brazil
e-mail: otavio.peres@ufpel.edu.br

C. Cacholas
Graduate Programme of Architectural and Urban Studies (PPGAU), Federal Fluminense University (UFF), Niterói, Brazil
e-mail: caiocacholas@id.uff.br

D. Rybski (ed.), *Compendium of Urban Complexity*, Understanding Complex Systems,
https://doi.org/10.1007/978-3-031-82666-5_12

12.1 Introduction

> Now, entropy is a very strange concept. Prigogine [1]

Entropy is seen everywhere—from the behavior of molecules to the evolution of the universe, from how life emerges to how hybrid complex systems like cities [2] come into being and continue to exist. Entropy is arguably one of the most powerful concepts to understand the world. Indeed, there are few concepts that have more widespread applicability in physical sciences than entropy and its irresistible quality in defining the structure and behavior of systems in different scientific fields [3]. Yet, it is also one of the most misunderstood concepts in science and common sense alike. It is so for several reasons, both in general and within the specific context of urban studies and the science of cities. It is a complex property rooted in thermodynamics, statistical mechanics, and information theory, which can be highly abstract and counter-intuitive, involving mathematical formulations and probabilistic reasoning. The fact that the concept often involves leaps from micro to macro scales poses challenges to understanding how the statistical behavior of individual parts contributes to the macroscopic behavior of a system, including cities, where the behavior of individual components contributes to emergent properties at the city level.

The frequent understanding of entropy as disorder can be misleading. While entropy is associated with the statistical measure of disorder in a thermodynamic system, this property does not directly translate into disorder in the everyday sense. The tremendous success of the concept in finding applications across a wide range of disciplines, from physics to urban studies, can make it difficult to reconcile its different interpretations and appreciate its universality. Finally, applying entropy to urban studies involves dealing with the multifaceted dynamics of cities, which are influenced by social, economic, and environmental factors. The complexity of urban systems can make it challenging to isolate and quantify the specific contributions of entropy, leading to potential misinterpretations. Entropy is applied in various fields of urban science, such as modelling spatial interactions, measuring diversity, assessing sustainability, and understanding self-organization. The diversity of applications can also contribute to confusion, as we may encounter entropy in different forms, materialities, and contexts.

This chapter seeks to demystify entropy and its main interpretations, along with some of its main explorations in the context of cities. It is intended as a means of understanding a famously hard-to-define concept while making sense of its plethora of applications. It aims to do so by identifying the main problems to which the concept has been put to use in research on cities, thus aiming to offer some semantic organization to an increasingly diverse area. It is intended as an introductory but comprehensive guide for those working in both qualitative and quantitative research in urban studies and the so-called "new science of cities". We begin by briefly elucidating the historical evolution of the concept, tracing its path from Clausius' [4] thermodynamic understanding and Boltzmann's [5] groundbreaking work on statistical mechanics in 19th-century to Claude Shannon's [6] information theory.

This foundation serves as a launching pad for its subsequent trajectories in biology and social sciences, filtering down into specialized fields of an emerging urban science that have gained momentum since the late 1960s. We shall unfold the multiple domains where the concept has been explored as a powerful lens to decipher the intricacies of cities—namely, in the following applications.

- Spatial interaction models of urban systems grounded in entropy-maximizing principles;
- The complexity and information latent in cities, i.e. how urban systems, like other dynamical systems, tend toward states of higher disorder and randomness associated with complexity and the amount of information in the built environment;
- Entropy as an approach to urban form, a lens to dissect the complexity inherent in the physical structures of cities, such as building footprints and street networks, as potential information signatures of distinct spatial cultures;
- Shannon entropy as a measure of urban diversity—a metric essential for comprehending the intricate fabric that defines cities;
- Entropy as a tool to measure spatial segregation in urban landscapes;
- Entropy as a means to understand urban densities and the forces that shape the growth and sprawl of cities over time;
- Entropy as a crucial property in the self-organization of cities and their role in social dynamics and cooperation, the intricate processes that allow cities to evolve in response to internal and external stimuli and fluctuations;
- Entropy as a means to assess urban sustainability—the complex interplay between human activities, environmental factors, and the resilience of urban systems.

We build such a panorama of multifaceted properties, domains, and applications of entropy, aiming for both an intuitive and nuanced understanding of this concept and how it has been used to unveil the dynamics of cities and their ever-evolving social and material fabric. Like other chapters in this compendium, we shall explore the measures associated with such domains, along with some of the main empirical findings in the fields. We shall conclude the chapter with some critical and open questions in the field of entropy research in urban science and a brief summary of the main ideas explored.

12.2 Historical Trajectories of the Concept

> The difficulties over the definition of entropy seem to originate from the fact that at one extreme, the concept is the basis of the second law of thermodynamics, which states that the entropy of a physical system must always increase, whereas, at the other extreme, the concept is used in defining the amount of information contained in a probability distribution. In fact, there is a somewhat tortuous route linking these two interpretations through arguments involving statistical physics. Yet the implications of the concept are still difficult to trace in any definitive sense. Batty [3]

How did the concept and measure of entropy come to be related to the city? The study of entropy certainly cuts across many different research domains, from its original context in physics through the pioneering definitions of information theory and the life sciences to views of society, economies, and the evolution of cities. We bring here a brief account of such a rich trajectory of the concept, pointing out their connections with and explorations in urban studies and the science of cities.

12.2.1 Origins in Thermodynamics and Statistical Mechanics: Clausius and Boltzmann Entropies

Clausius [4] introduced the concept of entropy in 1865 as a crucial idea in the Second Law of Thermodynamics. Etymologically, he derived the term in relation to the physical property of energy and the Greek word for change or transformation, "tropē". Clausius [4] defined entropy as the internal property that changes as heat energy moves around within a system—"the differential of a quantity which depends on the configuration of the system, but which is altogether independent of the manner in which that configuration has come about" [7]. Entropy is a way of understanding the arrow of time, the tendency of entropy to increase while energy tends to disperse and reach homogeneity. Ludwig Boltzmann [5] took the concept further in his kinetic theory of gases. Boltzmann's proposition is closely tied to the statistical interpretation of thermodynamics. He sought to connect the macroscopic behavior of a system, as described by classical thermodynamics, with the microscopic behavior of its constituent components, such as atoms or molecules. A microstate is the exact configuration or arrangement of positions of entities in the system. A macrostate, on the other hand, represents the macroscopic properties of the system, like temperature, pressure, and energy. A critical concept to think of entropy, particularly in connection with cities, is the definition of *phase space*. This mathematical space represents all possible configurations or positions of components in a system. Boltzmann's approach aims to explain macroscopic properties from the statistical distribution of microstates. Essentially, entropy is related to the number of ways the components in a system can be arranged while still producing the same macroscopic state.

Entropy is statistical in nature, i.e. for a given set of large-scale observable properties, every possible configuration of components that could give those properties is equally likely [8]. Systems with similar macroscopic properties can be characterized by their probability distribution in the phase space. The probability distribution describes the likelihood of finding the system in a given microstate. Basically, a macrostate gives the same or similar average distributions of microstates. The average distribution in the phase space is calculated by considering the probabilities of various microstates. Different distributions can lead to similar averages when they correspond to equivalent macrostates or satisfy certain statistical principles, like the ergodic hypothesis, i.e. over a sufficiently long time, a system will explore all accessible microstates (constrained by the laws of physics) with equal probability [5, 9].

The probability of a microstate is calculated through

$$S = \kappa \ln W ,$$

where S is the entropy, κ is the Boltzmann constant, and W is the number of possible microstates corresponding to a macrostate of the system. Entropy is proportional to the natural logarithm of the number of microstates associated with a particular macrostate. Even though the number of possible microstates may be exceedingly hard to calculate (imagine particles randomly moving within a room), there is no arbitrariness in deciding which states to count in, i.e. we equally count all states compatible with the macroscopic parameters [10]. In the absence of additional information, all microstates consistent with a given macrostate are equally probable, so these random distributions correspond to the same or equivalent microstates and satisfy the ergodic hypothesis. The goal is to explain certain macro-properties of the system to a desired degree of accuracy without having to explain the behavior of each individual component of configurations at the micro-level [11]. A vast number of possibilities of random configurations corresponding to the same average distribution or macrostate contain high entropy. In fact, the more microstates corresponding to a macrostate, the higher the entropy. The macrostate with the highest number of microstates has maximum entropy. This reflects the degree of disorder or uncertainty in the system.

Imagine a space with two *degrees of freedom* or coordinates to describe the position of cells in a 42x42 cellular grid (Fig. 12.1). This space contains 512 black cells whose different distributions represent different microstates or configurations of cells. Every specific configuration has the same probability of occurring, but there are many configurations that have roughly the same macro-properties. The first ones (cases 1-3) show different random microstates which give a similar average distribution and coarse grain into approximately the same macrostate. In fact, the vast majority of possible configurations correspond to such randomly mixed microstates, where cell distributions tend to contain none to low internal correlations.

In turn, different ordered configurations do not correspond to the same macrostate, like cases 5-9. Internally regular arrangements like spirals, clustered cells dispersing with the distance, checkerboard distributions and cellular rings, either deformed or orthogonal, correspond to rarer microstates in the set of possible microstates, so they seem like drops of order in a sea of disordered states. The larger the system, the less likely it is to find rare configurations like them—so rare they may never happen. At this point, we can only mention that biological structuring or human creation might be needed for such microstates to occur in real systems and become, say, organic bodies or cities (we will come back to this discussion in Sect. 12.3.3). Even when we consider rare distributions containing internal correlations, such as cases 4-9, only 6-8 resemble urban configurations—and, in a closer look, only cases 7-8 are capable of generating density through contiguity between cells while keeping permeability between blocks and could evolve into something like a city [12]. Of course, the very formation of such ring-shaped micro-structures is a highly unlikely event [13, 14].

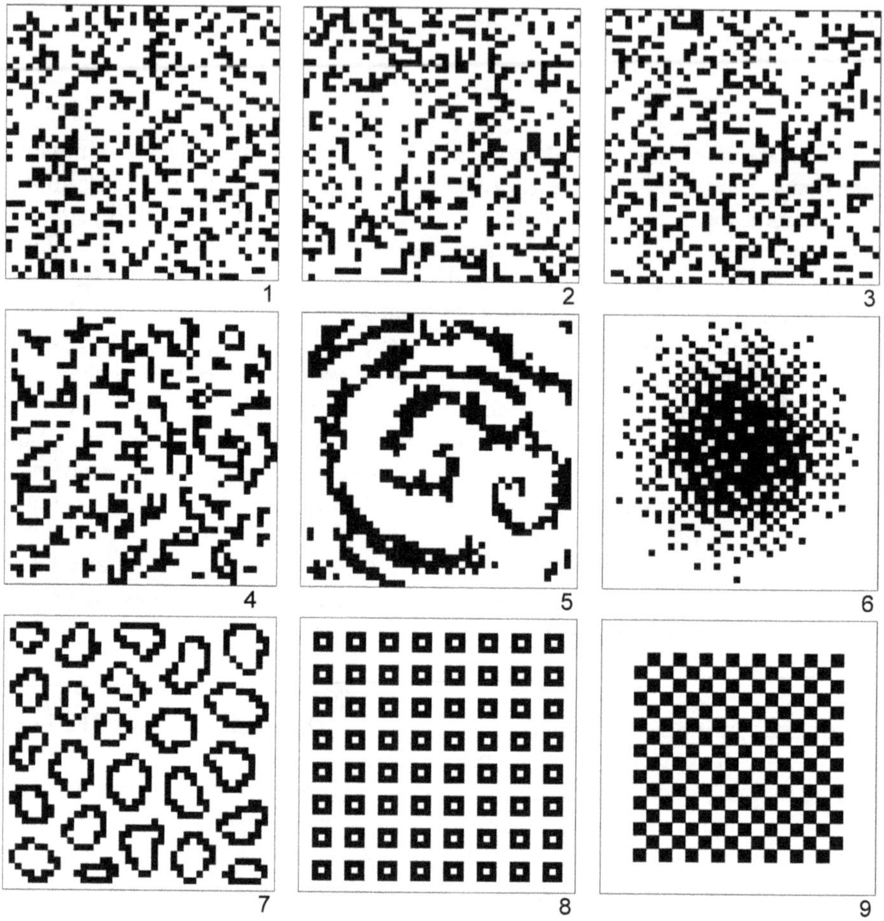

Fig. 12.1 Illustrative configurations in 42×42 grids. All configurations contain 512 black cells, so entropy varies only in relation to distribution. From top left to bottom right: cases 1-3 show different random microstates, which give a similar average distribution and may coarse-grain into the same macrostate. Microstate 4 would give a slightly lower entropy level. Microstates 5-9 show decreasing entropy levels and coarse-grain into different macrostates. Only configurations 6-8 resemble anything close to an urban distribution

12.2.2 Gibbs Entropy

Another way to look into entropy in contrast to Boltzmann's [5] microscopic understanding of the probability of different arrangements, coarse-graining into macrostates is Gibbs' [15] entropy, a function of the probability distribution over the phase space [16] (cf. [17, 18]). Gibbs wished "to determine how the whole number of systems will be distributed among the various conceivable configurations and velocities at any required time when the distribution has been given for some one time" [15]. For a system in equilibrium, entropy is related to the probabilities P_i of its microstates

$$S_G = -\kappa \sum_i P_i \ln(P_i) .$$

While the Boltzmann entropy is the entropy for a system in thermodynamical equilibrium, the Gibbs entropy is a generalization of the Boltzmann entropy holding for all systems, i.e. including non-equilibrium states. Both are a measure of the microstates available to a system, but the Gibbs entropy does not require the system to be in a single, well-defined macrostate. In equilibrium, all microstates belonging to the macrostate are equally likely, so, for N states, we obtain with $(P_i = 1/N)$

$$\begin{aligned} S_G &= -\kappa \sum_i \frac{1}{N} \ln\left(\frac{1}{N}\right) \\ &= -\kappa N \frac{1}{N} \ln(\frac{1}{N}) \\ &= -\kappa \ln N , \end{aligned}$$

where the latter term is the Boltzmann entropy for a system with N microstates [19]. If the probability distribution is very concentrated at one point in the phase space and very small everywhere else, the entropy state tends to zero [20]. When the probability of an event, position, or frequency of occurrence is evenly distributed over an area or throughout a space or unit of time, the level of uncertainty in the system dynamics and randomness in its distribution tends to be very high. It is very hard to predict the behavior of a system like this (cf. [21]). Differently from the arrangements in Fig. 12.1, which illustrated the probability of an arrangement as a whole, we are now dealing with the probability distribution of events or cells occurring within an arrangement (Fig. 12.2).

As a measure related to the probability distribution of events within a microstate, the Gibbs entropy may be particularly useful to understand systems with different gradients from mixed to clustered arrangements, such as the locational distribution of land uses (Sect. 12.3.4) and the clustering of social groups in spatial segregation (Sect. 12.3.5). As we shall see below, cities may not be examples of high entropy distributions but navigate towards low entropy states in varying degrees within the interaction of chaos and order [22].

12.2.3 *Information Theory: Shannon Entropy*

Shannon's [6] pioneering definition of information was proposed as a measure of uncertainty in the context of the problem of transmission of data, noise, and channel capacity from the point of view of engineering. The fundamental problem of communication is that of reproducing at one point a message selected at another

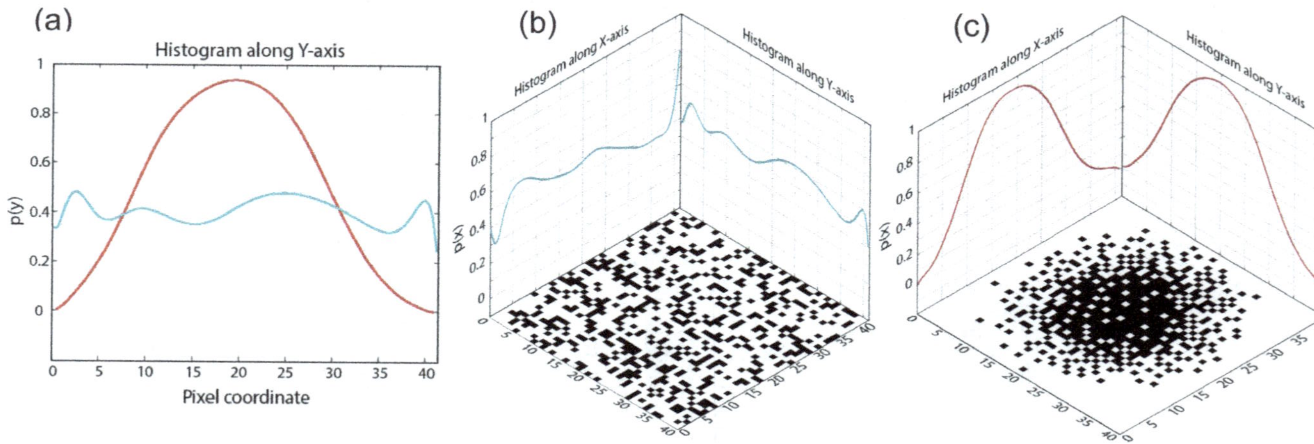

Fig. 12.2 **a** Histogram of the probability of events in a spatial distribution. One scenario with a high level of uncertainty or entropy (blue line) where every possible event (from 1 to 42) had a similar chance of occurring, and another with a low level of entropy (red line), where the probability of occurrence is concentrated around central values. These two probabilities may be illustrated by spatial distributions inspired by Gibbs' probability function. **b** A random distribution where cell positions find no pattern, implying that it would be hard to predict where the next cell would occur; and **c** a clustered distribution of cells with a dropping density according to the distance from the center ($x = 0$, $y = 0$). In such a distribution, the probability of position of new cells is given by the red curves

point, either exactly or approximately. If the number of messages in a set is finite, entropy is a measure of the distribution of probabilities that certain messages will occur. Similarly to Gibbs and Boltzmann, it defines entropy in terms of probabilities P_x of events. It quantifies the efficiency of a system, such as a code or language, in transmitting information. It is defined as the logarithm of the number of distinct messages that can be conveyed by selecting from the same set of symbols

$$H = -\sum P_x \log_2(P_x)\,.$$

This measurement gives the extent of initial uncertainty that can be resolved by a single message. It is a fine-grained approach to configurations such as the frequencies in which certain signs appear together in strings of symbols. Higher frequencies of certain co-occurrences amount to redundancies that reduce surprise (information) and uncertainty (entropy) in language. Shannon [6] illustrates this through a series of experiments, from random arrangements of symbols to restrictions to randomness based on the frequency of symbols found in words and sentences in English. In a zero-order approximation, independent symbols are listed randomly based on the same probability of occurrence.

XFOML RXKHRJFFJUJ ZLPWCFWKCYJ FFJEYVKCQSGHYD QPAAMKBZAACIBZLHJQD.

In a first-order approximation, independent symbols have the same probability of appearing that they have in English.

OCRO HLI RGWR NMIELWIS EU LL NBNESEBYA TH EEI ALHENHTTPA OOBTTVA NAH BRL.

In a second-order approximation, we have the frequency with which a symbol follows another in pairs in English.

ON IE ANTSOUTINYS ARE T INCTORE ST BE S DEAMY ACHIN D ILONASIVE TUCOOWE AT TEASONARE FUSO TIZIN ANDY TOBE SEACE CTISBE.

In a third-order approximation, symbols are randomly sequenced according to the likelihood of association with others in trigrams.

IN NO IST LAT WHEY CRATICT FROURE BIRS GROCID PONDENOME OF DEMONSTURES OF THE REPTAGIN IS REGOACTIONA OF CRE.

A first-order approximation to words shows symbols chosen independently from each other but with the same probability of appearing that they have in sentences in English.

REPRESENTING AND SPEEDILY IS AN GOOD APT OR COME CAN DIFFERENT NATURAL HERE HE THE A IN CAME THE TO OF TO EXPERT GRAY COME TO FURNISHES THE LINE MESSAGE HAD BE THESE.

In a second-order word approximation, Shannon comprises the probability that a word is followed by another specific word, with no further structure included,

> THE HEAD AND IN FRONTAL ATTACK ON AN ENGLISH WRITER THAT THE CHARACTER OF THIS POINT IS THEREFORE ANOTHER METHOD FOR THE LETTERS THAT THE TIME OF WHO EVER TOLD THE PROBLEM FOR AN UNEXPECTED.

A nearly intelligible message emerges even without syntactic rules or decoding the meanings of words.

Differences in frequency and sequence in symbols and words produce differences in the probability of occurrence and an increased probability of certain combinations. This means that the message components become more predictable and intelligible. The reduction of entropy allows communication [23]. This is also the limit to Shannon's approach without semantic information. Most importantly, as we have seen above, the entropy evaluation of a given probability distribution is a unique and unambiguous means of quantifying the intuitive difference between a broad distribution and a sharply peaked one [24, 25]. This notion can be naturally related to a general idea of uncertainty and randomness, where larger entropies correspond to broader distributions closer to equiprobability and randomness. Of course, Shannon proposed this formulation in a context where his problem was to understand sequences of information-carrying symbols [6]. Shannon's entropy quantifies the uncertainty associated with predicting a letter which follows a well-known portion of a string. Analogously, we shall see that increased frequencies of associations beyond one-dimensional strings, namely in three-dimensional arrangements of spatial entities like buildings, will create consistencies, correlations and structures in spatial formations fundamental to cities and their workings—and will set them apart from random distributions (see Sect. 12.3.3).

12.2.4 Entropy in Biological Systems

Another rich source of influence on the study of entropy in cities stems from the use of the concept in biology. Interpretations of the concept from Physics were used to understand life—from the thermodynamic conditions for the origin of life and the difference between living and non-living structures to the emergence of organized biological systems and their recursive adaptation in the face of environmental fluctuations and change. The second law of thermodynamics asserts that a closed system will inevitably experience an increase in entropy. However, living systems seem to challenge this law by maintaining low entropy. According to Prigogine [26], such systems will take on the configuration that minimizes its rate of entropy production [27]. Boltzmann himself first highlighted this concept, framing life as a struggle against entropy [8]. The thermodynamic understanding of life in the face of entropy posits that living systems diminish their internal entropy while elevating the entropy of their surroundings. They require energy to synthesise and maintain their highly ordered structures. This energy is often obtained through processes like metabolism, where complex molecules with a high degree of order (e.g. glucose) are broken down to release energy, also releasing simpler end products (e.g. carbon dioxide and water) and heat as a byproduct, increasing the disorder of the molecules in the environment [27]. This idea was closely interpreted in the study of urban metabolism and its entropic environmental effects (see Sect. 12.3.8).

Erwin Schrödinger [28] further explored this idea depicting life as a process that thrives on negative entropy. This is the case in the emergence of *order in the form of organic structures*, from molecules to cells to whole bodies. Living systems are characterized by highly specific and non-random structures. Atoms arrange themselves in organic molecules through physical and chemical combinatorial properties of atoms like carbon, which allow them to combine with other elements. Such far from random combinations form molecules like DNA. Schrödinger called these structures 'aperiodic crystals' [28, 29]. As microstates, such ordered configurations are statistically rare events in the space of possible configurations—except in the realm of biological systems, where combinations can achieve high frequency because such systems are able to synthesise and replicate themselves. DNA is an emblematic example. The molecular bases of DNA form four possible configurations connecting the double helix and amount to information-carrying sequences needed for coding other molecules and cellular structures. The micro-structures of DNA guide the *morphogenesis* of large structures emerging from the space of possible configurations, or what evolutionary biologists call *morphospace* [30, 31]. Exposed to random fluctuations in their external milieu, such self-assembling systems restrict themselves to occupying a limited number of states. Distilling structural regularities [32], they can form specialized subsystems like muscles and bones or circulatory and neural networks. In this *autopoietic formation of structures in space* [33] *across scales*, subsystems are connected and integrated into the functioning whole of the organism. Organisms are arrangements operating in coordination at different scales—a principle that came to pervade even the understanding of societies, from Durkheim [34] to Luhmann [35], and cities as systems of specialized, mutually-dependent subsystems (see Sect. 12.3.7). Of course, all these steps in the emergence of structures from smaller, nested structures are rare events in the horizon of possible events—even though actual organisms occupy a comparatively even smaller number of configurations in morphospace, in the non-ergodic universe above the level of atoms [36].

Uses of the concept of entropy in biology in ways appealing to urban metaphors were also applied to understand how living systems *survive*, how they respond to their environment and are able to improve and adapt to changing conditions. They have the ability to *resist a natural tendency to disorder* [32], i.e. their own degradation and decay as they metabolise and self-repair. Such resistance includes the production of specialized subsystems geared to protect and repair their structures, from immunological systems, as organisms are not closed systems in equilibrium, to the internal cleaning of entropic outputs of their metabolism. Such an ability is addressed as 'delayed entropy' by Schrödinger. "How would we express the marvellous faculty of a living organism, by which it delays the decay into thermodynamic equilibrium (death)?" [28]. Life is a property of a system to temporarily delay the process that leads the system toward maximum entropy [29]. Such biological interpretations of entropy influenced approaches to cities as metabolic systems drawing energy and self-structuring in the face of environmental fluctuations and randomness in their own morphogenesis while spilling effects of their growth and workings over their environment. Cities can self-repair and improve in the face of entropic forces—properties explored in views of urban self-organization and sustainability (Sects. 12.3.7 and 12.3.8).

12.2.5 Entropy in Social Systems

The concept of entropy has been explored in relation to societies as systems co-evolving with cities. Several theorists, within the framework of Social Entropy Theory (SET) originating in the 1960s, have specifically employed Shannon's information measure [37, 38]. Charvát et al. [39] introduced concepts like 'entropy of behavior' to assess homogeneity and interdependence. Horan [40] developed a measure of proportional reduction of uncertainty. Beyond internal entropy accumulation, threats to the social system also arise from the external environment. Entropy is viewed as the antithesis of information, controlled by system boundaries [41]. Galtung [42] analyzed entropy at micro- and macroscopic levels, identifying actor entropy and interaction entropy. Strong forces propel social systems into an oscillation between low- and high-entropy states. Action outcomes are not random, as agents operate within generalized roles, norms, and goals. Social norms, rules, culture, and language serve as constraints preventing maximum entropy. In turn, Bailey [43, 44] looked into how social systems increase organizational complexity and decrease internal entropy over time. The replication of recurring actions leads to order, resulting in entropy less than maximum. Removing action constraints could lead to increased entropy levels.

An insightful approach deals with counterfactuals and quasi-ergodic possibilities as entropic forces opened up by and surrounding people's daily actions. Luhmann [35] viewed societies as networks of interconnected subsystems where people face the challenge of selecting actions from an increasing array of possibilities [45]. In societies with growing information and agency, the complexity of possible interactions rises exponentially, leading to difficulties in selection. There is an increase in an unstructured type of complexity, an entropic informational complexity that may lead to organizational loss. The concept of *unstructured complexity* unraveled by possibilities that pose challenges to the selection and actualization of actions illuminates counterfactual risks that social systems recursively encounter.

Other approaches consider societies as collections of agents within geographic boundaries [45]. Bailey [46, 47] explored entropy in relation to city and population size. Social cybernetics concepts, such as context-dependency and agent self-reflexivity, were applied to issues like the division of labor, communication challenges, adaptation in complex societies, information's role in decision-making, and entropy reduction [48, 49]. However, these conceptualizations often lack spatial depth, treating space as a background rather than an integral part of entropy's resolution. Previously marginalized in such approaches, recently cities began to be seen as complex and active environments influencing and co-evolving with social systems [12, 23] (see Sect. 12.3.7).

We shall now see how these and other aspects of entropy have been interpreted and influencing the understanding of cities as complex systems.

12.3 Exploring the Concept of Entropy to Understand Cities: Domains, Measures and Applications

The use of the concept of entropy in urban studies draws upon thermodynamics, statistical physics, and information theory, along with their biological and sociological interpretations. As hinted above, applying the concept of entropy to cities may provide insights into how they emerged in different regions of the world and evolved into complex systems of an enormous diversity of shapes and forms. As interest in the concept of entropy grew in recent decades in urban science, several domains and properties have been explored.

12.3.1 Entropy-Maximizing Urban Systems and Spatial Entropy

The introduction of entropy concepts in relation to cities originated in the late 1950s. (i) Drawing analogies to statistical thermodynamics, it was motivated by concerns about how urban systems evolve toward a steady state (see [21]). Researchers in the early 1960s (such as [50, 51]) related entropy increase in closed systems to the dynamics of landscape ecologies and human systems. (ii) Other efforts focused on quantifying entropy as a measure of order to assess spatial object concentration or dispersion. (iii) A third line of approach culminates in the method of entropy maximizing focused on identifying constraints to the amount of entropy in an urban system. This method derives models of the distribution of energy in the system, introduced as a means to study the dynamics of urban systems by Wilson [11]. The concept was used to approach the most likely macro-distribution for micro-level interactions in a spatial economic system, namely through spatial interaction models able to capture the emergence of equilibrium solutions [52], in dynamics applied to transport flows and retailing, relating urban disaggregation to utilities and disutilities costs [11]. The entropy-maximizing paradigm was used to derive model formulations for spatial interactions and urban distributions, micro-economic behavior and input-output analysis [53]. Wilson [11] originally pursued an analogy with the statistical mechanics of entropy to find a way to reproduce the then-standard gravity model and a wide range of applications involving flows and locational structures related to the cost of transports and retail location models. Routes, flows and a disaggregation process emanate from the urban core.

The approach involves economic cost-benefit analysis in transport planning, studied through micro-economics of discrete choices [54]. The concept is used to generate interaction patterns of the journey to work subsystem, while indices of information theory entropy are used to unravel the nature and structure of these flows or movements of people [55]. For any flow pattern, a complete dispersion of inputs and outputs and, hence, economic linkages means maximum uncertainty in tracing the course of each flow. Furthermore, where there is a maximum concentration of inputs, there is little uncertainty in finding a place in the system for a randomly selected flow [55]. The most probable state of the elements of the system is achieved when its entropy is maximized subject to the constraints on the microstate of the system [11].

Harris and Wilson [56] extended the entropy-maximizing framework, embedding slow dynamics into spatial interactions. In a series of works in the 1970s, Batty [3, 57] explored a definition of spatial entropy as an evolving entropy-maximizing model introduced by Wilson, applied to the spatial aggregation problems, suggesting principles to partitioning spatial systems and uses of spatial entropy to cut off urban boundary to explore questions about optimal geometry of spatial systems. From then on, spatial interactions became widely related to Boltzmann's statistics methods, associating probabilities and maximization models of urban systems [11]. Entropy was linked to gravity models, rank-size rule (Chap. 1) and discrete choice models as a measure of interaction, uncertainty, diversity etc. [52]. For instance, Leibovici et al. [58, 59] consider as spatial information the distribution of co-occurrences between two or more observations with varying vicinities for the exploration of spatial patterns at different scales.

12.3.2 Entropy, Information and Complexity in Cities

Shannon's entropy certainly offers multiple domains encompassing order-disorder, certainty-uncertainty, minimum to maximum entropy, and zero to complete information [21, 60]. His definition blurred the boundaries between entropy, uncertainty, and information. Weaver [61] himself pointed out that the idea that "greater uncertainty and greater information go hand in hand" is deeply counter-intuitive. In fact, divergences between the statistical mechanic sense and information theory-based approaches on the relationship between entropy, uncertainty, and information were already felt in urban studies in the 1970s [3, 55]. More recently, the concepts of entropy and information have been more related through the appropriate lens of complexity. Batty et al. [21] propose a measure of complexity based on Shannon's information to grasp cities as they vary in scale, size, and spatial distribution of population. They interpret complexity in opposition to the idea of intricate systems able to hold order together in the face of unpredictable environments.

Batty chooses the interpretation that when something is entirely ordered, hence completely predictable, it is no longer complex. Of course, this conceptual choice renders the approach compatible with Shannon's view of entropy and information. The next step is to bring a view of the city as a distribution of events—namely, locations in relation to population size and area. Complexity is admitted by considering growth in the number of events (locations) and changes in their distribution. They adapt Shannon's measure partitioning information into two components, (i) a spatial entropy dealing with the distribution of information and (ii) an information density that deals with size. The idea is that the complexity of a system is inherently tied to the number of events it encompasses. Shannon's measure ensures that there is a trade-off between the distribution of information, the number of events, and their density as the urban system changes in size and scale, and the distribution evolves from highly ordered to less ordered, or vice versa.

Various methods in landscape ecology have explored adaptations of Shannon entropy and information-theoretical metrics to quantify and group similar two-dimensional spatial patterns. Some of these approaches aim for a universal classification of configuration types, arranged linearly based on increasing values [62–64]. In the context of cognitive studies, efforts to extract task-relevant information from the built environment, other spatial information approaches have used measures of entropy, distribution of spatial co-occurrences, or information density [65, 66, e.g.]. Additionally, Haken and Portugali [67, 68] have explored the embodiment of information in the built environment, quantifying Shannon's information in connection with Haken's synergetic qualitative approach to semantic information. Their work empirically assesses how basic cellular arrangements and categorizations of building facades convey varying amounts of information.

12.3.3 Entropy and Urban Form

There is a distinction between analyses of distributions of densities, locations, or activities and approaches geared to reveal correlations within distributions of urban form involving spatially well-defined entities such as buildings and streets. The latter aims to grasp structures at different scales, from detailed arrangements of buildings or streets at fine resolutions to large-scale structures in spatial morphologies. Approaches to entropy in urban form range from applications as measures of diversity as a proxy to disorder in street networks to explorations to capture information signatures in the built form distributions of cities in different world regions.

Entropy in street networks

The understanding of cities as topological structures was systematically introduced by Christopher Alexander in the mid-1960s [69, 70]. This graph-theoretical view ushered into approaches to configurations of buildings [13, 71, e.g.] and cities [72]. The space syntax approach offered the first topological analysis of cities as street networks to address the systemic structures and workings of cities as part of the social dynamics of people's movement and encounter systems. This line of research was popularized as (and reduced to) a purely morphological approach as 'street network analysis' in the 2000s [62, 73–76, e.g.]. Gudmundsson and Mohajeri [77] explored street networks to introduce a method based on Gibbs/Shannon entropy to measure angular and length variation between streets in 41 British cities

$$S = -\kappa \sum_{i=1}^{t} P_i \ln(P_i),$$

where S is entropy, κ is a positive constant, t is the number of bins with nonzero probabilities of streets, and P_i is the probability of streets falling in the i-th bin. Entropy values vary between zero (all the streets occupy a single bin) to the maximum value (a uniform distribution or lengths so that all bins have the same lengths or heights (Fig. 12.3, right). Findings show that entropy increases with street average length, distance from the city center and as cities group over time. Boeing [78] further developed the approach into an orientation-order indicator to quantify the extent to which a street network follows the spatial ordering logic of a single grid, and other indicators like street circuity applied to 100 cities worldwide and show consistent regional differences in values. Coutron et al. [79] explored such properties in a cognitive study to measure spatial navigation ability in more than 397,000 subjects from 38 countries through cognitive tasks embedded in a video game. Their findings revealed that people demonstrated better navigation performance in environments topologically resembling their upbringing. Growing up in cities with low street network entropy, such as Beijing (Fig. 12.3, left), resulted in better performance in digital environments with a regular layout. Conversely, those growing up outside cities or in cities with higher street network entropy, like Rio de Janeiro, had better performance in less ordered digital environments.

Note that Shannon entropy is more than a general diversity index. It is a well-defined measure of information that quantifies the degree of surprise the source of a sequence causes on the observer [6, 24]. Therefore, this interesting method focusing on entropy as a measure of variation in street orientations and lengths seems yet to be fully explored to uncover spatial information patterns. The entropy measure of the distribution of local traits like crossing angles or street orientations and lengths does not necessarily capture the global degree of order/disorder of street networks as long-range correlations. The apparent randomness at local scales frequently amounts to structure if looked at a larger scale (Fig. 12.4a). High angular variation may be found in ordered cities, of which concentric radial street networks are the simplest case (Fig. 12.4b). Furthermore, entropy-based diversity approaches applied to street networks can ignore the complexities of urban form latent in the distribution of buildings and the discrepancies between the order in street networks and in built form systems. Of course, cities can be spatially disordered even if their street networks are perfectly ordered: they may have low entropy in street orientation yet highly disordered morphologies [80] (Fig. 12.4d).

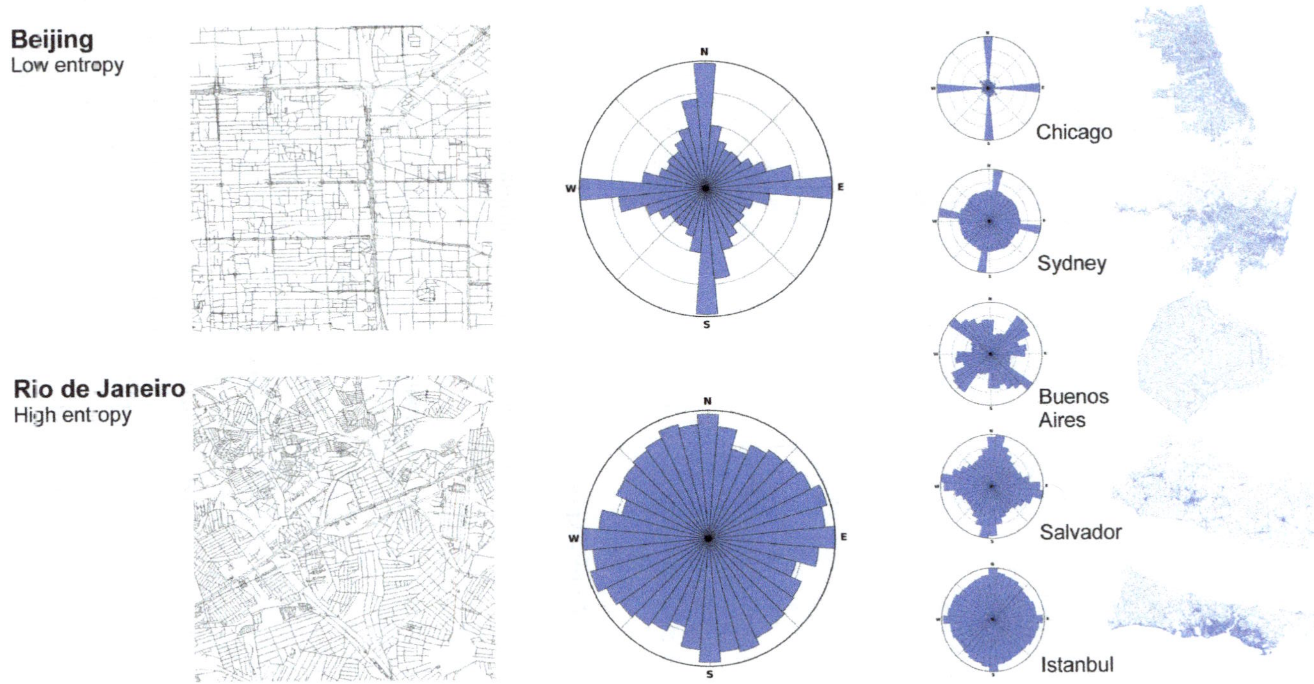

Fig. 12.3 Entropy measures can be applied to analyze variation in street orientations illustrated by polar histograms, e.g. Beijing and Rio de Janeiro (left and center), and Chicago, Sydney, Buenos Aires, Salvador and Istanbul (right)

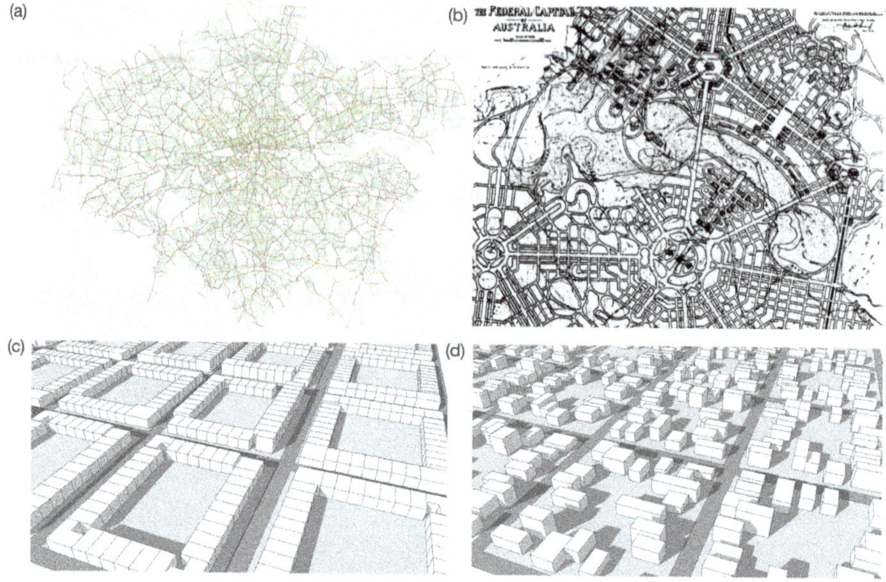

Fig. 12.4 High angular variation can be found in **a** the large-scale structures of 'organic' cities like London, UK, topologically identified through space syntax methods [81], and in **b** perfectly ordered radial street networks like the central area of Canberra, Australia, as designed by Walter Burley Griffin in 1911. The same street network in **c** and **d** can support very different built-form systems, e.g. ordered (bottom left) and random distributions of buildings (bottom right) [80]

Entropy in built form systems

A quite different situation corresponds to the built environment as a systemic whole, not selecting just a single local trait for characterizing it. Entropy measures can be applied to understand the ultimate feature defining urban form structures, i.e. buildings producing built-form configurations. Morphogenesis is a concept that originated in biology. In the context of cities, it addresses the process by which the form, structure, and organization arise during the development of cities. The emergence of structures in cities in the face of the uncertainty of possible morphogenetic paths is a crucial, if underestimated problem in urban science. The process may be approached as restrictions on randomness [13], echoing Shannon's [6] view of information as the amount of freedom of choices governed by probabilities. This is an open question in urban research. We may hypothesize that patterns emerge in the course of local societies actually solving problems of aggregating buildings into complexes and settlements. Such spatial solutions and formations seem to evolve in time through trial and error, conscious and unconscious spatial choices, contingencies and path dependence, among other potential factors [13, 14, 69]. The effect is the reduction of the uncertainty manifested as the increase in the probability of certain spatial distributions of buildings and open spaces. Higher frequencies bring regularities and

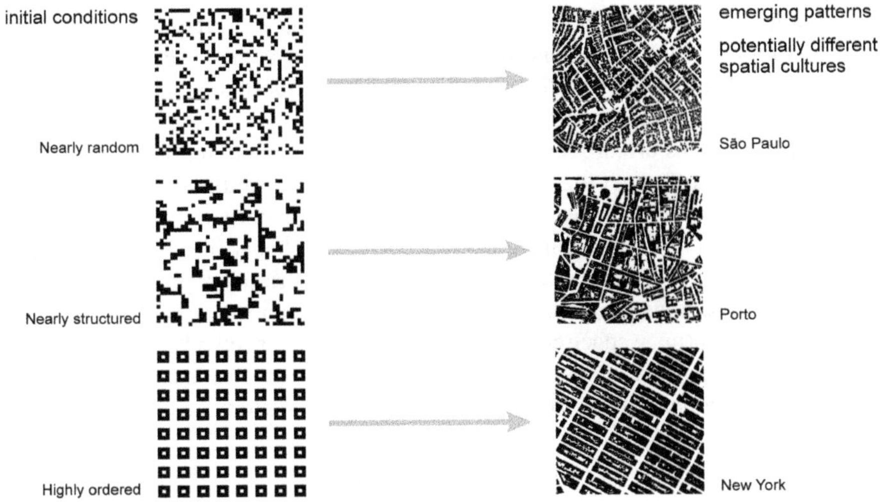

Fig. 12.5 Illustrations of possible archetypal morphogenetic paths. Initial conditions with different entropic levels (left) might evolve into distinct urban patterns and specific spatial cultures (right): São Paulo, Brazil; Porto, Portugal; New York, USA

internal correlations at different scales which, ultimately, may amount to large-scale structures in spatial formations.

Of course, these formations may take different morphogenetic trajectories across morphospace. Trajectories might be contingent and relate to local environmental conditions and societal codes—a plethora of factors that amount to cultural specificities. Here, the amount of randomness in initial morphogenetic conditions might matter. They might be related to culturally embedded choices bringing restrictions to the random creation of urban form, leading to different outputs which, over time, consolidate as patterns and, through path dependence, into distinct urban spatial cultures [80, 82]. Some spatial cultures might favor order and rules guiding additions to growing settlements. Others might be more relaxed when engaging in morphogenesis, allowing purely bottom-up processes to take over (Fig. 12.5).

Recent works [24, 80] have explored built form configurations to identify levels of entropy and disorder as information signatures of possible cultural traits. The approach uses Nolli maps of building footprints (representations of buildings and city blocks by black polygons or cells, and the open spaces of streets, squares and courtyards by white polygons or cells) as representations of the complete urban form of cities. Nolli maps are converted into cellular arrangements, a matrix of binary values, i.e. an information-carrying image where the symbols correspond to built or non-built spaces. The method allows the identification and counting of frequencies of cellular combinations in different urban contexts. Figure 12.6 shows the 1,000 most frequent cellular configurations from all possible combinations for a cell block with $n = 4 \times 4$, excluding configurations completely built or completely open, in 9 km^2 areas of two cities, i.e. Washington (USA) and Lagos (Nigeria). The total number of configurations found in Washington is 2,989; in Lagos, 6,825 (out of 65,534 possible cellular combinations). These distributions and numbers suggest substantial differences in how these distinct spatial cultures materialize morphogenetic possibilities and emphases on regularity and variety in urban form.

The method scans urban sections considered satisfactory to represent the spatial characteristics of different cities and to grasp correlations at sufficiently large scales, a necessary condition for correctly estimating entropy. Shannon entropy is applied to measure the amount of randomness and statistical complexity of the global system. To go beyond the local properties of the map, the distribution probability of the state of the entire image must be quantified. As the urban space is represented by matrices of 0 and 1 (open space cells and built form cells, respectively), this procedure theoretically corresponds to measuring the Shannon entropy H of a two-dimensional binary cellular sequence. The method defines the block entropy of order n through

$$H_n = -\sum_k P_n(\kappa) \log_2[(P_n)(\kappa)] ,$$

where a block of size n is a segment of n cells selected from the sequence under analysis, and the sum runs over all the k possible n-blocks. Two blocks with the same number of 0 and 1 but with a different spatial combination of them correspond to different configurations. The method counts the number of times each configuration k is found in the area, accounting for actual built form cellular configurations and not merely the density of cells (occupation density). Scanning cellular blocks from $n - 1$ to $n = 20$ allows for averaging entropy values across larger blocks of cells. The method considers features of longer-range order as it computes frequencies and identifies the same configurations in different locations, even quite far from each other, as correlations at a distance. Physical arrangements marked by higher levels of randomness and higher entropy levels are characterized by greater unpredictability. In contrast, patterns and regularities in urban structures correspond to lower entropy, which means a higher predictability. Results suggest some form of cultural idiosyncrasy consistently inherent to the built form of cities within their countries and regions, captured by a probabilistic measure of entropy [24, 80] (Fig. 12.7—cf. Chap. 10 on architectural design and self-organization).

12.3.4 Shannon Entropy as a Measure of Diversity

Entropy measures can be used as diversity indices, i.e. quantitative measures that may reveal how equiprobable specific features are in a distribution [24]. In general, spatial diversity measures focus on how a set of social variables, spatial characteristics, or urban attributes are evenly or unevenly distributed over landscapes and urban spaces. Applications in landscape ecology focus on patch mosaics related to size, density, fractal dimension, contagion, and nearest neighborhood distances [83], taking into account the relative percentage of land-use types within areas [84, 85]. In turn,

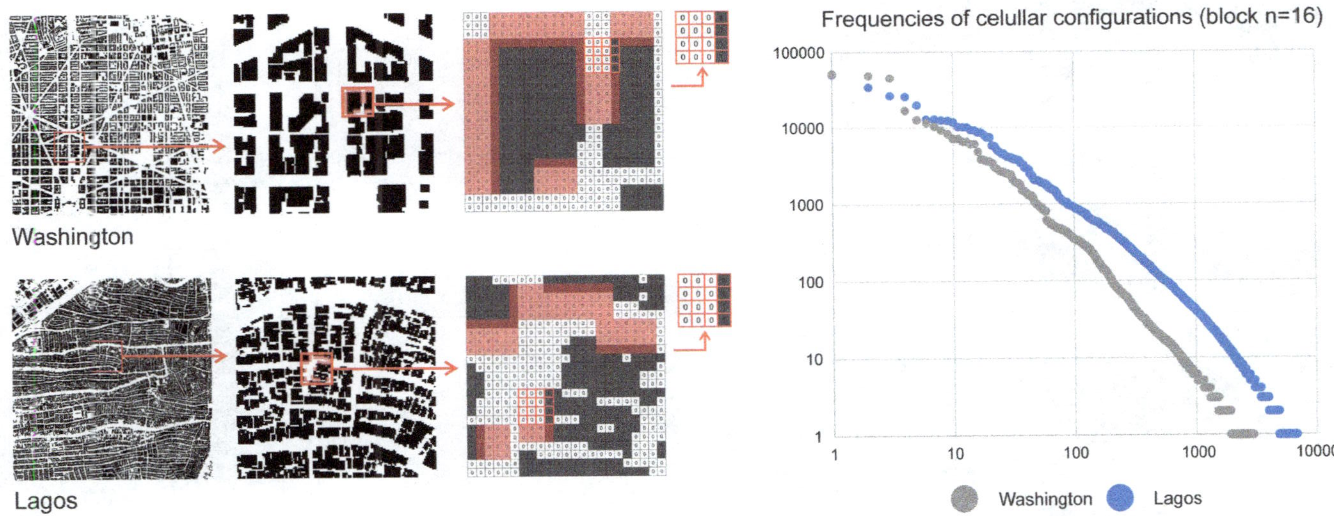

Fig. 12.6 Frequencies of cellular combinations in different urban contexts. The algorithm scans 9 km^2 areas in built form maps counting frequencies of cellular configurations for different block sizes (from $n = 1$ to $n = 20$). Amplified areas in red show a specific configuration (a cell block with $n = 16$) and where they are found. The total number of configurations $n = 16$ found in Washington (grey line) is 2,989 and 6,825 in Lagos (blue line) out of 65,534 possible combinations

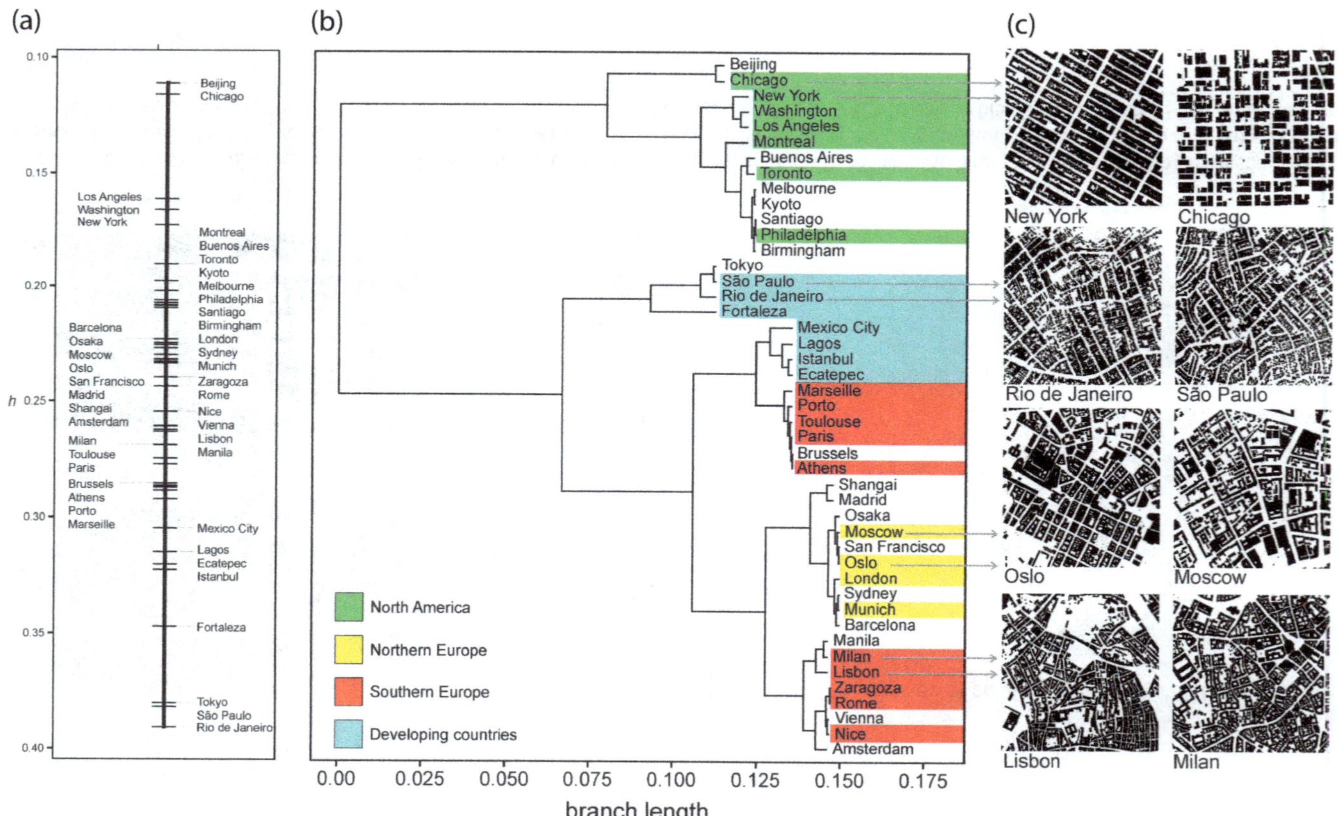

Fig. 12.7 Entropy levels as a proxy to differences in spatial cultures. **a** Estimated values of h for the 45 cities under analysis (see [80]). **b** Dendrogram of North American, European, Asian, Oceanian, African, and South American cities under analysis. Colors highlight the relative position of cities regarding world regions. **c** Subsections (1 km^2) show the morphologies of some of these cities, illustrating the method's potential to grasp distinct spatial cultures

entropy finds a broad range of applications as a diversity index in urban studies—from street orientations and lengths seen in Sect. 12.3.3 and urban parcel sizes [86] to the mixing of urban activities (land uses) or the segregation of social groups in urban spaces. We shall start this section by showing how Shannon entropy is applied to land-use mix (LUM).

Entropy measures of land-use mix

LUM entropy measures application started from a transportation perspective in the late 1980s, mainly focused on the potential influence of land-use diversity over transport choices such as walking, cycling, and public transport [87, 88]. LUM was extended into public health studies in the 2000s associated with walkability and health benefits [89]. Despite the variety of diversity measures for land uses, entropy-based LUM indices are likely the most often-used measure of land-use diversity [90, 91]. The concept of land-use mix implies that a distribution of a type of land use (e.g. retail or residential activities) across a limited spatial range is related to the distribution of other land uses. Cervero [87, 92] introduced Shannon entropy as a LUM measure, later employed by Frank and Pivo [89] to analyze the impacts of local diversity over trips and transportation choices (cf. [93, 94]). Frank et al. [89] proposed a standardization of the measure dividing by a natural logarithm of the number of classes J considered

$$H = -\sum_i \frac{P_i \ln(P_i)}{\ln(I)},$$

which reduces the influence of richness (the number of different land uses), emphasizing evenness (or unevenness) in the distribution of land use in an area. The standardization of Shannon entropy in land-use studies assumes that maximum entropy occurs when the percentages of each land-use distribution are exactly equal locally, reflecting a local best mix or balance of uses. Of course, the delimitation of areas of analysis according to administrative or analytic entities, such as radius-based buffers or street network distances, plays a relevant role in measurement. An apparently mixed local distribution of land uses can be dramatically different from distribution patterns over a whole city.

Song et al. [91] propose an adaptation of diversity measures considering both local and global compositions, computing an entropy index relative to a 'reference geography'. Local probabilities P_j expressed in the Shannon equation do not refer to local percentages but the ratio of local and global percentages, generating a local quotient for each land use (q) with values greater than 1. Local distributions can be compared with the global distribution, computing a local diversity relative to the diversity measured at a global scale. In such an approach, the local entropy-based LUM measure intrinsically relates to diversity unevenness at the global scale—something traditionally found in the dissimilarity index proposed by Duncan and Duncan [95] for estimating spatial segregation—our next subject.

12.3.5 Entropy and Spatial Segregation

Spatial segregation studies are based on the uneven locational distribution of homogeneous social groups (according to social characteristics, e.g. income, race, ethnicity and so on) across spatial areas, regions or places. Segregation is quite a diverse phenomenon that encompasses the separation of people in different situations of life, from institutions like schools to labor positions, but residential segregation is undoubtedly its most explored spatial domain. Similarly to land-use diversity, residential segregation was initially measured as the uneven locational distribution of population classified into social groups over a defined area. Since the 1950s, the favored measure was the dissimilarity index popularized by Duncan and Duncan [95]. A measure of entropy-based residential segregation only emerged two decades later, proposed by Theil and Finizza [96]—capturing interest in the field [97, 98]. The entropy index H measures evenness in residential location based on a weighted average deviation of the entropy in each spatially delimited area (e.g. census tract) from the entropy level of the city as a whole, therefore expressed as a fraction of the city's total entropy

$$H = -\sum_{i=1} \left(t_i \frac{\hat{H} - H_i}{\hat{H}} \right),$$

where $\hat{H}$ is the Shannon entropy of the distribution of social groups over the city as a whole, and H_i is the local entropy for the same set of social groups. H is related to the overall entropy of total area—a measure of how more socially homogeneous (and segregated) residential areas are on average than the total population of the city. H varies from 0 (residential areas with maximum social diversity or unevenness) to 1 (areas that have complete social homogeneity).

The entropy-based measure evolved since then through a set of adaptations as segregation measurement [99, 100] (see Chap. 7). Massey and Denton [101] identified five dimensions of measurement: evenness, exposure, concentration, centralization, and clustering. Reardon and O'Sullivan [102] readdressed the probabilistic isolation/exposure proposition by Bell [103] in 1954 to consider a simpler framework, i.e. spatial evenness and clustering (referring to the extent to which groups are segregated or not in urban space), spatial exposure (locations that favor encounters with members from another social group), and isolation (contact with members from one's own group) (Fig. 12.8). Note that these measures oversimplify the possibilities of interaction between groups and do not effectively handle encounters or contact since such social variables are dependent on people's movement, which is absent from analyses of purely locational distributions [14]. Locational exposure should be understood as a potential for actual social exposure.

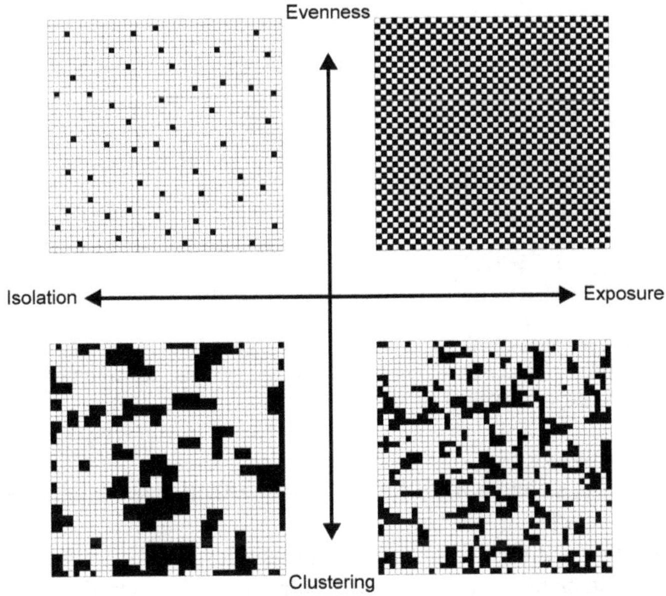

Fig. 12.8 Distributions of black and white cells according to locational properties, from highly segregated to less or non-segregated distributions in a two-dimensional space. Source: modified after [102]

12.3.6 Entropy as a Means to Understand Urban Size and Densities, Growth and Sprawl

A didactic way of seeing how entropy helps us understand urban densities is found in [21]. Batty shows that entropy varies not only as distribution changes within a configuration but also as the number of objects or occurrences in the distribution changes, reflecting a trade-off between number and distribution (Fig. 12.9a). The probabilities of discrete events occurring in a spatially extensional system like a city are not equally distributed across its space. Positions in geographical space are subject to differential potentials associated with their relative distance from other positions. The effect is a regular negative exponential distribution emanating from central positions—to be sure, a distribution found in different urban phenomena such as densities of population and retail activities or the rent curves fundamental in urban economics. The exponential function can be steeper or flatter according to a friction-of-distance parameter β. A steeper curve indicates higher energy costs to the movement (Fig. 12.9b—(see Chap. 4)). Such distributions have effects across virtually everything that happens in a city—from the probabilities of encountering certain activities or people on the streets to the distribution of built form, economic diversity and intensities of exchange. As a result, a comparatively smaller number of locations concentrate probabilities of occurrence of such events. The blue curve

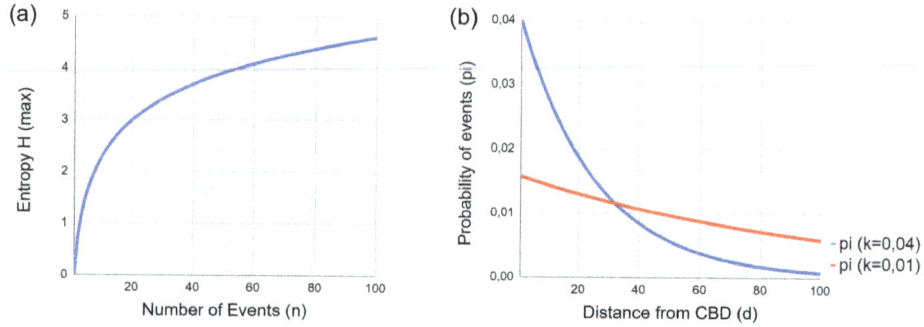

Fig. 12.9 Entropy and urban densities varying with distance. **a** Entropy increases with the number of events; **b** probabilities decrease with distance in urban systems according to different energy costs to movement. Source: modified after [21]

in Fig.12.9b reflects an internally differentiated distribution and contains more order than the flatter one (cf. Fig. 12.2).

Urban densities and their different yet intrinsic levels of order may be clarified as probability distributions and associated entropy levels—and they are inherently related to distance and therefore, city size. These properties are naturally useful to understand the implications of urban growth—and different forms of growth, with distinct regularities in the density of land occupation, as suggested by the curves in Fig.12.9b. An emblematic case of low-density growth is the so-called urban sprawl [104]. Urban sprawl has some key features, i.e. the amount of land that is built up, the dispersion of built-up areas over a landscape, and the density gradient of people living or working in these areas [105]. Methods for measuring urban sprawl cover a wide variety of aspects and landscape metrics, and have been used for its quantification [106, 107]. Confronting a lack of indicators to measure urban sprawl [108, 109] proposed an effective measurement of urban sprawl using entropy focused on two main domains. (i) distance from urban central places and (ii) spatial patterns. Shannon entropy is used as a measure of spatial concentration or dispersion of a geographical variable x among a number n of zones. A series of concentric rings from the city's central place can be defined, and the density of urbanized areas in each buffer zone is computed. Urban concentrations tend to obtain lower values of relative entropy, and at the end of the spectrum, higher levels of sprawl are associated with higher values of relative entropy.

Such measurement of sprawl as an entropic state of urban structures can account for time as the difference in entropy in t and $t + 1$ as an indicator of change, suggesting a city's trajectory towards either dispersion or compactness [109]. Shannon entropy has been widely used to understand sprawl dynamics in the last two decades [105, 110–114], as a measure of the degree of concentration or dispersion of spatial variables sensitive to the variations in the shapes and sizes of urban areas [109].

12.3.7 Entropy and the Self-Organization of Cities

The spatial production of cities through urbanization processes involves the emergence of structures. Similar to biological systems, cities are open, non-equilibrium systems functionally composed of different, specialized subsystems enacted by human and non-human agents, interacting and interchanging matter, information, and energy, cf. [115]. Cities as complex systems relate to entropy at different levels and moments of self-organization during their production and self-reproduction.

Cities resist a natural tendency to disorder in the form of degradation of their subsystems through agencies and mechanisms of self-repair, which is the subject of the following section.

The production of cities and urbanization processes

A thermodynamic interpretation of the emergence of organized urban systems means that cities diminish their internal entropy while elevating the entropy of their surroundings. This will have implications for the survival and sustainability of cities, societies, and the environment itself (see Sect. 12.3.8).

The autopoiesis of cities

A crucial question in the relationship of cities and entropy is how urbanization comes into being as a form of structuration countering tendencies towards entropy. Urban systems are characterized by specific and non-random structures. In the morphogenesis of structures emerging from possible configurations, from the micro to the macro structures, self-assembling urban systems restrict themselves to occupying a limited number of states. Through trial and error and other evolutionary processes (Sect. 12.3.3), urban systems may take on configurations that minimize entropy production, cf. [26]. Cities are formed as specialized, mutually dependent subsystems like activity systems materialized in buildings and interconnected by street networks. In such autopoietic structuration [33] across scales, subsystems become integrated into the functioning whole of the city.

The workings of cities, or how urban systems survive

Cities are systems that emerge as pockets of low entropy. Yet their own internal levels of entropy (endogenous entropy) can fluctuate while also exposed to random fluctuations in their external milieu (environmental or exogenous entropy). Entropy is produced inside and outside the system boundary [115] and relates to the loss of structure and organization. Cities require energy to synthesize and maintain their ordered structures. They respond to such losses and environmental fluctuations by adapting to changing conditions and improving their own structures.

12.3.8 Entropy and the Sustainability of Cities and Societies

The connection between entropy and urban sustainability as the ability of cities to persist over time without threatening adjacent systems has been mostly explored in relation to the thermodynamical understanding of entropy. Such an ability evokes Schrödinger's [28] notion of delayed entropy, a property of systems capable of temporarily delaying the natural process leading them toward maximum entropy. This delay is achieved through interactions with the environment, extracting negative entropy or free energy [32]. The interaction of complex urban systems with their environment, while essential for survival, is a 'double-edged sword'. It is both the source of the system's survival and the force responsible for its eventual decay into maximum entropy. Once a system can no longer extract negative entropy, it succumbs to disintegration by the environment [29]. In short, like other forms of human settlement, e.g. rural or non-urban spaces, the city needs a constant supply of energy to keep its entropy low. However, the amount of energy required in ever-growing economic systems materialized in the form of cities runs faster against the regenerative processes of the environment and its natural limits as an energy resource.

- *Energy use and entropic outputs during urban metabolism.* An open self-organized system needs to import available energy and material from their host environments in order to reproduce itself [115]. Processes of urbanization, such as urban growth and sprawl, and the daily workings of cities involve intensive use and transformation of resources, materials and energy, leading to negative externalities such as waste, effluents, and pollution. Such transformations are analogously found in biological systems (Sect. 12.2.4).
- *Cities as contributors to increasing environmental entropy.* Cities are recognized as significant contributors to the environmental crisis. Their exogenous consumption and externalities are directly linked with entropic environments. In urban growth and sprawl (Sect. 12.3.6), entropy must be kept within a range defined by a minimum value below which the system becomes vulnerable and a maximum value above which the system becomes unsustainable [116].

Entropy appears frequently described in the urban science literature as an opposite force to sustainability [115]. However, a direct opposition between entropy and sustainability is an oversimplification. Societies and cities also increase their internal entropies as an effect of the daily creation of new interactions, social relationships, groups and agencies, and through eventful cognitive, social, and technological innovations. Entropy increases as our everyday actions and interactions in the city go into momentary states of unpredictability and new possibilities are presented to us [23, 35]. These serendipitous processes may trigger positive entropic effects—even though they may be accompanied by initially negative ones such as the disruption of existing agencies, like the downfall of firms or whole industries. Furthermore, cities have become a crucial means to a broader sense of societal sustainability. Societies reproduce themselves subject to limited resources, fluctuating environments, external shocks and systemic crises, all contributing to uncertainty and hindering decision-making. The nature of these challenges can be clarified through the concept of entropy. Societies face entropy all the time (Sect. 12.2.5). While preparing for the negative effects of entropy is vital to how societies become resilient, fostering positive entropic effects triggered by innovation and serendipity is crucial for creative societies. Cities are key for both processes. Societies endure and manage different forms of entropy through three continuous processes.

1. *Creating physical and semantic urban structures.* To support volatile interaction systems, societies shape their built environment locally via autopoietic processes [33], self-organizing as they arrange physical space in the form of cities through morphogenesis and the semantization of the urban form, i.e. the creation of social contents, meaning and information associated with physical objects such as buildings and places [14].
2. *Exploring urban structures* as they encode information critical to people's lives, daily interactions and collective cooperation. People living in cities deal with daily activities and challenges through cognitive and practical mechanisms like retrieving and enacting information from their built environment [12, 68].
3. *Stressing and improving urban structures* in response to systemic fluctuations through properties like anti-fragility [117] and self-improvement mechanisms, like bottom-up collective learning and top-down monitoring capacities in planning. Such capacities trigger and guide recursive changes in the built environment, increasing urban resilience and keeping social entropy at bay through continuous works of adaptation in cities.

Cities can improve or hinder social dynamics, reducing or expanding societal demands and externalities because cities have structures that enable and shape such dynamics. Theoretically, two cities with the same population number and geographical size would perform differently if their internal structures were different—say, a fragmented urban structure that makes navigation difficult, disturbing interactions and triggering diseconomies, or a structure containing internal correlations capable of offering spatial legibility and fluency to daily movement and interaction. Urban structures, therefore, matter. They support the challenging passage from individual actions to large-scale interaction systems, a crucial recursive step in how societies manage entropy [118]. Cities are vital in that sense.

12.4 Criticism

The concept of entropy is often misunderstood both in general and within the specific context of urban science—for some possible reasons.

- *Complexity and abstraction.* Entropy is a highly abstract and counter-intuitive concept rooted in statistical mechanics and information theory. Its application involves mathematical formulations and probabilistic reasoning, making it challenging for researchers without a background in these fields to grasp its nuances. This is particularly the case in urban studies, which often involve researchers with backgrounds in fields such as sociology, geography, and planning. Without interdisciplinary training that includes concepts from physics and information theory, researchers may struggle to appreciate the full scope of the concept and its applications.
- *Urban dynamics are multifaceted.* Applying entropy in urban science involves deals with the multifaceted dynamics of cities, which are influenced by social, economic, and environmental factors. The complexity of urban systems can make it challenging to isolate and quantify the specific contributions of entropy, leading to potential misinterpretations.
- *Inconsistencies in the evolution and interpretations of the concepts over time.* The concept of entropy has evolved over time since its roots in classical thermodynamics. This evolution can lead to varying interpretations, especially when researchers are exposed to different formulations of the concept. The diversity of applications in urban studies can contribute to confusion, as researchers may find entropy in different forms and contexts [3]. For instance, an association with disorder in the everyday sense can lead to oversimplification.

While the concept of entropy has been applied in urban studies and city science, there are critiques associated with its use and the application of associated measures.

- *Assumption of equilibrium.* Entropy models often assume equilibrium or quasi-equilibrium conditions [3, 119]. However, cities are dynamic and rarely in a state of equilibrium. This assumption may not fully capture the ongoing evolution and adaptation of urban systems [120].
- *Single-domain approaches.* Some entropy measures focus on a single domain, such as land use or population distribution. This narrow focus may overlook the complexity of urban systems.
- *Overemphasis on physical structures.* Some entropy-based approaches emphasize the physical form of cities at the expense of its relations to semantic aspects of urban form.
- *Limited relations to social processes.* Entropy measures are frequently applied to spatial distributions. Connections to volatile dynamics of social interactions as intrinsic to the urban phenomenon and its complexity are frequently overlooked, along with the effects of top-down social agencies. Entropy-maximizing input-output frameworks [11, 119] are exceptions but do not encompass the richness of social interaction.

- *Overlooked temporal processes.* Many entropy-based approaches do not adequately consider temporal dynamics. Entropy measures applied in analyses of urban systems states may easily ignore changes and adaptations occurring within cities [59].
- *Exclusion of contextual information.* Context-specific factors, such as local policies, cultural nuances, and historical legacies, may not be adequately integrated into entropy calculations. Investigating the generalizability of entropy measures in various urban settings is an ongoing challenge.
- *Lack of standardization.* There is a lack of standardized methods for calculating entropy in urban studies. Different researchers may use different measures, making it challenging to compare findings and hindering the establishment of consistent benchmarks [52].
- *Data resolution issues.* The resolution of data used to calculate entropy can impact the results. Coarse-grained data may overlook fine-scale variations, while fine-grained data might introduce noise and overemphasize local patterns [24]. Choosing an appropriate level of resolution is a crucial but challenging task.
- *Data acquisition challenges.* Obtaining accurate and comprehensive urban data for entropy calculations can be challenging. Data gaps, inaccuracies, and the dynamic nature of cities also pose significant obstacles.
- *Limited predictive power.* Entropy models may have limited predictive power. Cities are influenced by unpredictable factors, and deterministic approaches might not account for the emergent properties of cities in response to such factors [120].

12.5 Open Questions

The use of the concept and measures of entropy in urban studies and city science has advanced our understanding of complex urban systems, but several open questions and challenges remain. Some of the key issues and problems yet to be fully addressed and explained include:

- *Multi-domain entropy.* The majority of entropy measures in urban studies focus on a single domain, such as land use or population distribution. Developing comprehensive multi-domain entropy measures that capture the complexity of urban systems in terms of social, economic, and environmental factors remains a challenge [52].
- *Dynamic interaction networks.* Cities operate as dynamic interaction networks [14], and current entropy measures may not fully capture the evolving relationships between different elements in the urban environment. Developing entropy measures that consider the multi-layered and hybrid network structures inherent in urban systems is a key step in research [21].
- *Multi-scale framework.* Entropy values can be scale-dependent, and there is a need to develop frameworks for interpreting entropy at various spatial and temporal

scales. Understanding how entropy values change based on the chosen scale is an area for further exploration [59, 121].

- *Integration of qualitative or semantic information.* Quantifying information in the context of cities is challenging. Unlike physical systems, information in urban settings is often qualitative, such as cultural richness, semantic information in space retrieved and enacted by agents [12, 23] and converting it into quantifiable measures for entropy calculations can be problematic. Promising insights may be gained from exploring methods to integrate qualitative and semantic information into entropy measures [68] and acknowledging the subjective aspects of the urban experience.
- *Policy and planning implications.* Bridging the gap between entropy measures and practical policy and planning decisions is an ongoing challenge. The mechanisms and thresholds that contribute to or hinder urban resilience [116], how entropy influences a city's ability to adapt and improve from environmental and internal fluctuations, and how entropic forces and their effects on different communities are fresh lines of inquiry. Exploring entropy measures to assess spatial scenarios as performance indicators to inform sustainable urban development and governance remains an important avenue for future research [115, 122, 123].

12.6 Summary

In this chapter, we followed the historical trajectory of the concept of entropy from its inception in thermodynamics and statistical mechanics in the last decades of the nineteenth century to some of its latest incarnations to understand cities as complex systems. We attempted to demystify Boltzmann's understanding of the probability of different molecular arrangements coarse-graining into macrostates and Gibb's function of the probability distribution over phase space. We visited Shannon's definition of entropy in information theory as a measure of uncertainty in the engineering of communication channels and the transmission of data in the mid-twentieth century. Evoking Prigogine [1] again, those pioneering visions of a very strange yet fundamental concept to understand the world transcended their original fields into a rich multidisciplinary tapestry of scientific applications.

Explorations in urban science were certainly enriched by how the concept was interpreted in biology—from the conditions for the origin of life as organized living systems and their metabolism to their recursive adaptation in the face of changing environments. The subsequent absorption of the concept in the social sciences threw light on how societies manage internal and external entropies. From the point of view of entropy and complexity, a key process is how microscopic actions are assembled into possible functional configurations. That is the passage from possibilities and uncertainties surrounding people's interactions to large-scale economic and political organization. Institutionalized structures like customs, rules, and culture have a role in generating such organized states [124, 125].

So have cities. Somewhere between structure and randomness, cities are more than an environment to the always ongoing emergence of social structures. Societies and cities are rather *co-evolving adjacent systems*. The city is what Parsons [124] called an 'action frame of reference'—a connective fabric that informs, constrains and enables interaction choices [23]. Thus, we explored the concept of entropy to understand cities, from the conditions of their emergence and evolution to the challenges in their continuity. We searched for approaches, measures and empirical applications of the concept to understand cities as entropy-maximizing interaction systems, along with urban form, the distribution and diversity of land uses, and patterns of spatial segregation and densities, growth and sprawl.

We looked into the self-organization of cities through the lens of entropy. From the space of possible configurations, the emergence of urban patterns might also minimize the rate of entropy in the workings of cities themselves, including the assembling of large-scale interaction systems out of individual actions [12]. Finally, we described entropy as an aspect of sustainability, seeing the role of cities in how societies navigate entropy by creating urban structures, exploring such structures in daily interactions, and stressing and improving them in response to internal and external fluctuations.

The concept and measures of entropy provide valuable insights into critical aspects of urban systems and urban life. Exploring such richness and addressing the challenges to integrate entropy approaches with broader perspectives considering the diverse and dynamic nature of cities outlined above seem essential steps for a more comprehensive understanding of urban complexity.

Acknowledgements Vinicius M. Netto acknowledges the support of the Research Centre for Territory, Transports and Environment | University of Porto | FEUP | CITTA, Porto, Portugal; the National Council for Scientific and Technological Development (Conselho Nacional de Desenvolvimento Científico e Tecnológico | CNPq, Brazil), grant number 315086/2020-3; and the Foundation for Research Support (Fundação de Amparo a Pesquisa do Rio de Janeiro | FAPERJ), grant number SEI-260003/015789/2021. Otavio M. Peres acknowledges the support of the National Council for Scientific and Technological Development (CNPq), Brazil.

References

1. I. Prigogine, What is entropy?. Die Naturwissenschaften **76**(1), 1–8 (1989). https://doi.org/10.1007/BF00368303
2. J. Portugali et al., Cities, complexity and beyond, in *Handbook on Cities and Complexity* (2021), p. 13. https://doi.org/10.4337/9781789900125.00008
3. M. Batty, Spatial entropy. Geogr. Anal. **6**(1), 1–31 (1974). ISSN: 0016-7363, 1538-4632. https://doi.org/10.1111/j.1538-4632.1974.tb01014.x. (Visited on 01/02/2024)
4. R. Clausius, *Ü ber verschiedene f ü r die Anwendung bequeme Formen der Hauptgleichungen der mechanischen W ä rmetheorie: vorgetragen in der naturforsch. Gesellschaft den 24. April 1865* (Verlag nicht ermittelbar, 1865)
5. L. Boltzmann, Weitere Studien über das Wärmegleichgewicht unter Gasmolekülen, in *Kinetische Theorie II* (1868/1970), pp. 115–225. https://doi.org/10.1007/978-3-322-84986-1_3

6. C.E. Shannon, A mathematical theory of communication. Bell Syst. Tech. J. **27**(3), 379–423 (1948). ISSN: 00058580. https://doi.org/10.1002/j.1538-7305.1948.tb01338.x. (Visited on 12/31/2023)
7. C.C. Gillispie, *The Edge of Objectivity: An Essay in the History of Scientific Ideas* (Princeton University Press, 2016). ISBN: 978-1-4008-8358-5 978-0-691-17252-1. https://doi.org/10.2307/j.ctvc779t6. JSTOR: http://www.jstor.org/stable/10.2307/j.ctvc779t6. (Visited on 12/31/2023)
8. R. O'Dowd, From telecollaboration to virtual exchange: state-of-the-art and the role of UNICollaboration in moving forward. J. Virtual Exch. **1**, 1–23 (2018). https://doi.org/10.14705/rpnet.2018.jve.1. (Visited on 01/02/2024)
9. J.C. Maxwell, Illustrations of the dynamical theory of gases, in *The Kinetic Theory of Gases: An Anthology of Classic Papers with Historical Commentary* (World Scientific, 1860), pp. 148–171. https://doi.org/10.1142/p281
10. M. Schwartz, Lecture 6: entropy, in *Physics 181, Statistical Mechanics* (Harvard University, 2021)
11. A.G. Wilson, *Entropy in Urban and Regional Modelling* (Pion Press, London, 1970)
12. V. Netto et al., Cities, from information to interaction. Entropy **20**(11), 834 (2018). ISSN 1099-4300. https://doi.org/10.3390/e20110834. (Visited on 01/07/2024)
13. B. Hillier, J. Hanson, *The Social Logic of Space*, 1st edn. (Cambridge University Press, 1984). ISBN: 978-0-521-23365-1 978-0-521-36784-4 978-0-511-59723-7. https://doi.org/10.1017/CBO9780511597237. (Visited on 01/07/2024)
14. V.M. Netto, *The Social Fabric of Cities* (Routledge, Taylor & Francis Group, London New York, NY, 2017). ISBN: 978-1-138-58864-6 978-1-4724-7066-9
15. J.W. Gibbs, *Elementary Principles of Statistical Mechanics* (Charles Scribner's Sons, New York, 1902)
16. S. Goldstein et al., Gibbs and Boltzmann entropy in classical and quantum mechanics, in *Statistical Mechanics and Scientific Explanation: Determinism, Indeterminism and Laws of Nature* (World Scientific, 2020), pp. 519–581. https://doi.org/10.1142/11591
17. R.L. Liboff, Gibbs vs. Shannon entropies. J. Stat. Phys. **11**(4), 343–357 (1974). ISSN: 0022-4715, 1572-9613. https://doi.org/10.1007/BF01009793. (Visited on 01/02/2024)
18. X. Gao, E. Gallicchio, A.E. Roitberg, The generalized Boltzmann distribution is the only distribution in which the Gibbs-Shannon entropy equals the thermodynamic entropy. J. Chem. Phys. **151**(3), 034113 (2019). https://doi.org/10.1063/1.5111333
19. J. Lalinský, *What Is the Conceptual Difference between Gibbs and Boltzmann Entropies?* (2014)
20. S.M. Carroll, Why Boltzmann brains are bad, in *Current Controversies in Philosophy of Science* (Routledge, 2020), pp. 7–20
21. M. Batty, Spatial entropy, information and complexity, in *Handbook on Entropy, Complexity and Spatial Dynamics*, ed. by A. Reggiani et al. (Edward Elgar Publishing, 2021). ISBN: 978-1-83910-059-8 978-1-83910-058-1. https://doi.org/10.4337/9781839100598.00015. (Visited on 01/02/2024)
22. J.P. Crutchfield, What lies between order and chaos?, in *Art and Complexity* (Elsevier, 2003), pp. 31–45. ISBN: 978-0-444-50944-4. https://doi.org/10.1016/B978-044450944-4/50004-9. (Visited on 12/31/2023)
23. V.M. Netto, J. Meirelles, F.L. Ribeiro, Social interaction and the city: the effect of space on the reduction of entropy. Complexity **2017**, 1–16 (2017). ISSN: 1076-2787, 1099-0526. https://doi.org/10.1155/2017/6182503. (Visited on 01/02/2024)
24. E. Brigatti et al., Entropy and hierarchical clustering: characterizing the morphology of the urban fabric in different spatial cultures. Chaos: Interdisc. J. Nonlinear Sci. **31**(11), 113138 (2021). ISSN: 1054-1500, 1089-7682. https://doi.org/10.1063/5.0069258. (Visited on 01/02/2024)
25. E.T. Jaynes, Information theory and statistical mechanics. Phys. Rev. **106**(4), 620–630 (1957). ISSN: 0031-899X. https://doi.org/10.1103/PhysRev.106.620. (Visited on 01/02/2024)

26. I. Prigogine, Modération et Transformation Irréversible Des Systèmes Ouverts, in *Bulletin de la Classe des sciences. Acad é mie royale de Belgique*, vol. 31 (1945), pp. 600–606
27. R.E. Ulanowicz, B.M. Hannon, Life and the production of entropy. Proc. R. Soc. London. Ser. B. Biol. Sci. **232**(1267), 181–192 (1987). https://doi.org/10.1098/rspb.1987.0067
28. E. Schrödinger, *What Is Life? The Physical Aspect of the Living Cell* (Cambridge University Press, 1944)
29. J. Portugali, Schrödinger's what is life?—complexity, cognition and the city. Entropy **25**(6), 872 (2023). ISSN: 1099-4300. https://doi.org/10.3390/e25060872. (Visited on 01/02/2024)
30. D.M. Raup, Geometric analysis of shell coiling: general problems. J. Paleontol. 1178–1190 (1966). ISSN: 00223360, 19372337
31. G.E. Budd, Morphospace. Current Biol. **31**(19), 1181–1185 (2021). ISSN: 09609822. https://doi.org/10.1016/j.cub.2021.08.040. (Visited on 01/07/2024)
32. K. Friston, A free energy principle for biological systems. Entropy **14**(11), 2100–2121 (2012). ISSN: 1099-4300. https://doi.org/10.3390/e14112100. (Visited on 01/02/2024)
33. H.R. Maturana, F.J. Varela, *Autopoiesis and Cognition: The Realization of the Living* (Springer, 1991). ISBN: 978-90-277-1016-1
34. E. Durkheim, *Rules of Sociological Method* (Simon and Schuster, 1895). ISBN: 978-0-02-907940-9
35. N. Luhmann, *Social Systems* (Stanford University Press, 1995). ISBN: 978-0-8047-2625-2
36. S.A. Kauffman, *A World beyond Physics: The Emergence and Evolution of Life* (Oxford University Press, 2019). ISBN: 0190871334, 978-0190871338
37. W. Buckley, *Sociology and Modern Systems Theory* (Prentice-Hall Inc, Englewood Clifs., 1967)
38. D.D. McFarland, Measuring the permeability of occupational structures: an information-theoretic approach. Am. J. Sociol. **75**(1), 41–61 (1969). ISSN: 0002-9602. https://doi.org/10.1086/224744. JSTOR: http://www.jstor.org/stable/2775612. (Visited on 01/02/2024)
39. F. Charvát, J. KuÄ?era, M. Soukup, Toward the system theory of dependence: further general theoretical remarks. Qual. Quant. **7**(1), 69–90 (1973). ISSN: 0033-5177, 1573-7845. https://doi.org/10.1007/BF00172369. (Visited on 01/02/2024)
40. P.M. Horan, Information-theoretic measures and the analysis of social structures. Sociol. Methods Res. **3**(3), 321–340 (1975). ISSN: 0049-1241, 1552-8294. https://doi.org/10.1177/004912417500300305. (Visited on 01/02/2024)
41. O.E. Klapp, Opening and closing in open systems. Behav. Sci. **20**(4), 251–257 (1975). ISSN: 00057940, 10991743. https://doi.org/10.1002/bs.3830200406. (Visited on 01/02/2024)
42. J. Galtung, Entropy and the general theory of peace, in *Essays in Peace Research,* ed. by J. Galtung, vol. L (Humanities Press, NJ, USA, Atlantic Highlands, 1975)
43. K.D. Bailey, Sociological entropy theory: toward a statistical and verbal congruence. Qual. Quant. **18**(1) (1983). ISSN: 0033-5177, 1573-7845. https://doi.org/10.1007/BF00221453. (Visited on 01/02/2024)
44. K.D. Bailey, Social entropy theory: an overview. Syst. Pract. **3**(4), 365–382 (1990). ISSN: 0894-9859, 1573-9295. https://doi.org/10.1007/BF01063441. (Visited on 01/02/2024)
45. J.G. Miller, *Living Systems* (University Press of Colorado, Niwot, Colo, 1995). ISBN: 978-0-87081-363-4
46. K.D. Bailey, *Sociology and the New Systems Theory: Toward a Theoretical Synthesis* (State University of New York Press, Albany, NY, 1994). ISBN: 978-0-7914-1744-7 978-0-7914-1743-0
47. K.D. Bailey, Sociocybernetics and Social Entropy Theory, in *Kybernetes*, vol. 35.3/4 (2006), ed. by B. Scott, pp. 375–384. ISSN: 0368-492X. https://doi.org/10.1108/03684920610653683. (Visited on 01/02/2024)
48. F. Geyer, The challenge of sociocybernetics, in *Kybernetes*, vol. 24.4 (1995), pp. 6–32. ISSN: 0368-492X. https://doi.org/10.1108/03684929510089321. (Visited on 01/02/2024)
49. F. Geyer, The march of self-reference, in *Kybernetes*, vol. 31.7/8 (2002), pp. 1021–1042. ISSN: 0368-492X. https://doi.org/10.1108/03684920210436318. (Visited on 01/02/2024)

50. L.B. Leopold, W.B. Langbein, *The Concept of Entropy in Landscape Evolution* (U.S. Government Printing Office, 1962). https://doi.org/10.3133/pp500A
51. L. Curry, The Random spatial economy: an exploration in settlement theory. Ann. Assoc. Am. Geogr. **54**(1), 138–146 (1964). ISSN: 00045608, 14678306. https://doi.org/10.1111/j.1467-8306.1964.tb00479.x. JSTOR: http://www.jstor.org/stable/2569469. (Visited on 01/07/2024)
52. A. Reggiani, L.A. Schintler, D. Czamanski (Eds.), *Handbook on Entropy, Complexity and Spatial Dynamics: A Rebirth of Theory?* (Edward Elgar Publishing, Cheltenham, UK; Northampton, MA, USA, 2021). ISBN: 978-1-83910-058-1. https://doi.org/10.4337/9781839100598
53. D.F. Batten, *Spatial Analysis of Interacting Economies: The Role of Entropy and Information Theory in Spatial Input-Output Modeling* (Springer Netherlands, 1982). https://doi.org/10.1007/978-94-017-3040-2
54. A.G. Wilson, Entropy in urban and regional modelling: retrospect and prospect. Geogr. Anal. **42**(4), 364–394 (2010). ISSN: 0016-7363, 1538-4632. https://doi.org/10.1111/j.1538-4632.2010.00799.x. (Visited on 01/02/2024)
55. M.A.O. Ayeni, The city system and the use of entropy in urban analysis. Urban Ecol. **2**(1), 33–53 (1976). ISSN: 03044009. https://doi.org/10.1016/0304-4009(76)90004-8. (Visited on 01/02/2024)
56. B. Harris, A.G. Wilson, Equilibrium values and dynamics of attractiveness terms in production-constrained spatial-interaction models. Environ. Plann. A: Econ. Space **10**(4), 371–388 (1978). ISSN: 0308-518X, 1472-3409. https://doi.org/10.1068/a100371. (Visited on 01/07/2024)
57. M. Batty, Entropy in spatial aggregation. Geogr. Anal. **8**(1), 1–21 (1976). ISSN: 0016-7363, 1538-4632. https://doi.org/10.1111/j.1538-4632.1976.tb00525.x. (Visited on 01/07/2024)
58. D.G Leibovici, Defining spatial entropy from multivariate distributions of co-occurrences, in *Spatial Information Theory: 9th International Conference, COSIT 2009 Aber Wrac'h, France, September 21–25, 2009 Proceedings 9* (Springer, 2009), pp. 392–404. https://doi.org/10.1007/978-3-642-03832-7_24
59. D.G. Leibovici et al., Local and global spatio-temporal entropy indices based on distance-ratios and co-occurrences distributions. Int. J. Geogr. Inf. Sci. **28**(5), 1061–1084 (2014). https://doi.org/10.1080/13658816.2013.871284
60. M. Batty et al., Entropy, complexity, and spatial information. J. Geogr. Syst. **16**(4), 363–385 (2014). ISSN: 1435-5930, 1435-5949. https://doi.org/10.1007/s10109-014-0202-2. (Visited on 01/08/2024)
61. W. Weaver, Recent contributions to the mathematical theory of communication, in *ETC: A Review of General Semantics* (1953), pp. 261–281. ISSN: 0014164X, 21689245
62. C. Claramunt, Towards a spatio-temporal form of entropy, in *Advances in Conceptual Modeling*, ed. by D. Hutchison et al., vol. 7518 (Springer, Berlin, Heidelberg, 2012), pp. 221–230. ISBN: 978-3-642-33998-1 978-3-642-33999-8. https://doi.org/10.1007/978-3-642-33999-8_28. (Visited on 01/07/2024)
63. L. Altieri, D. Cocchi, G. Roli, A new approach to spatial entropy measures. Environ. Ecol. Stat. **25**(1), 95–110 (2018). ISSN: 1352-8505, 1573-3009. https://doi.org/10.1007/s10651-017-0383-1. (Visited on 01/07/2024)
64. J. Nowosad, T.F. Stepinski, Stochastic, empirically informed model of landscape dynamics and its application to deforestation scenarios. Geophys. Res. Lett. **46**(23), 13845–13852 (2019). ISSN: 0094-8276, 1944-8007. https://doi.org/10.1029/2019GL085952. (Visited on 01/07/2024)
65. R. Rosenholtz, Y. Li, L. Nakano, Measuring visual clutter. J. Vis. **7**(2), 17 (2007). ISSN: 1534-7362. https://doi.org/10.1167/7.2.17. (Visited on 01/07/2024)
66. A. Woodruff, J. Landay, M. Stonebraker, Constant information density in zoomable interfaces, in *Proceedings of the Working Conference on Advanced Visual Interfaces* (ACM, L'Aquila Italy, 1998), pp. 57–65. ISBN: 978-1-4503-7435-4. https://doi.org/10.1145/948496.948505. (Visited on 01/07/2024)
67. H. Haken, J. Portugali, The face of the city is its information. J. Environ. Psychol. **23**(4), 385–408 (2003). ISSN: 0272-4944. https://doi.org/10.1016/S0272-4944(03)00003-3

68. H. Haken, J. Portugali, *Information Adaptation: The Interplay between Shannon Information and Semantic Information in Cognition* (Springer, 2015). https://doi.org/10.1007/978-3-319-11170-4
69. C. Alexander, *Notes on the Synthesis of Form*. 17. printing (Harvard University Press, Cambridge, Mass., 1964). ISBN: 978-0-674-62751-2
70. C. Alexander, *A City Is Not a Tree (1965)*, 50th Anniversary Edition (Sustasis Press, The Dalles, OR, 2015). ISBN: 9789463864817
71. P. Steadman, *Architectural Morphology: An Introduction to the Geometry of Building Plans* (Pion, London, 1983). ISBN: 978-0850860863
72. B. Hillier et al., Space syntax, a different urban perspective. Architects' J. **178**, 47–63 (1983)
73. B. Jiang, C. Claramunt, Topological analysis of urban street networks. Environ. Plan. B: Plan. Des. **31**(1), 151–162 (2004). ISSN: 0265-8135, 1472-3417. https://doi.org/10.1068/b306. (Visited on 01/07/2024)
74. P. Crucitti, V. Latora, S. Porta, Centrality measures in spatial networks of urban streets. Phys. Rev. E **73**(3), 036125 (2006). ISSN: 1539-3755, 1550-2376. https://doi.org/10.1103/PhysRevE.73.036125. (Visited on 01/07/2024)
75. S. Porta, P. Crucitti, V. Latora, The network analysis of urban streets: a dual approach. Physica A: Stat. Mech. Appl. **369**(2), 853–866 (2006). ISSN: 03784371. https://doi.org/10.1016/j.physa.2005.12.063. (Visited on 01/07/2024)
76. M. Barthélemy, A. Flammini, Modeling urban street patterns. Phys. Rev. Lett. **100**(13), 138702 (2008). ISSN: 0031-9007, 1079-7114. https://doi.org/10.1103/PhysRevLett.100.138702. (Visited on 01/07/2024)
77. A. Gudmundsson, N. Mohajeri, Entropy and order in urban street networks. Sci. Rep. **3**(1), 3324 (2013). ISSN: 2045-2322. https://doi.org/10.1038/srep03324. (Visited on 01/06/2024)
78. G. Boeing, Urban spatial order: street network orientation, configuration, and entropy. Appl. Netw. Sci. **4**(1), 67 (2019). ISSN: 2364-8228. https://doi.org/10.1007/s41109-019-0189-1. (Visited on 01/06/2024)
79. A. Coutrot et al., Entropy of city street networks linked to future spatial navigation ability. Nature **604**(7904), 104–110 (2022). ISSN: 0028-0836, 1476-4687. https://doi.org/10.1038/s41586-022-04486-7. (Visited on 01/07/2024)
80. V.M. Netto, E. Brigatti, C. Cacholas, From urban form to information: cellular configurations in different spatial cultures. Environ. Plan. B: Urban Anal. City Sci. **50**(1), 146–161 (2023). ISSN: 2399-8083, 2399-8091. https://doi.org/10.1177/23998083221107382. (Visited on 01/02/2024)
81. SpaceSyntaxLimited. *Space Syntax OpenMapping*. https://github.com/spacesyntax/OpenMapping. Accessed: March 2024. 2018
82. B. Hillier, The architecture of the urban object. Ekistics **56**(334/335), 5–21 (1989)
83. I. Vranken et al., A review on the use of entropy in landscape ecology: heterogeneity, unpredictability, scale dependence and their links with thermodynamics. Landscape Ecol. **30**(1), 51–65 (2015). ISSN: 0921-2973, 1572-9761. https://doi.org/10.1007/s10980-014-0105-0. (Visited on 01/06/2024)
84. M.G. Turner, R.H. Gardner, *Landscape Ecology in Theory and Practice: Pattern and Process* (Springer, New York, NY, 2015). ISBN: 978-1-4939-2793-7 978-1-4939-2794-4. https://doi.org/10.1007/978-1-4939-2794-4. (Visited on 01/06/2024)
85. B.B. Brown et al., Mixed land use and walkability: variations in land use measures and relationships with BMI, overweight, and obesity. Health Place **15**(4), 1130–1141 (2009). ISSN: 13538292. https://doi.org/10.1016/j.healthplace.2009.06.008. (Visited on 01/06/2024)
86. A. Bitner, M. Fialkowski, Entropy of the land parcel mosaic as a measure of the degree of urbanization. Entropy **23**(5), 543 (2021). ISSN: 1099-4300. https://doi.org/10.3390/e23050543. (Visited on 01/06/2024)
87. R. Cervero, *America's Suburban Centers: A Study of the Land Use -Transportation Link*. United States. Department of Transportation. Technology Sharing Program (1988). https://doi.org/10.21949/1527274. (Visited on 01/06/2024)

88. L.D. Frank, G. Pivo, Impacts of mixed used and density on utilization of three modes of travel: single-occupant vehicle, transit, walking. Transp. Res. Rec. **1466**, 44–52 (1994). ISSN: 0361-1981
89. L.D. Frank et al., Linking objectively measured physical activity with objectively measured urban form. Am. J. Prev. Med. **28**(2), 117–125 (2005). ISSN: 07493797. https://doi.org/10.1016/j.amepre.2004.11.001. (Visited on 01/06/2024)
90. R. Ewing, R. Cervero, Travel and the built environment: a meta-analysis. J. Am. Plan. Assoc. **76**(3), 265–294 (2010). ISSN: 0194-4363, 1939-0130. https://doi.org/10.1080/01944361003766766. (Visited on 01/06/2024)
91. Y. Song, L. Merlin, D. Rodriguez, Comparing measures of urban land use mix. Comput. Environ. Urban Syst. **42**, 1–13 (2013). ISSN: 01989715. https://doi.org/10.1016/j.compenvurbsys.2013.08.001. (Visited on 01/06/2024)
92. R. Cervero, Suburban employment centers: probing the influence of site features on the journey-to-work. J. Plan. Educ. Res. **8**(2), 75–85 (1989). ISSN: 0739-456X, 1552-6577. https://doi.org/10.1177/0739456X8900800203. (Visited on 01/06/2024)
93. K.M. Kochelman, Travel Behavior as a Function of Accessibility, Land Use Mixing, and Land Use Balance: Evidence from the San Francisco Bay Area. A Thesis Submitted in Partial Satisfaction of the Requirements for the Degree of Master of City Planning in City and Regional Planning (University of California, Berkeley, 1996). https://doi.org/10.3141/1607-16
94. R. Cervero, K. Kockelman, Travel demand and the 3Ds: density, diversity, and design. Transp. Res. Part D: Transp. Environ. **2**(3), 199–219 (1997). ISSN: 13619209. https://doi.org/10.1016/S1361-9209(97)00009-6. (Visited on 01/06/2024)
95. O.D. Duncan, B. Duncan, A methodological analysis of segregation indexes. Am. Sociol. Rev. **20**(2), 210 (1955). ISSN: 00031224. https://doi.org/10.2307/2088328. JSTOR: http://www.jstor.org/stable/2088328. (Visited on 01/06/2024)
96. H. Theil, *Statistical Decomposition Analysis* (North Holland Publishing Co, Amsterdan, 1972). ISBN: 9780720433142
97. D.R. James, K.E. Taeuber, Measures of segregation. Sociol. Methodol. **15**, 1 (1985). ISSN: 00811750. https://doi.org/10.2307/270845. JSTOR: http://www.jstor.org/stable/270845. (Visited on 01/06/2024)
98. M.J. White, Segregation and diversity measures in population distribution. Popul. Index **52**(2), 198 (1986). ISSN: 00324701. https://doi.org/10.2307/3644339. JSTOR: http://www.jstor.org/stable/3644339. (Visited on 01/06/2024)
99. M.J. White, The measurement of spatial segregation. Am. J. Sociol. **88**(5), 1008–1018 (1983). ISSN: 0002-9602, 1537-5390. https://doi.org/10.1086/227768. (Visited on 01/06/2024)
100. S. Openshaw, *The Modifiable Areal Unit Problem* (Geobooks, Norwich, 1984). ISBN: 0-86094-134-5
101. D.S. Massey, N.A. Denton, The dimensions of residential segregation. Soc. Forces **67**(2), 281 (1988). ISSN: 00377732. https://doi.org/10.2307/2579183. (Visited on 01/06/2024)
102. S.F. Reardon, D. O'Sullivan, 3. Measures of spatial segregation. Sociol. Methodol. **34**(1), 121–162 (2004). ISSN: 0081-1750, 1467-9531. https://doi.org/10.1111/j.0081-1750.2004.00150.x. (Visited on 01/06/2024)
103. W. Bell, A probability model for the measurement of ecological segregation. Soc. Forces **32**(4), 357–364 (1954). ISSN: 0037-7732, 1534-7605. https://doi.org/10.2307/2574118. (Visited on 01/06/2024)
104. J.R. Ottensmann, Urban sprawl, land values and the density of development. Land Econ. **53**(4), 389 (1977). ISSN: 00237639. https://doi.org/10.2307/3145984. JSTOR: http://www.jstor.org/stable/3145984. (Visited on 01/06/2024)
105. N. Nazarnia, C. Harding, J.A.G. Jaeger, How suitable is entropy as a measure of urban sprawl?. Landscape Urban Plan. **184**, 32–43 (2019). ISSN: 01692046. https://doi.org/10.1016/j.landurbplan.2018.09.025. (Visited on 01/06/2024)
106. G. Galster et al., Wrestling sprawl to the ground: defining and measuring an elusive concept. Hous. Policy Debate **12**(4), 681–717 (2001). ISSN: 1051-1482, 2152-050X. https://doi.org/10.1080/10511482.2001.9521426. (Visited on 01/06/2024)

107. Y.-H. Tsai, Quantifying urban form: compactness versus 'Sprawl'. Urban Studies **42**(1), 141–161 (2005). ISSN: 0042-0980, 1360-063X. https://doi.org/10.1080/0042098042000309748. (Visited on 01/06/2024)
108. P.M. Torrens, M. Alberti, Measuring sprawl, in *CASA Working Paper Series*, vol. 27 (2000). ISSN: 1467-1298
109. A. Gar-on Yeh, X. Li, Measurement and monitoring of urban sprawl in a rapidly growing region using entropy. Photogram. Eng. Remote Sens. **67**(1), 83–90 (2001). ISSN: 0099-1112
110. H.S. Sudhira, T.V. Ramachandra, K.S. Jagadish, Urban sprawl: metrics, dynamics and modelling using GIS. Int. J. Appl. Earth Obs. Geoinf. **5**(1), 29–39 (2004). ISSN: 15698432. https://doi.org/10.1016/j.jag.2003.08.002. (Visited on 01/06/2024)
111. J. Antony Vinoth Kumar, S.K. Pathan, R.J. Bhanderi, Spatio-temporal analysis for monitoring urban growth – a case study of Indore city. J. Indian Soc. Remote Sens. **35**(1), 11–20 (2007). ISSN: 0255-660X, 0974-3006. https://doi.org/10.1007/BF02991829. (Visited on 01/06/2024)
112. G. Borruso et al., (Eds.), Geographic information analysis for sustainable development and economic planning: new technologies. *Advances in Geospatial Technologies* (IGI Global, 2013). ISBN: 978-1-4666-1924-1 978-1-4666-1925-8. https://doi.org/10.4018/978-1-4666-1924-1. (Visited on 01/06/2024)
113. M.M. Aburas et al., Monitoring and assessment of urban growth patterns using spatio-temporal built-up area analysis. Environ. Monit. Assess. **190**(3), 156 (2018). ISSN: 0167-6369, 1573-2959. https://doi.org/10.1007/s10661-018-6522-9. (Visited on 01/06/2024)
114. S. Das, D.P. Angadi, Assessment of urban sprawl using landscape metrics and Shannon's entropy model approach in town level of Barrackpore sub-divisional region, India. Model. Earth Syst. Environ. **7**(2), 1071–1095 (2021). ISSN: 2363-6203, 2363-6211. https://doi.org/10.1007/s40808-020-00990-9. (Visited on 01/06/2024)
115. D.R. Rondinel-Oviedo, Construction and demolition waste management in developing countries: a diagnosis from 265 construction sites in the Lima metropolitan area. Int. J. Constr. Manag. **23**(2), 371–382 (2023). ISSN: 1562-3599, 2331-2327. https://doi.org/10.1080/15623599.2021.1874677. (Visited on 01/07/2024)
116. P. Cabral et al., Entropy in urban systems. Entropy **15**(12), 5223–5236 (2013). ISSN: 1099-4300. https://doi.org/10.3390/e15125223. (Visited on 01/07/2024)
117. N.N. Taleb, *Antifragile: Things That Gain from Disorder*. Random House Trade Paperback edition. (Random House Trade Paperbacks, New York, 2012). ISBN: 978-0-8129-7968-8
118. V.M. Netto, J. Meirelles, F.L. Ribeiro, Cities and entropy: assessing urban sustainability as a problem of coordination, in *Sustainability Assessment of Urban Systems*, ed. by C.R. Binder, R. Wyss, E. Massaro (Cambridge University Press, Cambridge, 2020), pp. 438–459. ISBN: 9781108574334. https://doi.org/10.1017/9781108574334
119. A. Wilson, *Entropy in Urban and Regional Modelling*, vol. 1 (Routledge, 1970)
120. M. Batty, S. Marshall, The origins of complexity theory in cities and planning, in *Complexity Theories of Cities have Come of Age: An Overview with Implications to Urban Planning and Design* (2012), pp. 21–45. https://doi.org/10.1007/978-3-642-24544-2_3
121. M. Batty, Space, scale, and scaling in entropy maximizing. Geogr. Anal. **42**(4), 395–421 (2010). ISSN: 0016-7363, 1538-4632. https://doi.org/10.1111/j.1538-4632.2010.00800.x. (Visited on 01/02/2024)
122. T. Ouyang et al., A new assessment method for urbanization environmental impact: urban environment entropy model and its application. Environ. Monit. Assess. **146**, 433–439 (1-3 2007). https://doi.org/10.1007/s10661-007-0089-1
123. Q. Wang et al., Research on the impact assessment of urbanization on air environment with urban environmental entropy model: a case study. Stochast. Environ. Res. Risk Assess. **26**, 443–450 (3 2011). https://doi.org/10.1007/s00477-011-0493-5
124. T. Parsons, *The System of Modern Societies*. 10. print. Foundations of Modern Sociology Series (Prentice-Hall, Englewood Cliffs (N.J.), 1971). ISBN: 978-0-13-881540-0 978-0-13-881557-8
125. A. Giddens, The constitution of society: outline of the theory of structuration: elements of the theory of structuration, in *Practicing History* (Routledge, 2004), pp. 121–142

Chapter 13
An Introduction to Mathematical Concepts of Power-Laws in Cities and Urban Systems

Diego Rybski

Abstract This chapter provides an introduction to how power-laws are used to characterize cities and urban systems. On the one hand, four power-law representations are used to describe city-size distributions. Here we detail how the respective exponents are related. For numerical simulations it might be necessary to study power-law distributed random numbers. The chapter includes some basics how to generate them. As another important aspect, the relation between power-law distributions and power-law correlations is presented. On the other hand, essentials on fractals and dimensions are provided. E.g. one section outlines why box-covering (aka box-counting) leads to an estimate of the fractal dimension.

13.1 Introduction

Two-dimensional graphs, i.e. plotting two variables of a function or data against each other, are common and fundamental to many scientific disciplines. Plotting represents the mapping of numbers onto space. In experiments, humans mapped numbers onto a logarithmic scale, in particular when the numbers are large [1]. This indicates that the logarithmic mapping of numbers onto space is a universal intuition that is also associated with "a logarithmic internal representation of sensation" [2]. In fact, the difference between 5 and 6 is noticeable but between 50 and 51 it might be negligible, although in both cases it is one. Logarithmic scaling takes relative differences into account by stretching small numbers and squeezing large numbers. Accordingly, in many contexts logarithmic scaling of axes is used complementary to linear axis.

If both axes of the initially mentioned graph are logarithmic, then a straight line corresponds to a power-law relation. Power-laws are common in nature and society, and scaling is fundamental to the theory of complex systems [3]. In the cities context,

D. Rybski (✉)
Leibniz Institute of Ecological Urban and Regional Development (IOER), Dresden, Germany
e-mail: ca-dr@rybski.de

Complexity Science Hub Vienna, Vienna, Austria

D. Rybski (ed.), *Compendium of Urban Complexity*, Understanding Complex Systems,
https://doi.org/10.1007/978-3-031-82666-5_13

power-laws and scaling appear in (i) distributions and (ii) fractals. In the following, respective mathematical concepts will be introduced.

13.2 The Power-Law Distribution

"When the probability of measuring a particular value of some quantity varies inversely as a power of that value, the quantity is said to follow a power law [...] distribution" [4]. Power-law distributions are interesting because they show scaling, i.e. they retain their shape when the considered quantity is multiplied by a constant which 'rescales' the distribution [3]. Power-law distributions are right-skewed and exhibit a long, heavy, or fat tail of non-zero likelihood of large values.

Not only city-sizes are analyzed in terms of power-law distributions but also many other quantities, including word-frequency [5], earthquakes [6], companies [7], sexual contacts [8], file sizes [9], vegetation patches [10], hurricanes [11], lakes [12], (tropical) forests [13], and many others [4]. Since power-law distributions are so omnipresent in nature, society, and technology, they can be considered as a normal distribution of a different kind, see also [14].

Despite the wide occurrence of power-law distributions, the processes behind them remain puzzling. Various mechanisms, models, and processes can generate power-law distributions, and different domains require their own processes. There is no universal mechanism. Here we give an introduction to power-law distributions, more details can be found in [4].

13.2.1 Four Representations

The *power-law distribution* can be traced back to Pareto [15] who found this sort of distribution in income. Consequently, they are also known as *Pareto distribution*. In words, the likelihood of a random variable decreases as a power-law with the value of the random variable.

In the literature, power-law distributions appear in four, in principle, equivalent forms. In order to compare the exponents, it is important to identify which representation has been use.

- The *probability density function* (pdf) describes the relative probability that a random variable has a given value.
- Alternatively, the size distribution can be expressed as *cumulative distribution function* (complementary cdf), describing the probability that a random variable will be found to have a value greater than a given value.
- Empirically, also the *rank-size representation* (rsr) is used, where the size is expressed as a function of the rank. This representation is sometimes called *Lotka form* [16].

- Often also a variant, i.e. the *inverse rank-size representation* (irs) is used, where the rank is plotted as a function of the size. This representation is sometimes called *Pareto form* [16].

For a historical overview in the context of cities we refer to [17]. A more thorough discussion can be found in Chap. 1.

We begin with the probability density function (pdf). For the power-law distribution it is

$$p(S) \sim S^{-\alpha_{\rm pdf}}, \tag{13.1}$$

where $p(S)$ is the probability that the random variable is found between S and $S + \mathrm{d}S$. The pdf is normalized, which means that its integral is unity. For simplicity we use the proportionality here and in the following three cases. In this representation the tail is on the right.

The complementary cumulative distribution function (complementary cdf) is given by

$$P(S) \sim S^{-\alpha_{\rm cdf}}. \tag{13.2}$$

It corresponds to the likelihood that the random variable is larger than S. Since the complementary cdf and the pdf are related via $P(S) = \int_S^\infty p(s)\,\mathrm{d}s$ it follows that $\alpha_{\rm pdf} \simeq \alpha_{\rm cdf} + 1$. In this representation the tail is also on the right.

In the rank-size representation (rsr), the power-law distribution takes the form

$$S \sim r^{-\alpha_{\rm rsr}}, \tag{13.3}$$

where r is the rank (the largest unit has rank 1). In this representation the tail is on the left.

Some authors also use the inverse, i.e. plotting the rank as a function of the size

$$r \sim S^{-\alpha_{\rm irs}}. \tag{13.4}$$

Since one is the inverse of the other, the exponents of the two rank-size representations are related via $\alpha_{\rm rsr} \simeq 1/\alpha_{\rm irs}$. Moreover, when dividing the rank by the number of observation, the inverse rank-size (irs) representation resembles the complementary cdf. Accordingly, $\alpha_{\rm irs} \simeq \alpha_{\rm cdf}$. In this representation the tail is also on the right.

As detailed in Table 13.1 all four exponents can be converted into any of the others. We use '$\simeq$' to relate the exponents to emphasize that the conversion is only theoretical and numerical estimates deviate naturally.

13.2.2 Moments and Characteristics

The power-law distribution is characterized by some interesting properties. Large events have larger probability compared e.g. to Gaußian or exponential distributions,

Table 13.1 Conversions between the exponents of different representation of the power-law distribution. These are (i) probability density function, pdf, Eq. (13.1), (ii) complementary cumulative distribution function, cdf, Eq. (13.2), (iii) rank-size representation, rsr, Eq. (13.3), and (iv) inverse rank-size, irs, Eq. (13.4)

	pdf	cdf	rsr	irs
pdf		$\alpha_{\text{pdf}} \simeq \alpha_{\text{cdf}} + 1$	$\alpha_{\text{pdf}} \simeq \frac{1}{\alpha_{\text{rsr}}} + 1$	$\alpha_{\text{pdf}} \simeq \alpha_{\text{irs}} + 1$
cdf	$\alpha_{\text{cdf}} \simeq \alpha_{\text{pdf}} - 1$		$\alpha_{\text{cdf}} \simeq \frac{1}{\alpha_{\text{rsr}}}$	$\alpha_{\text{cdf}} \simeq \alpha_{\text{irs}}$
rsr	$\alpha_{\text{rsr}} \simeq \frac{1}{\alpha_{\text{pdf}-1}}$	$\alpha_{\text{rsr}} \simeq \frac{1}{\alpha_{\text{cdf}}}$		$\alpha_{\text{rsr}} \simeq \frac{1}{\alpha_{\text{irs}}}$
irs	$\alpha_{\text{irs}} \simeq \alpha_{\text{pdf}} - 1$	$\alpha_{\text{irs}} \simeq \alpha_{\text{cdf}}$	$\alpha_{\text{irs}} \simeq \frac{1}{\alpha_{\text{rsr}}}$	

see Fig. 13.1. While exponential functions decrease quickly, the power-law decreases more moderately. Thus, depending on the exponent very large values can be drawn from a power-law distribution.

This property becomes more apparent when considering the moments. The q-th moment is given by

$$\begin{aligned} M_q &= \int_{S_{\text{min}}}^{\infty} S^q\, p(S)\, \mathrm{d}S \quad \text{with} \quad S_{\text{min}} > 0 \\ &\sim \lim_{s \to \infty} s^{q-\alpha_{\text{pdf}}+1} - S_{\text{min}}^{q-\alpha_{\text{pdf}}+1} . \end{aligned}$$

The limes diverges if $\alpha_{\text{pdf}} > q + 1$. This means, the first moment, associated with the mean, diverges when $\alpha_{\text{pdf}} < 2$ and the second moment, associated with the standard

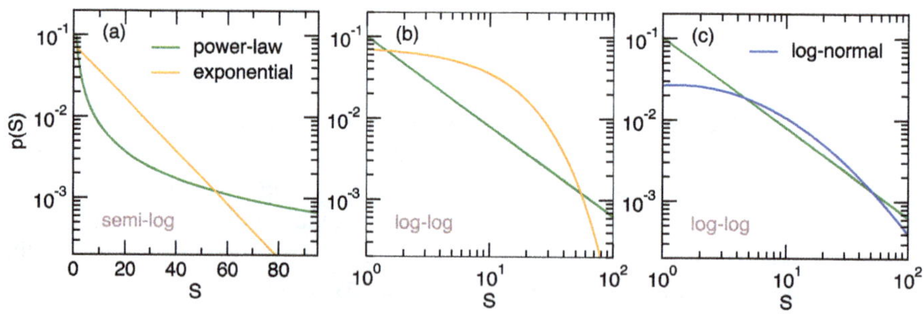

Fig. 13.1 Illustrative comparison of power-law, exponential, and log-normal probability density functions. **a** Power-law distribution $p(S) = (\alpha_{\text{pdf}} - 1)S^{-\alpha_{\text{pdf}}}$ with $\alpha_{\text{pdf}} = 1.1$ (green) and exponential distribution $p(S) = \lambda \exp(-\lambda S)$ with $\lambda = 0.075$ (orange) in semi-logarithmic representation. **b** The same as **a** but in double-logarithmic representation. **c** Power-law distribution as in **a** and **b** together with a log-normal distribution $p(S) = \frac{1}{2\pi S\sigma} \exp(-\frac{\ln S - \mu}{2\sigma^2})$ with $\mu = 1.2$, $\sigma = 1.5$ (blue) in double-logarithmic representation. The log-normal distribution corresponds to a parabola in log-space. One can see that for very large S the power-law is above the other distributions

deviation, diverges when $\alpha_{\text{pdf}} < 3$. For cities, typically size distributions with $\alpha_{\text{pdf}} \approx 2$ are reported [16].

Power-law distributions are said to scale [4], because multiplying the variable with a constant does not change the shape of the distribution, i.e. $p(c \cdot S) \sim (c \cdot S)^{-\alpha_{\text{pdf}}} \sim c^{-\alpha_{\text{pdf}}} \cdot S^{-\alpha_{\text{pdf}}} \sim S^{-\alpha_{\text{pdf}}} \sim p(S)$.

13.2.3 More on the Rank-Size Representation

While the pdf is normalized, see Eq. (13.7), in the case of the rank-size representation one can take advantage of the intercepts as illustrated in Fig. 13.2. Then, Eq. (13.3) becomes

$$S \approx S_{\text{max}}\, r^{-\alpha_{\text{rsr}}}\,. \tag{13.5}$$

where S_{max} is the size of the largest unit. The horizontal intercept requires $S(r{=}N) \approx 1$, where N is the rank of the last unit and the number of units, so that

$$S_{\text{max}} \approx N^{\alpha_{\text{rsr}}}\,. \tag{13.6}$$

We use '$\approx$' here because real-world data usually deviates from these relations. In particular, if the values are discrete, then usually a step function appears at the right end of the rank-size representation. When we replace the exponent in Eq. (13.6) with the respective pdf exponent, then we get $S_{\text{max}} \approx N^{1/(\alpha_{\text{pdf}}-1)}$ which is consistent with [4, Eq. (22)].

The number of units and the size of the largest one are related via Eq. (13.6). Moreover, the integral under the curve in Fig. 13.2 corresponds to the sum over all units. In the case of cities, S_i and r_i are the population size and rank of city i, respectively, N is the number of cities, e.g. in a country, and T is the total population in that country. The integral $T = \int_1^N S dr$ leads to

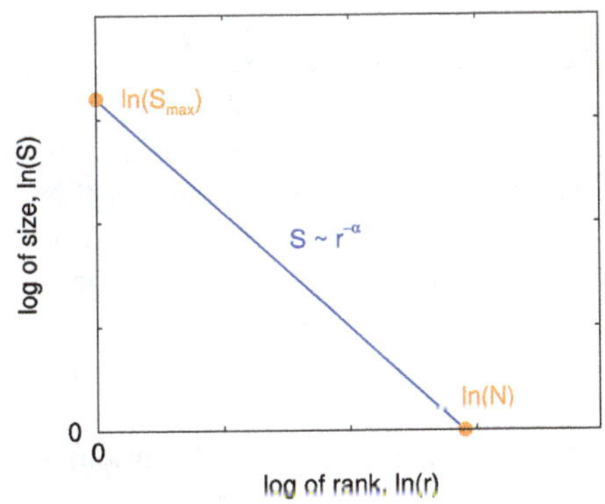

Fig. 13.2 In the rank-size representation, Eq. (13.5), the vertical and horizontal intercepts correspond to the size of the largest unit and the number of units, respectively. Moreover, the integral under the curve corresponds to the sum over all units. Please note, $\alpha = \alpha_{\text{rsr}}$ in this figure

$$T = \frac{S_{\max}}{1-\alpha_{\mathrm{rsr}}}\left(S_{\max}^{1/\alpha_{\mathrm{rsr}}-1}-1\right) \quad \text{for} \quad \alpha_{\mathrm{rsr}} \neq 1$$

making use of Eq. (13.5) and Eq. (13.6). For large $S_{\max}$ one can approximate $T \approx \frac{1}{\alpha_{\mathrm{rsr}}-1} S_{\max}$ for $\alpha_{\mathrm{rsr}} > 1$ and $T \approx \frac{1}{1-\alpha_{\mathrm{rsr}}} S_{\max}^{1/\alpha_{\mathrm{rsr}}}$ for $\alpha_{\mathrm{rsr}} < 1$. The latter case has been observed empirically for cities [18, therein Eq. (6) with $\theta \approx 0.79 < 1$]. For $\alpha_{\mathrm{rsr}} = 1$ we obtain

$$T = S_{\max} \ln S_{\max} .$$

Similar expressions can be obtained replacing $S_{\max}$ by N via Eq. (13.6).

In summary, the four quantities (size of largest unit, number of units, sum over all units, and the exponent) are related theoretically in case of an ideal power-law rank-size relation.

13.2.4 Generating Power-Law Distributions

For some problems it might be necessary or helpful to employ numerical simulations which require power-law distributed random numbers, i.e. following Eq. (13.1). Therefore, here we want to briefly sketch how such random numbers can be generated.

First we need to normalize to find the proportionality constant which we have omitted in Eq. (13.1). This means, we need to determine the proportionality constant b in $p(S) = bS^{-\alpha}$, i.e.

$$\int_{S_{\min}}^{S_{\max}} p(S)\,\mathrm{d}S = 1 \quad \text{and} \quad b = \left(\int_{S_{\min}}^{S_{\max}} S^{-\alpha}\,\mathrm{d}S\right)^{-1} , \tag{13.7}$$

where we assume that the distribution has minimum and maximum values $S_{\min}$ and $S_{\max}$, respectively. Integration leads to

$$b = (1-\alpha)\left(S_{\max}^{1-\alpha} - S_{\min}^{1-\alpha}\right)^{-1} . \tag{13.8}$$

Please note that in this section we only use pdfs so that we skip the subscript, i.e. $\alpha = \alpha_{\mathrm{pdf}}$, for better readability.

In order to generate power-law distributed random numbers, we need to transform pseudo random numbers as they come from the generator in the computer. Such pseudo random number generators usually provide uniformly distributed values between 0 and 1, i.e. $p'(x) = 1$. Therefore, a density transformation is necessary

$$p'(x)\,\mathrm{d}x = p(S)\,\mathrm{d}S .$$

Next we can specify the distributions and add the indefinite integrals

$$\int 1\,\mathrm{d}x = \int bS^{-\alpha}\,\mathrm{d}S ,$$

which leads to

$$S = \left[\frac{1-\alpha}{b}(x+c)\right]^{\frac{1}{1-\alpha}}, \tag{13.9}$$

where c is the integration constant. The integration constant can be obtained from the limits of the distribution. This means $S(x{=}0) = S_{\min}$ and $S(x{=}1) = S_{\max}$. Using the former, we get

$$S_{\min} = \left[\frac{1-\alpha}{b}\cdot c\right]^{\frac{1}{1-\alpha}}$$

and

$$c = \frac{b}{1-\alpha} s_{\min}^{1-\alpha}.$$

Introducing this expression in Eq. (13.9), we obtain

$$S = \left[\frac{1-\alpha}{b}x + s_{\min}^{1-\alpha}\right]^{\frac{1}{1-\alpha}}.$$

Finally, with the normalization constant from Eq. (13.8) we get

$$S = \left[(1-x)S_{\max}^{1-\alpha} + xS_{\min}^{1-\alpha}\right]^{\frac{1}{1-\alpha}}, \tag{13.10}$$

where we swapped $x \leftrightarrow (1-x)$ taking advantage of the fact that for uniformly distributed random numbers $p'(1-x) = p'(x)$. For $x = 1$ this results in $S_{\min}$ and for $x = 0$ it is $S_{\max}$. This means, when we transform random numbers x, that are uniformly distributed between 0 and 1, with Eq. (13.10), then the resulting random numbers S follow a power-law distribution with pdf-exponent α and are limited to the range between $S_{\min}$ and $S_{\max}$.

In some situations it is desired to generate random numbers that can get arbitrary large, i.e. without upper limit. The condition $S_{\max} \to \infty$ leads to an improper normalization integral $\int_{S_{\min}}^{\infty} S^{-\alpha}\,\mathrm{d}S \overset{!}{=} \frac{1}{b}$ which diverges for $\alpha < 1$. However, we obtain

$$S = x^{\frac{1}{1-\alpha}} S_{\min} \quad \text{for} \quad \alpha > 1 \tag{13.11}$$

when executing the analogous steps as before. For $x = 1$ we get $S_{\min}$ and for $x \to 0$ it is $S \to \infty$.

In summary, power-law distributed random numbers S can be obtained by transforming uniformly distributed random numbers x (e.g. `random()` in C) either via Eq. (13.10) or via Eq. (13.11).

13.2.5 Relating Power-Law Distributions and Correlations

Let us consider a quantity S that is power-law distributed according to Eq. (13.1), i.e. of the form $p(S) \sim S^{-\alpha_S}$ (all α-exponents in this section refer to the pdf and the subscript is omitted). If there is an approximate relation, yielding from S the most likely value of Y according to a power-law (aka *urban scaling*)

$$Y \sim S^{\beta}, \tag{13.12}$$

then Y also follows a power-law distribution $p(Y) \sim Y^{-\alpha_Y}$.

The corresponding complementary cdfs, Eq. (13.2), are related via

$$P_Y(Y) = P_S(t^{-1}(Y)),$$

where $t^{-1}(Y)$ is the inverse of the transformation $Y = t(S)$.

For Eq. (13.12) we have $t^{-1}(Y) = b\,Y^{1/\beta}$, where b is a prefactor, and we obtain

$$Y^{1-\alpha_Y} = (b\,Y^{1/\beta})^{1-\alpha_S}$$

from which follows that $1 - \alpha_Y = \frac{1-\alpha_S}{\beta}$ and

$$\beta = \frac{\alpha_S - 1}{\alpha_Y - 1}, \tag{13.13}$$

as also shown in more detail in [19].

However, Eq. (13.13) only represents an upper bound for perfect correlations. Consequently, weaker correlations between the two quantities can imply smaller β-values. This can be illustrated by a thought experiment. Let's imagine, we generate a set of numbers S following Eq. (13.1) with $\alpha_S \simeq 2$. Then we transform these numbers into Y following Eq. (13.12) with $\beta \simeq 1.1$. According to Eq. (13.13) we obtain power-law distributed Y with $\alpha_Y \approx 1.91$. In a cross-plot relating Y as a function of S we will recover $\beta \approx 1.1$. However, if we shuffle one of the sets, either Y or S, then any correlations will be destroyed and we expect $\beta \approx 0$, while $\alpha_S \approx 2$ and $\alpha_Y \approx 1.91$ still hold true. Accordingly, the β according to Eq. (13.13) only represents an upper bound, i.e. $\beta \leq \frac{\alpha_S - 1}{\alpha_Y - 1}$.

13.2.6 Further Aspects

Gini-coefficient

The Gini-coefficient G represents a common indicator to measure inequality (see also Chap. 5 and 7). The coefficient G is defined as the area under the Lorenz-curve divided by the area of the triangle given by the diagonal line. Both, the Lorenz-curve and the Gini-coefficient, have been introduced in the context of income inequality.

Interestingly, for power-law distributions there is a theoretical relation between the Gini-coefficient and the exponent of the distribution. Gastwirth [20] derives $G = \frac{1}{2\alpha_{\mathrm{cdf}}-1}$ which corresponds to

$$G = \frac{1}{2\alpha_{\mathrm{pdf}} - 3} \quad \text{for} \quad \alpha_{\mathrm{pdf}} > 2\,,$$

see also [21]. For an application of the Gini-coefficient in the context of carbon emissions, we refer to [22].

Related to this, Newman [4] derives the *80/20-rule*. If $P(x)$ is the share of population who's wealth exceeds x and $W(x)$ is the respective share of wealth, then

$$W = P^{(\alpha_{\mathrm{pdf}}-2)/(\alpha_{\mathrm{pdf}}-1)} \quad \text{for} \quad \alpha_{\mathrm{pdf}} > 2\,.$$

For $P = 0.2$ and $\alpha_{\mathrm{pdf}} = 2.15$ we obtain $W \simeq 0.81$. I.e. the approximately 20 % wealthiest own approximately 80 % of the wealth.

Log-normal and further distributions

There is a controversy around whether city-size distributions really follow the power-law distribution or rather the log-normal distribution, see [23, e.g.]. Thus, here we want to briefly discuss this alternative log-normal form, see also Chap. 1.

The log-normal distribution is interesting because it emerges from a multiplicative process, i.e. multiplying independent and identically distributed random variables leads, due to the Central Limit Theorem [24, e.g.], to a Gaußian distribution in log-space and accordingly to a log-normal distribution in linear space. Size-independent growth rates represent an assumption, which is also known as Gibrat's law [25, e.g.]. E.g. Eeckhout [26] reports both a log-normal distribution and proportionate growth.

Apparently, both forms, power-law and log-normal, represent broad distributions in the sense that they are skewed and the probability of occurrence extends over a wide range of scales. The log-normal distribution is *not* scale-invariant. However, very extreme events are more likely in the case of the power-law distribution and the likelihood still vanishes in the log-normal case, see Fig. 13.1c. The limited character of the log-normal distribution is also reflected in non-diverging mean and standard deviation. However, since the very extreme events are rare, it is difficult to statistically decide which functional form the tail of a given data set really follows. They become even more indistinguishable if the power-law distribution has an upper cutoff. For a historical overview also in other fields, we refer to [9].

A whole range of alternative, theoretical distributions has been proposed exhibiting power-law tails, see [27, e.g.]. Among them, the Double-Pareto log-normal (dPln) [28] distribution is interesting since it consists of different power-law tails in both extremes and a log-normal body in between. However, all of these forms require more parameters and e.g. dPln involves four parameters. Also the mathematical expressions are more complicated. It has been found that the dPln better describes city-size distributions than the log-normal distribution [29].

Last but not least, in extreme-value theory, the Fréchet distribution represents a sort of power-law distribution. It is one of three cases of the Generalized Extreme Value distribution and the moments can also diverge. However, the Fréchet distribution can be understood as a descendant of the power-law distribution so that not much might be gained.

13.3 Fractals

Next we want to introduce some basic concepts of fractal geometry. This part largely draws from [30], other relevant sources include [31–33]. Chapter 11 provides more details of fractals in the cities' context.

It makes sense to distinguish deterministic fractals and random ones. Deterministic fractals are generated according to a rule which is recognizable as a sort of pattern. The fractality is imposed top-down. Examples include the Koch curve or the Sierpiński carpet. Random fractals employ a random process according to which the fractality emerges bottom-up. The resulting fractals are irregular and exhibit disorder. Examples include percolation clusters (e.g. generated by the Leath Algorithm) or Diffusion Limited Aggregation [34–36]. Both types can be characterized by the fractal dimension.

The fractal dimension d_f can be defined via *mass scaling*, where the mass M of a structure scales with the length-scale L

$$M(L) \sim L^{d_f} . \tag{13.14}$$

In case of a plane, the mass inside a circle of radius L will scale following this relation with a dimension 2. For fractal structures, the exponent d_f will not be an integer number.[1] E.g. the Sierpiński carpet exhibits holes on all scales and has a dimension $d_f \approx 1.893$ (see Fig. 13.3). The fractal dimension is a measure of the compactness of the considered structure.

Generally, fractal structures are scale-invariant. This means, they have no characteristic scale. I.e. zooming in or out one cannot know at which scale one is looking. Formally, zooming in represents a change of length-scale by a factor. This means, we can introduce a factor in Eq. (13.14), i.e. $L \to bL$. Then

$$M(bL) \sim (bL)^{d_f} \sim b^{d_f} L^{d_f} \sim b^{d_f} M(L) \sim M(L) , \tag{13.15}$$

[1] However, fractals with integer dimension are also possible if it is smaller than the embedding dimension. E.g. the Sierpiński tetrahedron has a fractal dimension 2 but is embedded in 3 dimensions.

Fig. 13.3 Illustration of the Sierpiński carpet. From left to right the first iterations are displayed. In the asymptotic, a third of the length scale includes an eighth of the mass, implying a fractal dimension $d_f = \ln 8 / \ln 3 \approx 1.893$, see Eq. (13.16). In case of the left most square one third of the length scale includes a ninth of the mass, implying a dimension $d = \ln 9 / \ln 3 = 2$

where the last step is possible because b^{d_f} is constant (with respect to L). Accordingly, $M(bL) \sim M(L)$ implies that the considered structure is invariant under a change of scale, in the sense that the relation between mass and length-scale is not affected.

Equation (13.15) also includes

$$M(bL) = b^{d_f} M(L) \,. \tag{13.16}$$

It tells how the mass of a structure is related to the length-scale. As an example we can consider the Sierpiński carpet in Fig. 13.3. If we zoom in ($b < 1$) by a factor 3, i.e. $b = 1/3$, then the part we are looking at has an eighth of the original mass, i.e. $b^{d_f} = 1/8$. Combining both, leads to $(1/3)^{d_f} = 1/8$ and $d_f = \ln 8 / \ln 3 \approx 1.893$.

13.3.1 Why Does Box-Counting Provide an Estimate of the Fractal Dimension?

A simple and common approach to measure the fractal dimension of a structure is box-counting (aka box-covering). There are many variants and details to be considered around the method, but it essentially consists of counting how many boxes are necessary to cover the considered structure. By varying the size of the boxes, the fractal dimension is obtained from a double-logarithmic representation where the number of boxes is plotted as a function of the length-scale. E.g. in two dimensions one can use squares of size $l \times l$ and plot the number of boxes $N(l)$ versus l. This and similar methods are necessary for real-world data where we cannot make formal considerations as in the case of the Sierpiński carpet.

But why does box-counting provide an estimate of the fractal dimension? The boxes are similar to zooming in as discussed above and the scale of the boxes is $l = bL$. Then the number of boxes is simply given by

$$N(l) \simeq \frac{M(L)}{M(l)} = \frac{M(L)}{M(bL)} = b^{-d_f} = (l/L)^{-d_f} = L^{d_f} l^{-d_f} \sim l^{-d_f} ,$$

where the first equality is not exact, because the number of boxes $N(l)$ is an integer while $M(L)/M(l)$ is a rational number. Equation (13.16) has been used for the third equality. Again, the last step is possible because L^{d_f} is constant (with respect to l). In summary, we obtain $N(l) \sim l^{-d_f}$. I.e. when plotting the number of boxes versus the length scale of the boxes, the slope corresponds to the fractal dimension.

13.3.2 Every-Pair Interactions

An approach that directly relates to fractality is based on *every-pair interactions* (aka *pairwise-interaction model* [37]). Imagine a situation in which every part, component, or site interacts with every other of the system or structure. In the following we describe some theoretical considerations given that the interactions depend as a power-law on the distance and the considered structure can be described by fractal geometry.

We consider the Euclidean distance r_{ij} between sites i and j of a considered structure, transform it according to $r_{ij} \rightarrow r_{ij}^{-\gamma}$, and calculate an every-pair aggregate

$$I = \sum_{i=1}^{N} \sum_{j \neq i}^{N} r_{ij}^{-\gamma} ,$$

where N is the number of sites the considered structures consists of. The exponent describes how strongly the interactions decrease with the distance. Originally, $I_{ij} = r_{ij}^{-\gamma}$ were studied as pair interactions so that $I = \sum_{i=1}^{N} \sum_{j \neq i}^{N} I_{ij}$ was the interaction field [37]. Interactions usually decay with the distance ($\gamma > 0$) but for $\gamma = -1$ the quantity I relates to the average distance between all pairs of sites.

It has been shown in [37] that the every-pair aggregate I can be expressed theoretically by

$$I^{\text{theo}} = \frac{N_0}{1 - \frac{\gamma}{d_f}} \left[\left(\frac{N}{N_0} \right)^{1 - \frac{\gamma}{d_f}} - r_1^{d_f - \gamma} \right] , \tag{13.17}$$

where r_1 is the distance between first neighbors and N_0 is a constant (idealized, the average number of cells inside a circle of radius $r = 1$). Please note that this expression is restricted to structures which are connected via the four nearest neighbors (von Neumann neighborhood).

The every-pair aggregate and Eq. (13.17) have interesting properties as detailed in [37]. E.g. if $\gamma < d_f$ then $I^{\text{theo}} \sim N^{1-\frac{\gamma}{d_f}}$ for large N (long-range regime), for $\gamma > d_f$ it saturates $I^{\text{theo}} \rightarrow \frac{N_0 d_f}{\gamma - d_f} r_1^{d_f - \gamma}$ (short-range regime), and for $\gamma = d_f$ there is a logarithmic relation $I^{\text{theo}} \sim \ln N$. The d_f appearing in Eq. (13.17) suggests to use it to estimate the fractal dimension of unknown structures.

13.4 Summary and Discussion

In this chapter we have compiled mathematical properties around power-law distributions and fractal geometry that are relevant or interesting in the context of cities and urban systems. This is only an introduction, more details about power-law distributions can be found in [4]; more details about fractals can be found in [30, e.g.].

Here we would like to briefly mention two aspects which we did not discuss so far. First, measuring power-law distributions. Statistical testing of power-law distributions and estimating the respective exponents is a difficult undertaking. In the cities context, originally the rank-size relation has been used and the exponent has been estimated as slope in log-log representation. E.g. Zipf [5, Fg. 9-2] measures $\alpha_{\text{rsr}} \approx 1$, which is why power-law distributions with $\alpha_{\text{pdf}} \approx 2$ are also denoted *Zipf's law*, see also [17]. When estimating the exponent in the rank-size relation, at least three problems occur. (i) The bulk of small cities dominates the regression. (ii) Discrete small city sizes can cause steps in the curve. (iii) There is a lack of normalization (i.e. the sum of city population usually differs from the one implied by a regressions).

Various approaches have been proposed as improvement. A common but problematic practice is logarithmic binning. The idea is to approximate the pdf by a histogram where the bin size increases logarithmically. The exponent is then estimated via a linear regression in log-log representation, which does not represent a rigurous statistical treatment. Gabaix and Ibragimov [38] propose a slightly modified inverse rank-size representation $\ln(r - 1/2) = a - b \ln(S)$, where the parameters a and b stem from ordinary least squares regression. They argue that the shift of $1/2$ is optimal to reduce the bias due to the dominance of small cities. Although being an important issue, here we can only provide some references on the topic of testing or measuring the power-law distribution exponent [4, 39–43].

Second, power-law correlations and how to measure them. In Sec. 13.2.5 we have already mentioned urban scaling (see also [44]). It represents correlations between two variables which are characterized by a power-law relation, see Chap. 2 for more details. Measuring the respective exponent one faces similar problems as in the case of power-law distributions. The regression is dominated by the bulk of small cities, while the tail of large (but few) cities is usually more interesting. Applying a linear regression to the logarithmic quantities is lacking a rigorous statistical treatment. Again, although important, here we only provide related references on the topic [43, 45, 46].

To finalize, we would like to resume with a quote of caution.

> Power laws do have an interesting and possibly even important role to play, but one needs to be very cautious when interpreting them. The most productive use of power laws in the real world will therefore, we believe, come from recognizing their ubiquity (and perhaps

> exploiting them to simplify or even motivate subsequent analysis) rather than from imbuing them with a vague and mistakenly mystical sense of universality. Stumpf and Porter [14]

Acknowledgements I would like to thank A. Bunde for encouraging this chapter. I also appreciate useful comments from F.L. Ribeiro and J.W. Kantelhardt.

References

1. S. Dehaene et al., Log or linear? Distinct intuitions of the number scale in Western and Amazonian indigene cultures. Science **320**(5880), 1217–1220 (2008). https://doi.org/10.1126/science.1156540
2. S. Dehaene, The neural basis of the Weber-Fechner law: a logarithmic mental number line. Trends Cogn Sci. **7**(4), 145–147 (2003). https://doi.org/10.1016/S1364-6613(03)00055-X
3. M.E.J. Newman, Complex systems: a survey, in arXiv.org e-Print archive arXiv:1112.1440 [cond-mat.stat-mech] (2011). https://doi.org/10.48550/arXiv.1112.1440. https://arxiv.org/abs/1112.1440
4. M.E.J. Newman, Power laws, Pareto distributions and Zipf's law. Contemp. Phys. **46**(5), 323–351 (2005). https://doi.org/10.1080/00107510500052444
5. G.K. Zipf, Human Behavior, the Principle of Least Effort: An Introduction to Human Ecology (Reprint of, Edition) (Martino Publishing, Manfield Centre, CT, 1949), p. 2012
6. B. Gutenberg, C.F. Richter, *Seismicity of the Earth and Associated Phenomena* (Princeton University Press, Princeton, New Jersey, 1949)
7. R.L. Axtell, Zipf distribution of U.S. firm sizes. Science **293**(5536), 1818–1820 (2001). https://doi.org/10.1126/science.1062081
8. F. Liljeros et al., The web of human sexual contacts. Nature **411**(6840), 907–908 (2001). https://doi.org/10.1038/35082140
9. M. Mitzenmacher, A brief history of generative models for power law and lognormal distributions. Internet Math. **1**(2), 226–251 (2004). https://doi.org/10.1080/15427951.2004.10129088
10. T.M. Scanlon et al., Positive feedbacks promote power-law clustering of Kalahari vegetation. Nature **449**(7159), 209–212 (2007). https://doi.org/10.1038/nature06060
11. A. Corral, A. Ossó, J.E. Llebot, Scaling of tropical-cyclone dissipation. Nat. Phys. **6**(9), 693–696 (2010). https://doi.org/10.1038/nphys1725
12. C.P. McDonald et al., The regional abundance and size distribution of lakes and reservoirs in the United States and implications for estimates of global lake extent. Limnol. Oceanogr. **57**(2), 597–606 (2012). https://doi.org/10.4319/lo.2012.57.2.0597
13. F. Taubert et al., Global patterns of tropical forest fragmentation. Nature **554**, 519–522 (2018). https://doi.org/10.1038/nature25508
14. M.P.H. Stumpf, M.A. Porter, Critical truths about power laws. Science **335**(6069), 665–666 (2012). https://doi.org/10.1126/science.1216142
15. V. Pareto, *Cours d' é conomie politique: profess é à l'Universit é de Lausanne*, two volumes (F. Rouge, Librairie de l'Université, Lausanne, Paris, Leipzig, 1896, 1897)
16. C. Cottineau, MetaZipf. A dynamic meta-analysis of city size distributions. PLoS One **12**(8), e0183919 (2017). https://doi.org/10.1371/journal.pone.0183919
17. D. Rybski, A. Ciccone, Auerbach, Lotka, Zipf – pioneers of power-law city-size distributions. Arch. Hist. Exact Sci. (2023). https://doi.org/10.1007/s00407-023-00314-0
18. H.V. Ribeiro et al., Association between population distribution and urban GDP scaling. PLoS One **16**(1), e0245771 (2021). https://doi.org/10.1371/journal.pone.0245771

19. A. Gomez-Lievano, H. Youn, L.M.A. Bettencourt, The statistics of urban scaling and their connection to Zipf's law. PLoS ONE **7**(7), e40393 (2012). https://doi.org/10.1371/journal.pone.0040393
20. J.L. Gastwirth, The estimation of the Lorenz curve and Gini index. Rev. Econ. Stat. **54**(3), 306–316 (1972). https://doi.org/10.2307/1937992. https://www.jstor.org/stable/1937992
21. W. Pfähler, Relative Concentration Curve: Functional Form and Measures of Non-Proportionality. B. Econ. Res. **37**(3), 201–211 (1985). https://doi.org/10.1111/j.1467-8586.1985.tb00194.x
22. B. Zhou et al., A Gini approach to spatial CO2 emissions. PLoS One **15**(11), e0242479 (2020). https://doi.org/10.1371/journal.pone.0242479
23. B.J.L. Berry, A. Okulicz-Kozaryn, The city size distribution debate: resolution for US urban regions and megalopolitan areas. Cities **29**, S17–S23 (2012). https://doi.org/10.1016/j.cities.2011.11.007
24. E. Limpert, W.A. Stahel, M. Abbt, Log-normal distributions across the sciences: keys and clues. BioScience **51**(5), 341–352 (2001). https://doi.org/10.1641/0006-3568(2001)051[0341:LNDATS]2.0.CO;2
25. J. Sutton, Gibrat's legacy. J. Econ. Lit. **35**(1), 40–59 (1997). https://www.jstor.org/stable/2729692
26. J. Eeckhout, Gibrat's law for (all) cities. Am. Econ. Rev. **94**(5), 1429–1451 (2004). https://doi.org/10.1257/0002828043052303
27. M. Puente-Ajovín, A. Ramos, F. Sanz-Gracia, Is there a universal parametric city size distribution? Empirical evidence for 70 countries. Ann. Reg. Sci. **65**, 727–741 (2020). https://doi.org/10.1007/s00168-020-01001-6
28. W.J. Reed, M. Jorgensen, The double pareto-lognormal distribution—new parametric model for size distributions. Commun. Stat. **33**(8), 1733–1753 (2004). https://doi.org/10.1081/STA-120037438
29. K. Giesen, A. Zimmermann, J. Suedekum, The size distribution across all cities–double Pareto lognormal strikes. J. Urban Econ. **68**(2), 129–137 (2010). https://doi.org/10.1016/j.jue.2010.03.007
30. A. Bunde, S. Havlin, A brief introduction to fractal geometry, in *Fractals in Science*, ed. by A. Bunde, S. Havlin. Second Printing 1995 (Springer, Berlin, Heidelberg, New York, 1994). Chap. 1, pp. 1–25. ISBN: 3-540-56221-4
31. B.B. Mandelbrot, *The Fractal Geometry of Nature*. first published 1977 (W. H. Freeman an Company, New York, 1983). ISBN: 978-0-7167-1186-5
32. J. Feder, *Fractals* (Plenum Press, New York and London, 1988). ISBN: 0-306-42851-2
33. H.E. Stanley, Fractals and multifractals: the interplay of physics and geometry, in *Fractals in Disordered Systems*, ed. by A. Bunde, S. Havlin (Springer, Berlin, Heidelberg, New York, 1991). Chap. 1, pp. 1–49. ISBN: 3-540-54070-9
34. T.A. Witten Jr., L.M. Sander, Diffusion-limited aggregation, a kinetic critical phenomenon. Phys. Rev. Lett. **47**(19), 1400–1403 (1981). https://doi.org/10.1103/PhysRevLett.47.1400
35. A.S. Fotheringham, M. Batty, P.A. Longley, Diffusion-limited aggregation and the fractal nature of urban growth. Pap. Reg. Sci. Assoc. **67**(1), 55–69 (1989). https://doi.org/10.1007/BF01934667
36. M. Batty, P. Longley, S. Fotheringham, Urban growth and form: scaling, fractal geometry, and diffusion-limited aggregation. Environ. Plan. A **21**(11), 1447–1472 (1989). https://doi.org/10.1068/a211447
37. F.L. Ribeiro et al., Analytical solution for the long- and short-range every-pair-interactions system. Chaos Soliton. Fract. **183**, 114771 (2024). https://doi.org/10.1016/j.chaos.2024.114771
38. X. Gabaix, R. Ibragimov, Rank − 1/2: a simple way to improve the OLS estimation of tail exponents. J. Bus. Econ. Stat. **29**(1), 24–39 (2011). https://doi.org/10.1198/jbes.2009.06157
39. A. Clauset, C.R. Shalizi, M.E.J. Newman, Power-law distributions in empirical data. SIAM Rev. **51**(4), 661–703 (2009). https://doi.org/10.1137/070710111

40. R. Hanel et al., Fitting power-laws in empirical data with estimators that work for all exponents. PLoS One **12**(2), e0170920 (2017). https://doi.org/10.1371/journal.pone.0170920
41. M. Gerlach, E.G. Altmann, Testing statistical laws in complex systems. Phys. Rev. Lett. **122**(16), 168301 (2019). https://doi.org/10.1103/PhysRevLett.122.168301
42. J.M. Moore, G. Yan, E.G. Altmann, Nonparametric power-law surrogates. Phys. Rev. X **12**(2), 021056 (2022). https://doi.org/10.1103/PhysRevX.12.021056
43. E.G. Altmann, *Statistical Laws in Complex Systems*. Draft 0.1 – June 28. 2025. https://link.springer.com/book/10.1007/978-3-031-73164-8
44. L.S. D'Acci (Ed.), *Urban Scaling – Allometry in Urban Studies and Spatial Science*. Advances in Regional Economics, Science and Policy (Routledge, 2024). ISBN: 9781032264400
45. J.C. Leitão et al., Is this scaling nonlinear?. Roy. Soc. Open Sci. **3**(7), 150649 (2016). https://doi.org/10.1098/rsos.150649
46. Y. Xiao, P. Gong, Removing spatial autocorrelation in urban scaling analysis. Cities **124**, 103600 (2022). https://doi.org/10.1016/j.cities.2022.103600

Index

D. Rybski (ed.), *Compendium of Urban Complexity*, Understanding Complex Systems,
https://doi.org/10.1007/978-3-031-82666-5

GPSR Compliance
The European Union's (EU) General Product Safety Regulation (GPSR) is a set of rules that requires consumer products to be safe and our obligations to ensure this.

If you have any concerns about our products, you can contact us on

ProductSafety@springernature.com

In case Publisher is established outside the EU, the EU authorized representative is:

Springer Nature Customer Service Center GmbH
Europaplatz 3
69115 Heidelberg, Germany

www.ingramcontent.com/pod-product-compliance
Lightning Source LLC
Chambersburg PA
CBHW071749250825
31638CB00006B/619

* 9 7 8 3 0 3 1 8 2 6 6 5 8 *